THE GLOBAL GOVERNAN[ ]
CLIMATE CHANGE

Global Environmental Governance

Series Editors: John J. Kirton, University of Toronto, Canada and
Miranda Schreurs, Freie Universität Berlin, Germany

Global Environmental Governance addresses the new generation of twenty-first
century environmental problems and the challenges they pose for management and
governance at the local, national, and global levels. Centred on the relationships
among environmental change, economic forces, and political governance, the
series explores the role of international institutions and instruments, national and
sub-federal governments, private sector firms, scientists, and civil society, and
provides a comprehensive body of progressive analyses on one of the world's most
contentious international issues.

The Global Governance of Climate Change

G7, G20, and UN Leadership

JOHN J. KIRTON
University of Toronto, Canada

ELLA KOKOTSIS
G7 Research Group, Canada

Routledge
Taylor & Francis Group

LONDON AND NEW YORK

First published 2015 by Ashgate Publishing

2 Park Square, Milton Park, Abingdon, Oxfordshire OX14 4RN
711 Third Avenue, New York, NY 10017

Routledge is an imprint of the Taylor & Francis Group, an informa business

First issued in paperback 2017

British Library Cataloguing in Publication Data
A catalogue record for this book is available from the British Library.

The Library of Congress has cataloged the printed edition as follows:
Kirton, John J.
 The global governance of climate change : G7, G20, and UN leadership / by John J. Kirton and Ella Kokotsis.
    pages cm. -- (Global environmental governance)
 Includes bibliographical references and index.
 ISBN 978-0-7546-7584-6 (hardback)
 1. Climatic changes--International cooperation. 2. Climatic changes--Government policy.
 3. Environmental policy--International cooperation. I. Kokotsis, Ella. II. Title.

 QC903.K5618 2015
 363.738'7456--dc23

                                                                    2015014525

ISBN 978-0-7546-7584-6 (hbk)
ISBN 978-0-8153-8041-2 (pbk)

Contents

PART V   CONCLUSION

List of Figures

List of Tables

About the Authors

John J. Kirton is a professor of political science and the director and founder of the G8 Research Group, co-director and founder of the G20 Research Group, as well as the co-director of the BRICS Research Group and the Global Health Development Program, at the Munk School of Global Affairs at Trinity College in the University of Toronto. He is also a Non-Resident Senior Fellow at the Chongyang Institute for Financial Studies in Renmin University of China. He is the author of *G20 Governance for a Globalized World* and *Canadian Foreign Policy in a Changing World*, as well as many articles and chapters, and editor and co-editor of numerous books in the Global Governance series and Global Environmental Governance series published by Ashgate.

Ella Kokotsis, PhD, is the Director of Accountability for the G8 and G20 Research Groups at the Munk School of Global Affairs at Trinity College in the University of Toronto. An expert on summit accountability and compliance, she has consulted with the Canadian government's National Round Table on the Environment and the Economy, the Council on Foreign Relations on their African development agenda, with the Russian government on global health issues in the lead-up to the 2006 St. Petersburg Summit, and with the Government of Canada on numerous summit-related issues during the 2010 Canadian G8 and G20 Summits. Her scholarly research methodology for assessing compliance continues as the basis for the annual accountability reports produced by the G8 and G20 Research Groups. She is author of *Keeping International Commitments: Compliance, Credibility and the G7 Summits*, as well as many articles and chapters.

Preface

This book has been a long time in the making. It has required the talents and dedication of many individuals to get it done. For John J. Kirton, it constitutes the convergence of two longstanding areas of research—on the Group of Seven (G7), the Group of Eight (G8), and the Group of 20 (G20) and on global environmental governance. His work on the G7/8 began in 1987–1988, when he founded the G7 Research Group, which expanded to become the G8 Research Group in 1998, and then added the G20 Research Group, which he founded in 2008. Much of his scholarly work on the G7/8 and related plurilateral summit institutions was published by Ashgate Publishing in series that he co-edited, first on G8 Governance, subsequently on Global Finance, and now on Global Governance.

Kirton's work on global environmental governance began in 1989 when he became a member of the Foreign Policy Committee of the National Roundtable on the Environment and the Economy, an advisory body to the Canadian prime minister on sustainable development. One of its first major projects was to prepare advice for the United Nations Conference on the Environment and Development (UNCED) at Rio de Janeiro in June 1992. He subsequently worked on projects with the North American Commission for Environmental Co-operation and served as principal investigator of the scholarly research program on "Strengthening Canada's Environmental Cohesion through International Regime Reform" (Envireform). He is grateful for the financial support provided by the Social Sciences and Humanities Research Council of Canada for the Envireform project, and to his colleagues at the University of Toronto and beyond who participated in its work. Some of the work it produced was published in another series that he co-edits for Ashgate Publishing on Global Environmental Governance.

Ella Kokotsis's work on the G7/8 and global environmental issues began in the mid 1990s with her doctoral dissertation and subsequent book on compliance with summit decisions. One of her case studies was the compliance of G7 members with their commitments on climate change, along with biodiversity, development, and aid to the former Soviet Union. She has continued her work as director of accountability for the G8 Research Group and G20 Research Group, and has published extensively on their compliance and accountability. Her research on climate change has resulted in several scholarly publications as well as presentations at academic conferences and symposiums in the United States, Europe, Russia, and Asia.

In producing this book, we are grateful to the contribution of many individuals who provided essential and impressive assistance in so many ways along the way. Jenilee Guebert, who served as research director of the G8 and G20 Research Groups, and Caroline Bracht and Julia Kulik, who are both senior researchers, have contributed to many chapters and participated in many summits with us. They have overseen the changing team of researchers and produced much of the data contained in the appendices. Zaria Shaw played an important part in researching and editing portions of the work. Madeline Boyce and Ana Zotovic provided extensive documentary research and participant observation at the Copenhagen Conference of the Parties to the United Nations Framework Convention on Climate Change in 2009. The research and initial drafting of several sections was skillfully done by Leanne Rasmussen, Kevin Moraes, and Rozalind Theriault. Takumi Shibaike, Olga Milkina, and Alecsandra Dragus made critical contributions. Many individuals contributed to the painstaking but essential research for producing the compliance assessments that underpin the analysis in this book. These include Rebeca Ramirez, Aurora Hudson, the many members of the G8 Research Group and G20 Research Group in their annual compliance assessments since 1996 and 2008 respectively, and students in the University of Toronto's POL 456/2256Y course since 2009. Maria Marchyshyn was a loyal contributor throughout.

The book was magnificently edited by Madeline Koch, as the managing director of the G8 Research Group, G20 Research Group, and the Envireform project. Kirstin Howgate, Margaret Younger, Brenda Sharp, and their colleagues at Ashgate Publishing displayed their usual skill and unusual but much appreciated patience all along the way.

We are especially grateful to our spouses, Mary and Rob, and our children, Michael and his wife Linda, Joanna and her husband Dennis, and Max and Zara, for their constant support, understanding, and patience over many years, as we relied on them to bear more than their fair share of household duties while we focused fully on the seemingly endless task of getting this book designed, researched, written and finally published. A special heartfelt thanks also go to Ella's parents, Barbie and Peter, and sister Chippy, who regularly took care of Max and Zara while she researched and co-authored this book.

As this book reveals, the creation of global leadership on climate change governance came from the initiative of three individuals in 1979—Helmut Schmidt, Jimmy Carter, and Joe Clark. The description and explanation in this book of their work and all that followed depended critically on the dedicated, talented work of the students in the G7/8 Research Group and G20 Research Group, starting with their participation in their first summit in 1988. It is to all of them, as visionary statesmen and dedicated students, that we dedicate this book.

John J. Kirton
Ella Kokotsis
Toronto, Canada
May 2015

Abbreviations and Acronyms

APEC	Asia-Pacific Economic Cooperation
ASEM	Asia–Europe Meeting
BRIC	Brazil, Russia, India, and China
BRICS	Brazil, Russia, India, China, and South Africa
CCAC	Climate and Clean Air Coalition to Reduce Short-Lived Climate Pollutants
CCS	carbon capture and storage
CDM	Clean Development Mechanism
CFC	chlorofluorocarbon
CIS	Commonwealth of Independent States
COP	Conference of the Parties to the United Nations Framework Convention on Climate Change
DFAIT	Department of Foreign Affairs and International Trade, Canada
ENB	Earth Negotiations Bulletin
EPA	United States Environmental Protection Agency
ESWG	Energy Sustainability Working Group
FASS	foreign affairs sous sherpa(s)
FTA	free trade agreement
FTT	financial transaction tax
G5	Group of Five (Brazil, China, India, Mexico, and South Africa, which were invited to the G8 summits from 2005 to 2009)
G7	Group of Seven (Canada, France, Germany, Italy, Japan, the United Kingdom, the United States, and the European Union)
G8	Group of Eight (G7 plus Russia)
G20	Group of 20 (G8 plus Australia, Argentina, Brazil, China, India, Indonesia, Korea, Mexico, Saudi Arabia, South Africa, and Turkey)
G77	Group of 77 developing countries
GATT	General Agreement on Tariffs and Trade
GDP	gross domestic product
GEF	Global Environment Facility
GEOSS	Global Earth Observation System of Systems
HIPC	heavily indebted poor country
IAEA	International Atomic Energy Agency
ICAO	International Civil Aviation Organization
ICT	information and communication technologies
IEA	International Energy Agency
IFI	international financial institution
IMF	International Monetary Fund
IMO	International Maritime Organization
IPCC	Intergovernmental Panel on Climate Change
JODI	Joint Oil Data Initiative

LDC	least developed country
MDG	Millennium Development Goal
MEF	Major Economies Forum on Energy Security and Climate Change
MEM	Major Economies Meeting on Energy Security and Climate Change
MOP	Meeting of the Parties to the Kyoto Protocol
MtCO2e	million metric tons of carbon dioxide equivalent
NATO	North Atlantic Treaty Organization
NAFTA	North American Free Trade Agreement
NEPAD	New Partnership for Africa's Development
NGO	nongovernmental organization
NIEO	New International Economic Order
NRTEE	National Round Table on the Environment and the Economy
O5	Outreach Five (see Group of Five)
OECD	Organisation for Economic Co-operation and Development
OPEC	Organization of Petroleum Exporting Countries
ppm	parts per million
PSI	plurilateral summit institution
REDD	Reducing Emissions from Deforestation and Forest Degradation
SARS	severe acute respiratory syndrome
SME	small and medium-sized enterprise
Three Rs	reduce, reuse, recycle
UN	United Nations
UNCED	United Nations Conference on Environment and Development
UNCLOS	United Nations Convention on the Law of the Sea
UNCSD	United Nations Commission on Sustainable Development
UNESCO	United Nations Educational, Scientific, and Cultural Organization
UNEP	United Nations Environment Programme
UNFCCC	United Nations Framework Convention on Climate Change
UNGA	United Nations General Assembly
UNSC	United Nations Security Council
WMD	weapons of mass destruction
WMO	World Meteorological Organization
WSSD	World Summit on Sustainable Development
WTO	World Trade Organization

PART I
Introduction

Chapter 1
The Global Challenge of Climate Change

Climate change is arguably the most compelling global issue of our time. Along with nuclear war, it is the world's only challenge that could conceivably threaten the continuation of human life on the planet as a whole. Its negative effects are already evident in the increasing frequency, severity, and unpredictability of deadly heat waves, fires, droughts, floods, hurricanes, tsunamis, dying coral reefs and other deteriorating ecosystems, melting ice, and infectious disease from insects invading long immune domains. Growing evidence suggests it could submerge the coastal cities of major powers, or even extinguish some small island states beneath the warming and thus expanding and rising seas. Because climate is a complex adaptive system, driven by only partly understood, path-dependent processes, cumulative effects, non-linear dynamics, multiple feedback mechanisms, and abrupt system shifts, it is especially difficult to predict its precise course and consequences. It is equally difficult to control through the galaxy of intergovernmental organizations inherited from the 1944–1945 order with their selective, siloed, slow-moving structures and the mechanistic management models and approaches at the core.

Yet the global community showed long ago that, through visionary leadership, it could take the prescient, preventive, ambitious steps to control climate change before its major and perhaps irreversible harmful effects arrived and moved beyond human control. On June 28, 1979, at the conclusion of their Tokyo Summit, the leaders of the Group of Seven (G7) major market democracies called for "alternative sources of energy" that would "help to prevent further pollution" caused by the "increases of carbon dioxide and sulphur oxides in the atmosphere" (G7 1979). They thus acknowledged through this voluntary consensus the need to halt immediately, at 1979 levels, the concentration of carbon dioxide in the world's climate. In the following five years, they and the other countries in the Organisation for Economic Co-operation and Development (OECD) moved in this desired direction, as their emissions into the atmosphere declared (Sustainable Energy Development Center 2006, 48; Barnes 1994, 42).

In acting so boldly in 1979, G7 leaders were forwarding the more general carbon-controlling environmental principles embedded in their group from its very start. At the conclusion of their first summit at Rambouillet, France, on November 15–17, 1975, the six leaders of France, the United States, the United Kingdom, Germany, Japan, and Italy declared that "our common interests require that we continue to cooperate in order to reduce our dependence on imported energy through conservation and the development of alternative sources" (G7 1975). In 1976, now with Canada's leader present, they noted the need for the rational use of energy resources (G7 1976). In 1977, with the European Community added, they affirmed the principle of "more efficient energy use" (G7 1977). At the first summit Germany hosted, at Bonn in 1978, they directly declared: "in energy development, the environment and human safety of the population must be safeguarded with greatest care" (G7 1978). And then, in Tokyo in 1979, they took up the issue of carbon emissions directly and concluded that its concentration in the atmosphere must stabilize right away.

The central role of the G7, and the Group of Eight (G8) with Russia added after 1998, in global climate governance stands in sharp contrast to the historic absence of any powerful broadly multilateral intergovernmental organization dedicated to the control of climate change.[1] The charter of the United Nations is silent about the existence, let alone the value, of the natural environment. The UN system lacks any specialized functional organization to deal with either energy or the environment overall, instead relying on the fragile United Nations Environment Programme (UNEP) created in 1972 (Kirton 2004a; Biermann and Bauer 2005; Lesage et al. 2010). The Atlantic system of international organizations, centred in the North Atlantic Treaty Organization (NATO) and the OECD, generated only the International Energy Agency (IEA) in 1974, from which France—one of the G8's original members—at first stood aloof. The global community thus was institutionally defenceless as the oil shocks of 1973 and 1979 from the Arab members of the Organization of Petroleum Exporting Countries (OPEC) assaulted the world economy, and as the trees dying from acid rain in North America and Europe showed that the increasing use of coal and other hydrocarbons polluted the atmosphere and killed living things. When George Kennan (1970), one of the world's pioneers of the post-World War II order, called for a new powerful plurilateral international institution to meet these ecological challenges, only the G7 responded to the call. The UN followed a different approach, establishing the weak UNEP in 1972, adding a few issue-specific conventions with their small, fragmented secretariats in 1992, and periodically holding summits on sustainable development (Strong 1973; Kirton 2004a). At the dawn of the twenty-first century, at its development-oriented Millennium Summit in New York in September 2000, the UN recognized at the highest level in a comprehensive context the existence and value of the natural environment in general terms (Kirton, Kulik, Bracht, et al. 2014). But it still took no major actions to put its new principle into effect through the Millennium Development Goals launched there (Kirton, Kulik, Bracht, et al. 2014). Nor did it do so with the required ambition at its Rio+20 Summit in June 2012.

The Debate among Competing Schools of Thought

It thus matters how, how well, and why the G8 and its related plurilateral summit institutions (PSIs) led by the Group of 20 (G20) since 2008 have led global climate governance during the 40 years in which they have been engaged in this task. To date, however, accounts of the G8's involvement and leadership in climate governance have been inconclusive, whether referring to that involvement directly or as part of the G8's broader environmental performance, and its relationship with the broader G20 and the UN. They range widely from judgements of harmful through irrelevant to effective and beneficial, and differ on why the G8 and G20 performed as they did and what their proper role should be.

The first school of thought considers G8 climate governance to be little more than domestically oriented deliberation and delegation (Bayne 2000b, 65). In this view, G8 leaders discuss pressing, high-profile environmental issues such as climate change for domestic political management, to satisfy attentive audiences back home. But they are otherwise anxious to delegate direction setting, decision making, and delivery to lower levels of the

---

1   Throughout this book, the term G8 is used to refer to the entire history of the G7 and G8, beginning in 1975 up until 2014. However, G7 is used in reference to events that took place when Russia became a member in 1998.

G8 system, including to the ministerial, official, and multistakeholder bodies they created for this task. A variant of this school adds that "though G7 environment ministers began meeting regularly each year, this was not due to instructions from the summit, and it was not clear that the leaders paid much attention to their findings" (66).

Other schools see the G8 as far more effective, in more far-reaching ways, but with vastly different results. A second school, focusing on principled and normative direction setting, views the G8 as the centre of ecologically destructive disciplinary neo-liberalism. Stephen Gill (2000, 18, 1) argues that the G7 acts in a "deliberate and strategic manner" at the centre of a disciplinary neo-liberalism that "tends to atomise human communities and destroys the integrity of the ecological structures that support all life forms," thereby generating a "crisis of social reproduction on a world scale, a crisis that is ecological as well as social."

A third school, led by Steven Bernstein (2000, 2001), argues in a similar if less far-reaching fashion that the G8 has been complicit in converting the ecological values that flourished at the birth of the UNEP in 1972 into a new doctrine of "liberal environmentalism" in which economic and market values took pride of place. Bernstein asserts that the G7 began in 1985 with a legitimation of the market norms of liberal environmentalism, then moved by 1988–1989 to the more state interventionist, Keynesian approach of the Brundtland Commission Report, but shifted in 1990 to market mechanisms once again. The cause was the preference of Margaret Thatcher's United Kingdom and Ronald Reagan's United States, with an implantation of the latter's policies coming when the administration of George H. Bush arrived in 1989.

A fourth school argues, in contrast, that the G8 continually asserted and progressively developed the principles and norms of "embedded ecologism" (Kirton 2002a; Hattori 2007). Here the values of environmental enhancement, trade liberalization, and social cohesion occupied an equal and integrated place, even though market-oriented neo-liberalism flourished in the early 1980s and trade values dominated from 1994 to 1998. The latter was due to a G8 deference and delegation to the new, heavily legalized and organized, "hard law" but ecologically unfriendly World Trade Organization (WTO) (Kirton 2002d, 2004a; Abbott et al. 2000). A very strong version of this school from Takashi Hattori (2007, 78–9, 91) argues that the G8 at its 1988 and 2005 summits developed ever expanding and interlinked environmental norms within Japan's public and its government, led to pro-environmental institutional innovation in the latter, and educated a future Japanese prime minister whose government led on climate change at the third Conference of the Parties (COP) to the United Nations Framework Convention on Climate Change (UNFCCC). The G8's domestic political and direction-setting performance was thus powerful, in part because of Japan's emerging domestic policy by 2005 of integrating environment, economy, and international cooperation.

A fifth school emphasizes institutionalized international interdependence as driving the G8 to ever greater performance on critical issues such as climate change and biodiversity, through its decisional, delivery, and development of global governance tasks (Kokotsis 1995, 1999). Here higher levels of G8 commitment and compliance came from the availability of the multilateral UN institutions after its Rio summit in 1992, the growing array of G8 environmental institutions, and more powerful environmental bureaucracies in G8 member governments to help those governments transform their collective commitments into implementing action. This school suggests that G8 governments were also inspired to do so due to the personal commitment and political capital of individual G8 leaders, perhaps as they presciently and responsibly recognized that international interdependence required collective action to address emerging ecological problems that were inherently global, even if these problems currently harmed outsiders far more than G8 members and citizens themselves.

A sixth school claims, in contrast, that the G8 was a growing environment and climate governor but an implementation failure, due to its lack of a secretariat, the changing agenda of each host, and its trusting, unmonitored delegation of environmental issues to the World Bank and International Monetary Fund (IMF). James Barnes (1994, 1–2) writes:

> Not until environmental organizations began to put forward initiatives to the G-7 in the mid-1980s were environmental topics addressed seriously. Since 1985, environmental and development issues have been featured with increasing prominence. Yet, as with many other G-7 issues, there has been little follow-up by the member Governments. This is partially because there was no "secretariat" for the Summits, and the tendency of each host Government to focus upon the current year's agenda rather than reviewing the commitments and concerns of prior years ... [and] the continual referral of responsibilities by the G-7 to the World Bank and International Monetary Fund without simultaneously developing accountability mechanisms or reviewing how the Bretton Woods institutions have carried out their duties.

A seventh school sees the G8 more generally as a UN supporter, with varying effects. Joseph Aldy and Robert Stavins (2008) view the 2007 G8 Heiligendamm Summit, with its five invited guests of Brazil, China, India, Mexico, and South Africa (Group of Five, or G5, also sometimes referred to as "Plus Five"), as supporting the UN's climate governance by calling for a continuation of the UN approach in the post-Kyoto years. Others agree but judge the G8 too ambitious, due to its inappropriate adoption of the Kyoto approach of setting long-term targets and pressing to produce a similar, post-Kyoto regime. Andrew Green (2008) thinks that the G8 was wrong in 2007 to set a target of reducing emissions by 50 percent by 2050, as it should have focused on actions that could be done in the near term. Similarly José Goldemberg (2007) notes that the 2007 summit endorsed a Kyoto-like cap and trade approach, which was potentially very effective. Soledad Aguilar (2007) also notes that the G8 simply supported the UNFCCC as the appropriate forum for post-Kyoto negotiations, and did not do much by itself on climate change.

An eighth school presents the G8 as a UN supporter first, and fall-back governor should the UN fail. James Sebenius (1991, 111) notes that the 1990 G7 Houston Summit "endorsed expeditious framework negotiations" to produce the convention on climate change called for by the UN General Assembly in a unanimous resolution in December 1988. But he concludes that if this convention was unsuccessful, a small group of industrial states that had unilaterally committed to greenhouse gas control—the EU, European Free Trade Area, Japan, Australia, Canada, and others with stabilization or reduction targets—could produce a smaller-scale convention among themselves.

A ninth school, in contrast, points to the G8's informal governance as being appropriate from the start, due to the need for commitments to be voluntary and the refusal of the United States and other G8 partners to accept the sanctions that hard law regimes bring. Thomas Schelling (2002) argues that "neither the United States nor the other major developed countries will likely accept serious sanctions for missing emissions targets." A variant from George Kennan (1970) argues that a small group of 10 or so powerful and committed states, including the Soviet Union, had to act plurilaterally from the start to produce a badly needed global climate regime. Another variant from Peter Haas (2002, 77–80) suggests that G8 summits are effective on environmental and climate issues due to their small, ongoing, private nature, the nesting of G7 finance and trade ministers within international organizations such as OECD, IMF, and Bank for International Settlements, their lack of "profound political schisms," and the presence of popular concerns and perceived links to national interests. Similarly Manfred Milinski and his colleagues (2008) see the G8 as potentially creating climate cooperation because it embodies the small, non-anonymous groups that best preserve the common good.

A tenth school sees the G8 summit providing promising political leadership due to its members' mobilization of political will, in contrast to the pessimism that experts bring. Ruth Greenspan Bell (2006) notes that "many politicians are more optimistic. In July 2005, leaders of the group of eight highly industrialized states (G-8) pledged to put themselves 'on a path to slow and, as the science justifies, stop and then reverse the growth of greenhouse gases'."

The arrival of the G20 as a forum of finance ministers and central bank governors in 1999 and then as a leaders' summit in 2008 led to a focus on this broader plurilateral body and gave rise to a new debate about how and why the G20 did and could govern climate change.

Here the first school sees G20 failure. Jing Huang (2009, 439) argues that the G20 as an ad hoc, crisis-driven institution for networking and consensus formation has "yet to come up with any long-term plans or strategies for major global issues such as climate change, poverty eradication, water crisis, energy security, and food supply."

The second school sees potential effectiveness due to the G20's size, capability, leaders' prescience, issue linkage, and transparency. David Victor concluded that the G8 would not induce effective climate cooperation unless it included all the major carbon-producing powers (Victor et al. 2006; Victor 2006b). Victor (2006a, 101) suggested that Paul Martin's proposal for a leaders' twenty (L20) was "the most interesting idea" to cope with climate change because it offered the required small group, non-binding institution, and bottom-up approach. In the same spirit of broadening beyond the G8, from the political world Hillary Clinton (2007) held that "we must create formal links between the International Energy Agency and China and India and create an 'E-8' international forum modeled on the G8. This group would be comprised of the world's major carbon-emitting nations and hold an annual summit devoted to international ecological and resource issues." Peter Haas (2008b) saw the G8 plus China successfully creating technology cooperation for climate change mitigation. Once the G20 summit had arrived, Barry Carin and Alan Mehlenbacher (Carin and Mehlenbacher 2010a, 33) argued that the G20 is "as good as it gets" as a group with the potential to take effective action on climate change as it has the right number of members and amount of capability, and can be farsighted, create issue linkage, and fully disclose the values and interests of its members.

The third school sees G20 promise crowded out by financial crisis. David Victor and Linda Yueh (2010) assert that the G20, which "played the pivotal role in crafting new financial regulations after the Asian financial crisis, seemed to be a promising forum for addressing energy and climate issues as well, but topics such as the global economic meltdown of 2008 have crowded them out at the top of the agenda."

The fourth school sees the G20 as a proven performer in particular places. Joy Kim and Suh-Yong Chung (2012) argue that the G20 has effectively delivered on financing, phasing out fossil fuel subsidies, and engaging business stakeholders and has much potential to facilitate UN climate change negotiations. This potential is due to the G20's ability to build issue-specific coalitions, arising from its institutional flexibility, informal high-level dialogue, and ability to address diverse issues. Thijs Van de Graaf and Kirsten Westphal (2011, 28) agree that the G20's actions in phasing out fossil fuel subsidies show that it has great potential, based on its more representative membership than a G8 with a "good record" as an agenda setter and deliberative forum on climate change in the past.

The fifth school, arising more recently, sees the G20 producing results of broader importance in shaping a new multilateral regime. Sander Happaerts (2015) argues that discussions with rising powers in clubs such as the Major Economies Forum on Energy Security and Climate Change (MEF) and the G20 made it possible for those powers to consider making their own contributions to emissions reductions. The uncertainty of climate change and the diverse interests enhanced by the rise of big emerging powers in the failing UN regime led the major climate powers to "forum shop" to minilateral clubs such as the G20 and MEF. Starting in 2005, the G8 Plus Five led a strategy to separate the rising powers from the G77 and China to bring down the Kyoto "firewall," and develop a new regime based on a bottom-up, "pledge and review" approach. Within the UN, the Copenhagen Accord produced at the 2009 COP heralded the success of that approach (Happaerts 2015). Similarly, Sylvia Karlsson-Vinkhuyzen and Jeffrey McGee (2013, 74) conclude that such minilateral forums "appear to have exercised significant influence over global climate governance" by allowing powerful countries "to advocate a discourse supporting a more exclusive negotiation process and voluntary approach to mitigation commitments that appears to have flowed back into the UNFCCC process." Anthony Brenton (2013, 542) also argues that since 2005 the "Great Powers" began to meet more closely and that their cooperation contributed significantly to the Copenhagen Accord, "making their involvement crucial to the achievement of any worthwhile agreement." Maximillian Terhalle and Joanna Depledge (2013) agree on the need for the powers to cooperate but highlight the need for a multi-issue forum.

The sixth school emphasizes even more expansively the G20's past and potential central success in generating a new, effective climate control regime. Ross Garnaut (2014b, 223) argues that the G20 has a "track record" of leadership action on climate change with positive as well as "pending" outcomes. He argues that the G20 played an important role in setting the target of limiting temperature rise to no more than 2°C above pre-industrial levels at Copenhagen in 2009, in assisting domestic reform to reduce fossil fuel subsidies in some countries, and, at its 2013 St. Petersburg Summit, in reducing hydrofluorocarbon emissions. He thus concludes that the G20 is well suited to be the main forum to guide collective action on climate change, as it "contains the most influential developed and developing countries, can stand outside the entrenched and stereotypical divisions that have become barriers to effective action within the UNFCCC, with its huge and unwieldy membership and traditions of symbolic posturing" (2014a). Andrew Light (2009) similarly feels that the G20 had performed better than the UNFCCC in producing multilateral and bilateral success.

Puzzles

While each of these schools offers important insights, none adequately explains the particular pattern of how and why the G8 and then the G20 shaped global climate governance and thereby controlled climate change. In particular, none accounts for the failure of the G8, working with or without the UN, and reinforced by the Major Economies Meeting on Energy Security and Climate Change (MEM; later the MEF) and the G20 summits, to adequately control climate change after 2008, even as the world moved cumulatively toward a harmful temperature rise of more than 2°C.

With regard to the early years of G7 leadership, none fully accounts for the particular pattern of a reluctant America, especially under Republican presidents, adjusting through the G7/8 to the enthusiastic environmental leadership of Germany, Japan, Italy, and Canada, and of a Russia joining them in the Kyoto Protocol while leaving America alone on the outside. While Nicholas Bayne's collective management model of general G8 governance accurately accounts for the equal influence of the Europeans in inducing the Americans to adjust in the environmental field, it does not account for the critical role played in 2001 by the unlikely coalition of Japan, Canada, and Russia acting together, against and without George W. Bush's America, to bring the Kyoto Protocol to life. Nor does it account for the calculation of Tony Blair as host to the G8 plus the G5 at Gleneagles in 2005 that he could mobilize a newly re-elected Bush as an ally in the G8-led climate change control cause.

Such puzzles point to the need to conceive of the G8, always in potential and at key moments in practice, as a concert of equals in which any country can lead and align with any other, even against a strongly opposed and most powerful America, and thereby induce the US, and later countries outside the globally predominant club to adjust to produce a collective G8 and ultimately global result (Kirton 1999b). It further points to the need to conceive of the G8—with first another five members, then a full 17 in the form of the MEF, and, finally, as the G20—as an expanding club of equals acting as the hub of a global governance network for a globalized world (Kirton 2013a).

The value of such moves are reinforced by the puzzles presented by the case of climate change for all the traditional major explanatory models of G8 performance. The initial American leadership model developed by Robert Putnam and Nicholas Bayne (1987) cannot easily explain why effective G8 global environmental action flourished under Ronald Reagan and George H. Bush, but diminished in the latter Bill Clinton–Al Gore years and virtually disappeared during the early Barack Obama ones when the US had a committed president whose party controlled both houses of Congress. C. Fred Bergsten and C. Randall Henning's (1996) false new consensus model, with its portrait of impotent governments in a market-driven, globalizing world, cannot account for those periods and places of effective G8 environmental governance, in a world where ecological change is the ultimate global, domestically driven process, with complex, cumulative, interactive, and delayed effects that are difficult to anticipate, understand, or control. The model of democratic institutionalism developed by Ella Kokotsis (1995, 1999) has difficulty explaining why the G8's environmental performance and leadership appear to have declined after 1992, when the United Nations Conference on Environment and Development (UNCED) created more multilateral organizational capacity for implementation and when the G8 established its own annual environment ministers' forum. John J. Kirton's (1993) concert equality model has difficulty explaining the 1998–2004 decline in the climate change performance of an institution where Russia's inclusion had increased the club's global preponderance and ultimately internal equality both overall and in issue-specific ecological and energy capability, and where ecological interdependencies and inter-vulnerabilities had tightened in a finite world under increasing ecological stress.

These puzzles are only partly resolved by accounts of G20 summit governance since 2008 and by the systemic hub model that explains this performance (Kirton 2013a). On the surface, the G20 combines—as institutional equals in a single forum—the world's two leading climate polluters, China and the United States, and the world's major powers and polluters that cause and can thus control climate change. The first eight G20 summits from 2008 to 2013 dealt increasingly with climate change and started to shift the leadership of the world's evolving climate change control regime from the prevailing UN-led, divided, development-first one to a new, inclusive, equal, environment-first one. But as its ninth summit in 2014 dramatically revealed, the G20 faced severe internal divisions and remained unable to meet the challenge by creating a full control regime that worked, either outside or inside the UN.

Purpose

In the face of such puzzles arising from the existing accounts, this book has three primary purposes. The first is to chart the course of leadership in global climate governance in a complete way by adding the work of consequentially involved informal PSIs to a literature that overwhelmingly emphasizes the formal multilateral organizations of the UN system and the formal legal conventions and protocols they produced. This account thus focuses on the contribution of the G8 and the G20. In so doing, it seeks to build upon and broaden, rather than repeat, the rich analyses in the many important books already produced on UN-centred global governance of climate change (see Bodansky 2001; Hoffmann 2005; Weiss and Jacobson 2000; Rowlands 1995; Haas 2008b; Harris 2001; Bernstein 2001).

The second purpose of this book is to examine systematically the cadence and causes of the G8- and G20-centred contributions to climate change control. It identifies and describes, in as much detail as possible, the G8's initiation, shaping, and support of global climate governance, along with the many times G8 leaders and their G20 colleagues withdrew from the playing field to focus on other things. To chart G8 performance carefully over these phases and to identify the key trends, this book applies to G8-centred climate governance the six dimensions of performance highlighted by the concert equality model of general G8 governance and by the systemic hub model of G20 governance: domestic political management, deliberation, direction setting, decision making, delivery, and the development of global governance (Kirton 2013a). To explain such performance, it both conducts a detailed process tracing of the diplomacy ahead of, at, and after key G8 and G20 summits, especially in relationship to the UN, and fully applies the six causes highlighted by the concert equality model of successful G8 governance: shock-activated vulnerability; multilateral organizational failure; predominant equalizing capability; common democratic purpose; political cohesion based on leaders' domestic control, capital, continuity, commitment, competence, and public support; and the constricted, controlled participation of a cherished club. It adds the new cause contained in the systemic hub model of G20 governance—the G20 as an institutional and emerging interpersonal club that serves as the hub of a growing network of global summit governance for a globalized world (Kirton 2013a).

This study extends the concert equality and systemic hub models in several ways. It explores how visible shocks from the same or similar fields can catalyze G8 action in an issue area whose problem is characterized by chronic, cumulative, incremental and often invisible change and how shocks in different, less directly related areas can have a diversionary or general galvanizing effect. It also examines the competence of leaders to comprehend and act on the inevitably incomplete and contested scientific consensus about a phenomenon characterized by complexity, uncertainty, and the other attributes of complex adaptive systems. It further highlights individual leadership—how a single leader acquires the initial awareness of the problem and consequent willingness to act, and does so successfully, in highly proactive and prescient ways, long before subsequent shocks or shock sequences in the same issue area force slow, cybernetically learning leaders to adapt and long before a clear, scientific consensus forms.

The third purpose of this book, and in some ways the most important one, is to explore how informal PSIs, led by the G8 and G20, and the formal multilateral ones, above all the UN, interact to produce global governance on an issue as complex and challenging as climate change. In keeping with the anti-bureaucratic bias of G7 leaders, Kirton (1989, 1993) specified in the initial and revised concert equality model that the absence or failure of multilateral organizations in governing a particular problem generated G8 governance success (see also Putnam and Bayne 1987; Karlsson et al. 2012). In sharp contrast, in her book on G7 compliance with commitments made at G7 summits on the issues of climate change, biodiversity, debt relief to the poorest, and assistance to the former Soviet Union, Kokotsis (1995) argued that the presence of a powerful, dedicated, multilateral organization, such as the IMF in the field of finance, helped generate high G7 performance in the key performance dimension of delivery, while the absence of such an organization—as in climate change, especially prior to 1992—caused low compliance. However, subsequent work by Kirton and colleagues on the causal catalysts of compliance showed, in keeping with Kokotsis's democratic institutionalist model, that when G8 leaders referred in their commitment to the core, dedicated, multilateral organization on the subject, G8 members' compliance with that commitment generally rose (Kirton 2006a). A subsequent effort to construct and apply a full-scale framework to analyze the relationship between the G8 and multilateral organizations found that under different conditions, the G8 could work to produce a successful governance result, with or without the relevant organization (Kirton et al. 2010). A similar conclusion arises from recent work on the relationship between the G8 and G20 (Larionova and Kirton 2015).

This book thus offers the first analysis of the G8/G20-UN relationship in the field of climate change. It conducts a comprehensive, detailed exploration of the dynamics of how the G8/G20 and the UN interrelate. It does so in a changing context where new international institutions—PSIs as well as multilateral organizations—enter the field. It also highlights and develops an understanding of the dynamics of institutionally embedded regimes operating in cooperation and competition in an uncertain world to determine which will prevail (Raustiala and Victor 2004; Kirton 2013a). This book thus provides the first systematic account of leadership in global climate governance that properly puts the G8 and G20, in its relationship with the UN, at the centre.

The Argument

This book argues that the G8 created, retreated from, and returned with the new G20 to lead global climate change governance. It did so by inventing and later reinvigorating an effective, inclusive, equal, environment-first control regime that was very different than the failing, divided, inequitable, development-first UN one dominating from 1989 to 2004. This cadence of the creation, retreat, and return of G8 leadership in global climate governance was caused primarily by the shocks from energy and acid rain in the 1970s and extreme weather events since 2005, by the absence of UN climate governance in the 1970s and its increasing ineffectiveness since 2005, and by the growing globally predominant, internally equalizing capability of the G7 from 1975 to 1988 and the more inclusive G8 plus G5 since 2005 and G20 since 2008. Also important were the slow convergence of G8, G5, and G20 members on democratic principles, the political control, capital, commitment, competence, and civil society support of key G7 leaders, and the growth of the G8, G5, and G20 as cherished clubs at the hub of a network of global summit governance needed to protect all from their mounting vulnerability to the consequences of uncontrolled climate change.

The Phases of G8 Leadership

G8 leadership in global climate governance unfolded quite unevenly, through the three major phases of creation, retreat, and return, with distinct periods within each phase.

Creation, 1979–1988

The first phase, of creation from 1979 to 1988, saw the G7 invent an effective, inclusive, ambitious, environment-first control regime through its own direct, informal, voluntary action. It put the environment first by targeting climate change as a key condition in support of energy security, along with energy conservation, efficiency, and alternatives as key principles. It ambitiously set a carbon control target and timetable as zero increases immediately in carbon concentrations in the atmosphere. It included all consequential carbon-polluting powers in its actions, if not its institution, to meet this goal. It did so through informal but politically obligatory oil-import quotas and more comprehensive, bottom-up, sectoral actions by G7 members, and by bold economic and security measures to defeat the outside carbon powers of the Soviet Union and bloc and the major oil-producing emerging powers of the time. It proved effective in moving toward its goal, as the emissions of OECD members declined from 1979 to 1985, the emerging oil powers with their development-first demands for a New International Economic Order (NIEO) were defeated, and the Soviet Union and bloc were driven to the brink of collapse by 1988.

This phase of creation encompassed three distinct periods of G7 governance. The first was a period of invention from 1979 to 1984. During this time German chancellor Helmut Kohl, supported by America's Jimmy Carter and Canada's Joe Clark, led their colleagues at the 1979 Tokyo Summit to the consensus on stabilizing atmospheric carbon and sulphur concentrations immediately. They implemented this consensus at subsequent summits through more direct sectoral controls on G7 members in 1980–1981 and then resistance to the demands of the consequential carbon-intensive communist and emerging oil-rich countries from 1982 to 1984.

The second period saw a rediscovery from 1985 to 1986. At the second German-hosted Bonn Summit in 1985, the G7 returned to govern climate change directly and made its first precise, politically obligatory commitment on the subject, which members complied with well during the following year.

The third period featured a revival from 1987 to 1988. At their second Venice Summit in 1987, G7 leaders started their continuing deliberations on climate change. At Toronto in 1988, they expanded their climate agenda and called for the creation of the Intergovernmental Panel on Climate Change (IPCC).

Retreat, 1989–2004

The second phase, of retreat from 1989 to 1984, saw the G7 step back from direct global climate leadership of its own in favour of shaping and then supporting the emerging UN-centred regime. As it developed through the UNFCCC created in 1992 and its Kyoto Protocol produced in 1997, this alternative UN-centred regime increasingly became a divided, development-first, ineffective one. Despite the apparent equality for the environment and the economy contained in the name of the UN Conference on Environment and Development at Rio de Janeiro in June 1992, development came first, under the defining principle of sustainable development. Here environmentalists were given the conditioning adjective of ecologically "sustainable" while the developing countries and community were given the noun economic "development" as the central goal.

This UN regime offered a top-down approach featuring an overall global goal but initially set virtually no specific targets or timetables, let alone any goals sufficiently ambitious to control the steadily compounding carbon emissions and concentrations and the global warning they brought. Under the principle of common but differentiated responsibilities, it divided the world into a small group of developed countries that were obliged to control their emissions and a much larger group of developing countries that were not. The latter were thus left free and incentivized to develop in their prevailing, preferred carbon-intensive way. Employing a formal, legal, broadly multilateral convention-protocol approach, the UN regime began with generalities that were to be progressively but slowly developed into more specific measures at annual meetings of the members who had consented to join—a group that excluded from the precise Kyoto Protocol produced in 1997 the world's largest carbon polluter and power, then the United States as an operational carbon constrainer, and a China that assumed this position less than a decade hence. The result was an increasing failure to take the needed action, especially as the little group of countries that had agreed in 1997 to begin controlling their emissions by a little largely did not do so by 2004.

This phase of G7 retreat in favour of the UN and its alternative approach contained three distinct periods. The first was a reinvention from 1989 to 1992. Starting at its Paris Summit in 1989, where environmental subjects took up one third of the communiqué, the G7 did much through its increasing climate commitments to reinvent the regime for global climate governance by initially shaping and subsequently decisively supporting the creation of the compromised but acceptable alternative UN regime at UNCED in 1992.

The second period saw a reinforcement of the Rio regime from 1993 to 1997. The G7 reduced its attention to climate change but improved its members' compliance with their G7 climate commitments and increased its support for implementing the divided UN regime. This culminated at the June 1997 Denver Summit of the Eight, which helped create the UN's Kyoto Protocol in December of that year.

The third period featured a retreat from 1998 to 2004. This retreat came in two ways. First, G7 attention and commitments on climate change dropped to very low levels after the Kyoto Protocol arrived. Second, the G7 started to shift from simply supporting the UN regime to increasingly call for an inclusive one in which all carbon-producing and thus polluting powers controlled their carbon.

Return, 2005–2014

The third phase, of a return to direct G8-centred climate leadership from 2005 to 2014, saw the G8 shift from supporting the increasingly ineffective UN regime toward an enhanced version of its initial, inclusive, ambitious, environment-first, effective, G8-led regime. That regime now centred on informally agreed action by the G8, with its new, increasingly equal G5, MEM/MEF, and G20 partners in direct support. It put climate change as the central focus, whose control was backed by a broad array of environmental, energy, economic, and other measures. It ambitiously set a specific target and timetable of a temperature rise no greater than 2°C above pre-industrial levels and a 50 percent reduction in G8 greenhouse gases by 2050. Above all, it secured a consensus and partial commitment from all G5, MEM, and G20 powers that they would also control their carbon, with China in the lead. It again did so through informal but politically obligatory sectoral actions by all partners, contentiously but consensually agreed to at G8-centred summits expanding to eventually include as equals all the major carbon powers combined in the G20.

This evolving new regime promised to be effective. G8-wide and US greenhouse gas emissions declined each year since 2005. There was now hope that the newly constrained major emerging carbon powers and polluters would eventually follow this G8 and now G20 lead. In 2014, global carbon emissions dropped for the first time in 40 years without an economic downturn to help.

This return to the principles of a G8-led, inclusive regime, now including all G20 members, unfolded through four distinct periods. The first saw a restoration from 2005 to 2007 of the basic principles of the initial G7-led regime of 1979–1988, with the G5 powers of China, India, Brazil, South Africa, and Mexico now included at the G8 summits, starting at Gleneagles in 2005 and continuing through to Heiligendamm in 2007. These three years featured soaring summit attention to, priority for, and commitments on climate change, and the related development of global governance inside and outside the G8.

The second period involved reaching out from 2008 to 2009 beyond the G5 to bring the emerging powers of Korea, Indonesia, and Australia into the summit and regime. They were included along with the G5 powers as members equal to the G8 in the MEM, which was elevated to a summit in 2008 and renamed the MEF in 2009, and then in the new G20 summit, which started actively governing climate change in 2009. The G8's 2009 L'Aquila Summit saw the G8 give climate change the greatest attention, priority, and link to the G8's foundational democratic and human rights principles.

The third period saw the realization from 2010 to 2011 that the UN regime was failing. This was dramatically displayed by the shock of the disappointing COP-15 in Copenhagen in December 2009, the arrival of the ad hoc Copenhagen Accord cobbled together at the last minute there by the US, China, India, South Africa, and Brazil, and the entry of the G20 summit into global climate governance in a sustained, expanding way. The economically oriented G20 brought the vision of a green economy into essential support of climate change control. The UN's stunning Copenhagen failure brought reduced but still robust G8 climate governance at Muskoka in 2010 and Deauville in 2011, while that of the G20 rose amidst a new financial crisis erupting in Europe at Toronto in June 2010, Seoul in November 2010, and Cannes in November 2011.

The fourth period featured the increasing replacement from 2012 to 2014 of the principles and leadership of the old divided UN regime by the new inclusive, expanded, equal, G8- and G20-centred one as the dominant centre of global governance on climate change. This process of replacement first saw the G20 emerge alongside and then ahead of the G8 as the PSI leader on climate change control. It then saw the G8 replaced by the G7 alone at its summit in 2014, as Russia's invasion and annexation of the Crimean region of Ukraine disrupted but did not delay the G7's revival of climate leadership, arising on the way to the landmark COP meeting to replace the Kyoto regime in Paris in December 2015. As 2015 began, the inherited divided UN regime was steadily being replaced in principle and practice by the inclusive, expanded, equal G8- and G20-led one.

The Causes of G8 Leadership

This three-phase cadence of G8 and then G20 climate governance—from creation through retreat to return—was driven largely by the six causes highlighted by the concert equality model of G8 governance and the one cause added by the systemic hub model of G20 governance (Kirton 1989, 1993, 2013a).

According to the concert equality model, awareness of the vulnerability of major powers is activated by intensifying new nonstate shocks that cannot be controlled through unilateral actions taken by the major powers or by the old multilateral organizations that they control. Their leaders thus mobilize their globally predominant powers through balanced bargains among equals to respond, on the basis of shared core principles. They can also secure the consent of their domestic polities to invent and agree on appropriate responses with their powerful colleagues in their increasingly cherished, compact summit clubs. The expanded model of systemic club governance adds that the leaders' personal summit club increasingly acts as the hub of a network of global systemic governance that spreads their influence to produce effective, legitimate global governance as a whole.

Shock-Activated Vulnerability
The first cause was shock-activated vulnerability, specifically the cluster, sequence, and size of shocks on the same and similar subjects that showed all G8 and then G20 members, especially the most powerful United States and the major emerging powers led by China, their equalizing vulnerability to the destructive, deadly, and even possibly existential new threat of climate change (Kirton 2004a, 1993). Such shocks, defined generally by surprise, sudden eruption, and threat, create the new issues, crises, and uncertainty that lead states to choose informal intergovernmental organizations to respond and lead these organizations to perform well (Vabulas and Snidal 2013). These shocks came in three categories.

The first category contained shocks with the same content as climate change and its major component of global warming. These most clearly took the form of abnormally severe and persistent heat waves and droughts that produced warming felt directly by many humans and caused visible destruction, particularly to agriculture and food, and death to humans through malnutrition and dehydration and other direct health impacts.

The second category comprised shocks similar in content and closely causally connected to climate change, but requiring cognitive, social, and political construction by G8 leaders and publics to forge the causal link between the content of the shock and its climate change cause, consequence, and required control response. Such shocks included visible, destructive, and deadly extreme weather events such as hurricanes, tsunamis, typhoons, floods, storm surges, and other sudden sea-level rises. They extended to oil-price spikes and abrupt supply cut-offs, oil spills from tankers and drilling rigs, nuclear reactor accidents that suddenly transmitted deadly particles through the atmosphere around the world, and the acid rain from coal burning and transportation that produced visibly denuded forests and dead lakes.

The third category consisted of shocks quite different in content from climate change, at least before the causal complexities were comprehended and the range of uncertainty reduced. They included war, revolution, terrorism, financial and economic crisis, and acute outbreak events in health. Such shocks tended to divert G8 leaders' immediate attention from climate change toward these other visibly urgent concerns. But they could also enhance the G8 and G20 clubs' general cumulative institutional and interpersonal solidarity in ways that made agreement easier in all issue areas, including climate change. They could also bring semi-intended and recognized benefits for climate change control, such as the widely applauded swift, large-scale end of the Soviet Union, Soviet bloc, Soviet system, and Soviet ideology after 1989, and the widely reduced economic growth after the economic and financial crises in 1979, 1997, and 2008. Such crises could be mobilized by influential agents through new cognitive constructions for climate change control, such as with the G20's post-2008 emphasis on creating green growth.

Such dramatic shocks politically activated in politicians, elites, and publics the awareness of the vulnerability of people, property, nature, and other valued things. This activation was more likely to come from subsequent shocks with the same content or similar content to the initial or previous ones, even if they might now be smaller in size, scale, or scope. The memory of the death and destruction caused by the earlier shock, and the failure of the traditional policy approaches in response, meant that subsequent shocks spurred increasingly sensitized, usually slow-learning G8 leaders and supportive publics to take faster, more ambitious, innovative action. This was especially the case if the shocks struck their own, highly capable countries with the destruction and death that demonstrated how vulnerable they now were to such new, nonstate security threats—those unintended, untargeted threats brought by nonstate or even non-human actors that flowed often unpredictably from anywhere to anywhere through pathways that no one fully understood. Climate change was a classic case (Kirton 1993).

Such vulnerability was usually a chronic and less visible but still potentially costly condition, as was the growing sensitivity to globalization and interdependence on which this vulnerability partly rested (Keohane and Nye 1977). Among the leading same- or similar-subject vulnerabilities spurring global climate governance were the acid rain that was transmitted through the atmosphere across international borders to create visibly denuded and dead trees and dead fish, and the smog that visibly choked cities and killed citizens in capitals such as Beijing. Beyond lay the full array of greenhouse gas emissions, sequestrations, and concentrations largely invisible to the naked eye and thus reliant on scientific instruments, satellites, authoritative consensual knowledge, and visual media such as television to be rendered credibly real.

Thus the first phase of G8 climate leadership—the creation of its inclusive regime from 1979 to 1988—was driven by the oil supply, price, and tanker spill shocks of 1979, following those from 1973, the famine in Africa, the nuclear explosion at Three Mile Island in the US in 1979, and the visible vulnerability of European and North American trees and lakes to acid rain. The shock of the Soviet invasion of Afghanistan at the end of 1979 and the ensuing new cold war soon diverted G7 attention but brought carbon-reducing benefits from a global economy in decline, until the 1985 Bonn Summit linked the two subjects. Then the deadly heat wave afflicting the US in the summer of 1988 led the G7's Toronto Summit to have the G8 govern climate change ever since.

The following 15 years from 1989 to 2004 brought a G8 retreat to support the UN regime. This was due to the diversionary if positive shocks of the Cold War victory bringing a democratizing Russia into the G8, the absence of serious oil shocks with sustained effects, and effective action on acid rain. Also important were the diversionary shocks of the Asian-turned-global financial crisis from 1997 to 2002 and the al Qaeda terrorist attack on the US on September 11, 2001. The short-lived oil spike brought by the 1990–1991 Gulf War and the 1986 Chernobyl nuclear explosion were insufficient to overcome G8 confidence that the UN had the formula and time to slowly produce an effective, if divided, development-first climate change control regime.

From 2005 to 2014, the return of inclusive, now expanded G8-led and then G20-led climate governance was driven by the cadence of increasingly severe and widespread climate shocks. These had begun with deadly heat waves in France, elsewhere in Europe, and in India, as well as the Asian tsunami in late December 2004, rising oil prices, Hurricane Katrina afflicting the US in late August 2005, and, as solidarity-creating diversionary shocks, the rising post-September 11 terrorist attacks including the attack in London as the UK hosted the Gleneagles Summit on July 7, 2005.

Multilateral Organizational Failure
The second cause of the G8-centred creation, retreat, and return of global leadership on climate change was the recurrent failure of the multilateral organizations emerging from the 1940s UN system to control a climate change problem that was becoming cumulatively worse and approaching critical thresholds by 2005.

Constant core constraints came from the presence of the economically devoted IMF and World Bank as the pre-eminent multilateral organizations of the post-World War II order, the absence of any recognition of the existence or value of the natural environment in the 1945 UN charter, and the lack of a functional UN organization dedicated to environmental protection or energy as the World Health Organization was for health (Biermann and Bauer 2005; Lesage et al. 2010) The failure of the World Meteorological Organization from 1950 and UNEP born in 1972 to take up climate change control at the political or leaders' level left the G7 to create global climate governance from 1979 to 1988. A G7 that trusted the UN then helped create, shape, and support the IPCC after 1989, the UNFCCC's COP in 1992, the Kyoto Protocol's Meeting of the Parties (MOP) in 1997, the UNFCCC Secretariat in Bonn, and the UN sustainable development summits in 1992, 1997, and 2002. This UN action led the G8 to retreat to support the UN and its divided, development-first regime, shaped by international institutions where the developing country majority was decisive. Only with the evident failure of this UN regime by 2005, peaking at the Copenhagen COP in 2009, did the G8 return to lead with its own expanded, inclusive, equal regime. It did so first with the Gleneagles Plus Five in 2005, then the MEM/MEF summits in 2008–2009 and finally the G20 summits in 2009–2013. These new PSIs containing all the major carbon-producing powers, ultimately as equal members, allowed for the emergence and effectiveness of the new G8-led, expanded, inclusive, equal regime. They also provided direct solutions, as on phasing out fossil fuel subsidies and shaping the new UN-beyond-Kyoto regime due to arrive in December 2015.

Predominant Equalizing Capability
The third cause of the G8-centred creation, retreat, and return of global climate leadership was the changing configuration of globally predominant and internal equality in capability of a G7 and then G8, expanding into a G20. During the creation of its own inclusive climate regime from 1979 to 1988, the G7 had considerable global predominance and substantial internal equality among its members, especially with the outside communist and developing world in relative decline after 1979. From 1989 to 2004, when the G8 retreated to let the UN lead, the G8's global predominance and internal equality slowly declined, as the strong, sustained rise in relative capability of the G5 powers increasingly overwhelmed the G7's addition of Russia in 1998, the earlier unification of Germany, and the periodic expansion of the European Union (Kirton 2013a). From 2005 to 2014, that strong relative capability rise of the G5, as well as other MEM/MEF and G20 powers, and the G8's growing success in incorporating them as peers in an enlarged summit group allowed the G8, now with the

G20, to return to global climate leadership with its expanded, inclusive, equal, environment-first regime.

Common Principles
The fourth cause of the creation, retreat, and return of G8 climate leadership was the changing democratic convergence of the G7 and G8 plus members. The all-democratic G7 pioneered global climate governance from 1979 to 1988. Then, with post-Cold War globalization creating a more politically open global community and the G7 slowly incorporating a reforming Russia into a now slightly less democratic G8, an optimistic G7 was content to leave global climate leadership to the more multilateral UN, now more democratically like-minded than in the earlier confrontational Cold War–NIEO years. After 2005, the democratic dominance of the expanded G8 plus G5, MEM/MEF, and G20 allowed the G8 to return to expanded, inclusive, equal global leadership, even with the constraints of a still communist China, authoritarian Saudi Arabia, and recidivist Russia in 2014.

Political Cohesion
The fifth cause of the G8-centred creation, retreat, and return of climate leadership was the changing domestic political cohesion of members. This cohesion was composed of the leaders' political control, capital, and public support or pressure, and their personal continuity, competence, and convictions about climate change. The creation from 1979 to 1988 was brought by the domestic political dominance, personal convictions, public support, and G7 experience of co-founder Helmut Schmidt of Germany, and continued by his successor Helmut Kohl at Bonn in 1985 and beyond. Necessary support came from Canada's Joe Clark, America's Jimmy Carter, and Britain's Margaret Thatcher, with expertise in chemistry, and Europeans pushed by Green parties, in Germany above all. During the G8's retreat from 1989 to 2004, new leaders with limited convictions about the need to control climate change arrived in Britain under John Major, Russia under Boris Yeltsin and Vladimir Putin, Japan under Junichiro Koizumi, and the US under George H. Bush from 1989 to 1992 and, above all, George W. Bush from 2001 to 2004. From 2005 to 2014, the return of climate leadership was fuelled by Britain's re-elected Tony Blair in 2005, Germany's Angela Merkel (a physicist who had served as Kohl's environment minister before assuming the chancellorship in 2005), Barack Obama as US president since 2009, his supportive G20 colleagues from Korea and Mexico, and China's new president Xi Jinping.

Constricted, Controlled Participation in a Club at the Network Hub
The sixth cause of this cadence of G8 creation, retreat, and return was the changing constricted, controlled participation within the G8 and then G20 club, operating at the hub of a growing network of global summit governance for a globalized world. The G7 created global climate governance from 1979 to 1988 when it was a compact club of initially six members that added an environmentally committed and competent Canada in 1976 and the EU in 1977. Its leaders stood at the hub of a network of global PSI governance including the summits of the Commonwealth, NATO, and La Francophonie after 1985. The retreat in favour of UN leadership from 1988 to 2004 coincided with the G7's increasing inclusion of the Soviet Union and then Russia, followed by a growing, changing array of invited countries and multilateral organizations in 1996 and from 2001 on, even if the surrounding network of PSIs rapidly increased in those post-Cold War years. The G8's return to climate

leadership was caused by its continuing but carefully controlled expansion to the G5, then MEM/MEF, and, finally, the still constricted G20 with a constant membership from 2008 to 2014.

The Outline

To develop this argument, this book proceeds as follows. Part II, "Creating the Exclusive G7 Regime," covers the first phase of creation from 1979 to 1988. Chapter 2, "Invention, 1979–1984," examines the G7's invention of global climate governance in 1979, when it called for an end to increases of carbon concentrations in the atmosphere, and its follow-up in the years to 1984. Chapter 3, "Rediscovery, 1985–1986," focuses on the rediscovery of climate change in 1985 and its follow-up in 1986. Chapter 4, "Revival, 1987–1988," explores the surge of G7 global climate leadership that started to create what became the successor and ultimately different UN-led regime.

Part III, "Retreating to Support the Divided UN Regime," covers phase two from 1989 to 2004. Chapter 5, "Reinvention, 1989–1992" examines the G7's initial post-Cold War surge to create the UN regime at UNCED in Rio de Janeiro in 1992. Chapter 6, "Reinforcement, 1993–1997," charts the G7's full retreat and shift to support the global climate leadership of the UN, its UNFCCC, and its Kyoto Protocol in 1997. Chapter 7, "Retreat, 1998–2004," examines this retreat through to the near death of G8 climate governance at George W. Bush's Sea Island Summit in 2004.

Part IV, "Pioneering the Inclusive Global Regime," covers phase three from 2005 to 2014. Chapter 8, "Restoration, 2005–07," explores the increasing return of the basic principles of the initial G7-led regime, now with the G5 powers included as regular participants at the expanded G8 summits, Gleneagles in 2005, St. Petersburg in 2006, and Heiligendamm in 2007. Chapter 9, "Reaching Out, 2008–2009," examines how the G8 brought first the G5, then the emerging powers of Korea, Indonesia, and Australia, and, finally, Saudi Arabia, Turkey, and Argentina into the expanded regime as equals, under the new MEM and G20 summits in 2008. Chapter 10, "Realization, 2009–10," shows how the shock of the failure of the UN and the ad hoc Copenhagen Accord in December 2009 was followed by reduced but still robust climate governance by the G8 at Muskoka in 2010 and Deauville in 2011, and increasingly by the G20 at Toronto in June 2010, Seoul in November 2010, and Deauville in November 2011. Chapter 11, "Replacement 2012–14," shows how the old, divided UN regime was slowly but steadily being superseded by the new inclusive, environment-first, G8-centred one, coming from the G20 as the emerging dominant centre of global governance on climate change.

Part V, "Conclusion," contains Chapter 12, which summarizes the key arguments in this analysis and connects the three-phase cadence of G8 climate leadership to the six causes of the revised concert equality model. It does so by quantitatively relating changes in the causes to those in the six dimensions of G8 and G20 summit performance. It also suggests the prospects and possibilities for global climate governance in the years ahead.

Methods and Materials

This analysis of G8- and G20-centred leadership on climate change follows a multi-method approach. For each G8 and G20 summit it engages, insofar as possible, in detailed process tracing to show how the host planned and prepared the summit and how the members and participants arrived at, advanced, and adjusted their positions in the lead-up to and at the summit, as well as the major achievements that the summit produced. It also employs input-output matching through a more systematic quantitative analysis of the six major dimensions of performance at each summit: domestic political management, deliberation, direction setting, decision making, delivery through compliance with and implementation of the commitments, and the development of global governance both inside the G8-G20 and outside, above all at the UN and multilateral organizations as a whole.

In general, in evaluating a summit's overall performance, as the years covered in this study unfold, greater emphasis is placed on the subsequent sequence of six dimensions of performance, with delivery privileged throughout. The performance on these dimensions and the outcomes of the process tracing are related to the condition at the time of the six major causes of the concert equality model of G8 governance, with its supplement from the systemic hub model of G20 governance. The causes shared by both models—starting with shock-activated vulnerability—are considered the most salient, but with compact club cohesion acquiring more salience in the later years, and individual leadership a critical component throughout. The result is a compact model of six effects (the performance dimensions) and six causes (of the concert equality/systemic hub models) assessed against 49 G7/8 and G20 summits over 40 years. It provides the best available account of how the political complex adaptive system of G8-G20 summitry governed the physical complex adaptive system of climate change.

To identify the impact of G7/8 and G20 governance on the broader UN-based multilateral system and regime, and the reverse influence flows that followed, this study relies on input-output matching. It shows in particular how principles and proposals first adopted at the leaders' level in the G7/8 and G20 were subsequently accepted collectively by leaders and ministers in their climate-relevant meetings at the UN, if often with some delay, in less detail, and in a distinctive form. It largely does not trace the process of how G7/8 and G20 members individually and collectively operated within the UN to produce this broader change. Other scholars have now begun to do so in the case of the G20 (Brenton 2013; Happaerts 2015; Karlsson-Vinkhuyzen and McGee 2013; Terhalle and Depledge 2013). Nonetheless, this initial process-tracing scholarship and the input-output matching conducted for this book are sufficient to support the bold claims that it makes. But there is much left for scholars and others to do, to close this important causal gap.

The materials for this study, particularly the detailed process tracing of G7/8 and G20 summitry, come from scholarly works on the G8, G20, and global governance of climate change, publications by and autobiographies and biographies of G8 and G20 participants, policy commentaries, newspaper and other media accounts, and the documents in the archives of the G8 Research Group and G20 Research Group at the John Graham Library at Trinity College in the University of Toronto. In addition, this study draws on extensive private interviews, conversations, participation in policy dialogues, and discussions of drafts with key G8 and G20 participants, from the leaders' level on down, embracing those from all members of the G8 and almost all members of the G20. John J. Kirton has had several hundred such encounters with G8 participants since 1977 and more than a hundred with G20 participants since 1999. Ella Kokotsis has conducted dozens more, primarily from 1995 to 1997. Kirton and Kokotsis have attended virtually all G8 and G20 summits since 1988, serving as accredited members of the media to cover the event, attend the briefings and scrums, and access the materials available only at the event. The quantitative evidence on the dimensions and causes of summit performance, including the extensive data set on compliance, have been largely assembled since 1988 by the researchers, research assistants, and members of the G8 Research Group and G20 Research Group. These continually improving data sets are available at the G8 Information Centre.[2]

The Issue Area of Climate Change

For this study, focused on the international institutional context of the G8, G20, and UN, the issue area of climate change includes climate change itself, greenhouse gases, global warming, the UNFCCC and Kyoto Protocol, and carbon emissions. It includes the Global Environment Facility (because one of its six main functions is to provide grants for climate change projects in developing countries). It includes sustainable development (because the UN defines climate change as falling within the scope of sustainable development). It usually excludes energy, energy efficiency, clean energy, nuclear energy, or alternative energy in general, unless they are linked to climate change adaptation, mitigation, or risk.

The concept of leadership used in this volume is the classic one advanced by Oran Young (1991), embracing structural, directional, and idea-based leadership (Underdal 1994; Carin and Mehlenbacher 2010b; Karlsson et al. 2012). The concept of regime employed, expressed in the six dimensions of performance, is based on the classic one codified by Stephen Krasner (1983, 2) as "implicit or explicit principles, norms, rules, and decision-making procedures around which actors' expectations converge in a given area." It adds the subsequent recognition that issue areas such as climate change are constructed, not given, that expectations matter primarily when they lead to action, and that decision-making procedures include the creation of international institutions within which regimes are embedded, operated, and creatively changed.

References

Abbott, Kenneth W., Robert Keohane, Andrew Moravcsik, et al. (2000). "The Concept of Legalization." *International Organization* 54(3): 401–20.

Aguilar, Soledad (2007). "Elements for a Robust Climate Regime Post-2012: Options for Mitigation." *Review of European Community and International Environmental Law* 16(3): 356–67.

Aldy, Joseph E. and Robert N. Stavins (2008). "Climate Policy Architectures for the Post-Kyoto World." *Environment* 50(3): 7–17.

Barnes, James (1994). *Promises, Promises! A Review: G7 Economic Summit Declarations on Environment and Development.* Washington DC: Friends of the Earth.

Bayne, Nicholas (2000). *Hanging In There: The G7 and G8 Summit in Maturity and Renewal.* Aldershot: Ashgate.

Bell, Ruth Greenspan (2006). "What to Do about Climate Change?" *Foreign Affairs* 85: 105–13. http://www.foreignaffairs.com/articles/61710/ruth-greenspan-bell/what-to-do-about-climate-change (January 2015).

---

2 Data sets on the G8 and G20 are available at the G8 Information Centre at http://www.g8.utoronto.ca, which also contains an index of the G8 and G20 archives housed at the John Graham Library at Trinity College in the University of Toronto.

Bergsten, C. Fred and C. Randall Henning (1996). *Global Economic Leadership and the Group of Seven*. Washington DC: Institute for International Economics.

Bernstein, Steven (2000). "Ideas, Social Structure, and the Compromise of Liberal Environmentalism." *European Journal of International Relations* 6(4): 464–512. doi: 10.1177/1354066100006004002.

Bernstein, Steven (2001). *The Compromise of Liberal Environmentalism*. New York: Columbia University Press.

Biermann, Frank and Steffen Bauer, eds. (2005). *A World Environmental Organization: Solution or Threat for Effective International Environmental Governance*. Aldershot: Ashgate.

Bodansky, Daniel (2001). "The History of the Global Climate Change Regime." In *International Relations and Global Climate Change*, Urs Luterbacher and Detlef F. Sprinz, eds. Cambridge MA: MIT Press, pp. 23–39.

Brenton, Anthony (2013). "'Great Powers' in Climate Politics." *Climate Policy* 13(5): 541–46.

Carin, Barry and Alan Mehlenbacher (2010a). "Constituting Global Leadership: Which Countries Need to Be Around the Summit Table for Climate Change and Energy Security?" *Global Governance* 16(1): 21–38.

Carin, Barry and Alan Mehlenbacher (2010b). "Constituting Global Leadership: Which Countries Need to Be Around the Summit Table for Climate Change and Energy Security?" *Global Governance* 16(1): 21–37.

Clinton, Hillary Rodham (2007). "Security and Opportunity for the Twenty-First Century." *Foreign Affairs*, November/December. http://www.foreignaffairs.com/articles/63005/hillary-rodham-clinton/security-and-opportunity-for-the-twenty-first-century (January 2015).

G7 (1975). "Declaration of Rambouillet." Rambouillet, November 17. http://www.g8.utoronto.ca/summit/1975rambouillet/communique.html (January 2015).

G7 (1976). "Joint Declaration of the International Conference." San Juan, June 28. http://www.g8.utoronto.ca/summit/1976sanjuan/communique.html (January 2015).

G7 (1977). "Appendix to Downing Street Summit Conference." London, May 8. http://www.g8.utoronto.ca/summit/1977london/appendix.html (January 2015).

G7 (1978). "Declaration." Bonn, July 17. http://www.g8.utoronto.ca/summit/1978bonn/communique (January 2015).

G7 (1979). "Declaration." Tokyo, June 29. http://www.g8.utoronto.ca/summit/1979tokyo/communique.html (January 2015).

Garnaut, Ross (2014a). "G20 Should Facilitate International Cooperation on Climate Change." *East Asia Forum*, April–June pp. 33–35. http://press.anu.edu.au/wp-content/uploads/2014/06/whole.pdf (January 2015).

Garnaut, Ross (2014b). "The G-20 and International Cooperation on Climate Change." In *The G-20 Summit at Five: Time for Strategic Leadership*, Kemal Dervis and Peter Drysdale, eds. Washington DC: Brookings Institution Press, pp. 223–45.

Gill, Stephen (2000). "The Constitution of Global Capitalism." Paper presented at the annual convention of the International Studies Association, March 15, Los Angeles.

Goldemberg, José (2007). "Energy Choices Toward a Sustainable Future." *Environment* 49(10): 7–11.

Green, Andrew (2008). "Bringing Institutions and Individuals into a Climate Policy for Canada." In *A Globally Integrated Climate Policy for Canada*, Stephen Bernstein, Jutta Brunnée, David Duff, et al., eds. Toronto: University of Toronto Press, pp. 246–57.

Haas, Peter M. (2002). "UN Conferences and Constructivist Governance of the Environment." *Global Governance* 8(1): 73–91.

Haas, Peter M., ed. (2008). *International Environmental Governance*. Aldershot: Ashgate.

Happaerts, Sander (2015). "Rising Powers in Global Climate Governance: Negotiating Inside and Outside the UNFCCC." In *Rising Powers and Multilateral Institutions*, Dries Lesage and Thijs Van de Graaf, eds. New York: Palgrave Macmillan, pp. 317–42.

Harris, Paul (2001). *International Equity and Global Environmental Politics: Power and Principles in US Foreign Policy*. Aldershot: Ashgate.

Hattori, Takashi (2007). "The Rise of Japanese Climate Change Policy: Balancing the Norms of Economic Growth, Energy Efficiency, International Contribution, and Environmental Protection." In *The Social Construction of Climate Change: Power, Knowledge, Norms, Discourses*, Mary Pettenger, ed. Aldershot: Ashgate, pp. 75–97.

Hoffmann, Matthew (2005). *Ozone Depletion and Climate Change: Constructing a Global Response*. Albany: SUNY Press.

Huang, Jing (2009). "A Leadership of Twenty (L20) Within the UNFCCC: Establishing a Legitimate and Effective Regime to Improve Our Climate System." *Global Governance* 15(4): 435–41.

Karlsson, Christer, Mattias Hjerpe, Charles Parker, et al. (2012). "The Legitimacy of Leadership in International Climate Change Negotiations." *Ambio* 41(1): 46–55. doi: 10.1007/s13280-011-0240-7.

Karlsson-Vinkhuyzen, Sylvia I. and Jeffrey McGee (2013). "Legitimacy in an Era of Fragmentation: The Case of Global Climate Governance." *Global Environmental Politics* 13(3): 56–78.

Kennan, George (1970). "To Prevent a World Wasteland: A Proposal." *Foreign Affairs* 48 (April): 401–13. http://www.foreignaffairs.com/articles/24149/george-f-kennan/to-prevent-a-world-wasteland (January 2015).

Keohane, Robert O. and Joseph S. Nye (1977). *Power and Interdependence: World Politics in Transition*. Boston: Little, Brown.

Kim, Joy and Suh-Yong Chung (2012). "The Role of the G20 in Governing the Climate Change Regime." *International Environmental Agreements: Politics, Law, and Economics* 12(4): 361–74.

Kirton, John J. (1989). "Contemporary Concert Diplomacy: The Seven-Power Summit and the Management of International Order." Paper presented at the annual convention of the International Studies Association, 29 March–1 April, London. http://www.g8.utoronto.ca/scholar/kirton198901 (January 2015).

Kirton, John J. (1993). "The Seven Power Summits as a New Security Institution." In *Building a New Global Order: Emerging Trends in International Security*, David Dewitt, David Haglund, and John J. Kirton, eds. Toronto: Oxford University Press, pp. 335–57.

Kirton, John J. (1999). "Explaining G8 Effectiveness." In *The G8's Role in the New Millennium*, Michael R. Hodges, John J. Kirton, and Joseph P. Daniels, eds. Aldershot: Ashgate, pp. 45–68.

Kirton, John J. (2002a). "Embedded Ecologism and Institutional Inequality: Linking Trade, Environment, and Social Cohesion in the G8." In *Linking Trade, Environment, and Social Cohesion: NAFTA Experiences, Global Challenges*, John J. Kirton and Virginia W. Maclaren, eds. Aldershot: Ashgate, pp. 45–72.

Kirton, John J. (2002b). "Winning Together: The NAFTA Trade-Environment Record." In *Linking Trade, Environment, and Social Cohesion: NAFTA Experiences, Global*

Challenges, John J. Kirton and Virginia W. Maclaren, eds. Aldershot: Ashgate, pp. 79–99.

Kirton, John J. (2004). "Generating Effective Global Environmental Governance: The North's Need for a World Environmental Organization." In *A World Environmental Organization: Solution or Threat for Effective International Environmental Governance?*, Frank Biermann and Steffen Bauer, eds. Aldershot: Ashgate, pp. 145–74.

Kirton, John J. (2006). "Explaining Compliance with G8 Finance Commitments: Agency, Institutionalization, and Structure." *Open Economies Review* 17(4): 459–75.

Kirton, John J. (2013). *G20 Governance for a Globalized World*. Farnham: Ashgate.

Kirton, John J., Julia Kulik, Caroline Bracht, et al. (2014). "Connecting Climate Change and Health Through Global Summitry." *World Medical and Health Policy* 6(1): 73-100. doi: 10.1002/wmh3.83.

Kirton, John J., Marina V. Larionova, and Paolo Savona, eds. (2010). *Making Global Economic Governance Effective: Hard and Soft Law Institutions in a Crowded World*. Farnham: Ashgate.

Kokotsis, Eleanore (1995). "Keeping Sustainable Development Commitments: The Recent G7 Record." http://www.g8.utoronto.ca/scholar/kirton199503/kokotsis/index.html (January 2015).

Kokotsis, Eleanore (1999). *Keeping International Commitments: Compliance, Credibility, and the G7, 1988–1995*. New York: Garland.

Krasner, Stephen (1983). *International Regimes*. Ithaca, NY: Cornell University Press.

Larionova, Marina V. and John J. Kirton, eds. (2015). *The G8-G20 Relationship in Global Governance*. Farnham: Ashgate.

Lesage, Dries, Thijs Van de Graaf, and Kirsten Westphal (2010). *Global Energy Governance in a Multipolar World*. Farnham: Ashgate.

Light, Andrew (2009). "Showdown among the Leaders at Copenhagen." December 18, Center for American Progress. https://www.americanprogress.org/issues/green/news/2009/12/18/7049/showdown-among-the-leaders-at-copenhagen/ (January 2015).

Milinski, Manfred, Ralf D. Sommerfeld, Hans-Jürgen Krambeck, et al. (2008). "The Collective-Risk Social Dilemma and the Prevention of Simulated Dangerous Climate Change." *Proceedings of the National Academy of Sciences* 105(7): 2291–94. doi: 10.1073/pnas.0709546105.

Putnam, Robert and Nicholas Bayne (1987). *Hanging Together: Co-operation and Conflict in the Seven-Power Summit*. 2nd ed. London: Sage Publications.

Raustiala, Kal and David G. Victor (2004, April). "The Regime Complex for Plant Genetic Resources." *International Organization* 58(2): 277–309.

Rowlands, Ian H. (1995). "The Climate Change Negotiations: Berlin and Beyond." *Journal of Environment & Development* 4(2): 145–63. doi: 10.1177/107049659500400207.

Schelling, Thomas C. (2002). "What Makes Greenhouse Sense?" *Foreign Affairs*, May/June. http://www.foreignaffairs.com/articles/58002/thomas-c-schelling/what-makes-greenhouse-sense (January 2015).

Sebenius, James K. (1991). "Designing Negotiations Toward a New Regime: The Case of Global Warming." *International Security* 15(4): 110–48.

Strong, Maurice (1973). "One Year After Stockholm: An Ecological Approach to Management." *Foreign Affairs*, July, pp. 690–707. http://www.mauricestrong.net/images/docs/one%20year%20after%20stockholm%281973%29.pdf (January 2015).

Sustainable Energy Development Center (2006). *Global Energy Security: Summary of Russia's G8 Presidency*. Moscow: Sustainable Energy Development Center.

Terhalle, Maximilian and Joanna Depledge (2013). "Great-Power Politics, Order Transition, and Climate Governance: Insights from International Relations Theory." *Climate Policy* 13(5): 572–88.

Underdal, Arild (1994). "Leadership Theory: Rediscovering the Arts of Management." In *International Multilateral Negotiation: Approaches to the Management of Complexity*, I. William Zartman, ed. San Francisco: Jossey-Bass, pp. 178–97.

Vabulas, Felicity and Duncan Snidal (2013). "Organization Without Delegation: Informal Intergovernmental Organizations (IIGOs) and the Spectrum of Intergovernmental Arrangements." *Review of International Organizations* 8(2): 193–220.

Van de Graaf, Thijs and Kirsten Westphal (2011). "The G8 and G20 as Global Steering Committees for Energy: Opportunities and Constraints." *Global Policy* 2: 19-30. doi: 10.1111/j.1758-5899.2011.00121.x.

Victor, David G. (2006a). "Recovering Sustainable Development." *Foreign Affairs* 85(1): 91–103.

Victor, David G. (2006b). "Toward Effective International Cooperation on Climate Change: Numbers, Interests, and Institutions." *Global Environmental Politics* 6(3): 90–103.

Victor, David G., Amy M. Jaffe, and Mark H. Hayes, eds. (2006). *Natural Gas and Geopolitics: From 1970 to 2040*. Cambridge: Cambridge University Press.

Victor, David G. and Linda Yueh (2010). "The New Energy Order." *Foreign Affairs* 89(1): 61–73.

Weiss, Edith Brown and Harold K. Jacobson, eds. (2000). *Engaging Countries: Strengthening Compliance with International Environmental Accords*. Cambridge MA: MIT Press.

Young, Oran R. (1991). "Political Leadership and Regime Formation: On the Development of Institutions in International Society." *International Organization* 45(3): 281–308. doi: 10.1017/S0020818300033117.

PART II
Creating the Exclusive G7 Regime

Chapter 2
Invention, 1979–1984

Introduction

The Challenge

As the 1970s unfolded, there arose growing scientific concern, and initial policy interest, in the harmful consequences of human-induced climate change. Previous prevailing scientific estimates that the earth was due for a new phase of global cooling slowly gave way to an evolving consensus that human-created greenhouse gas emissions would lead to harmful global warming. Prominent policy figures such as George Kennan in 1970 and Crispin Tickell in 1977 pointed to the dangers. Yet, with considerable scientific uncertainty and only initial, sporadic policy-related attention at the international level, there was no interest in creating an intergovernmental regime to control climate change.

Since its start in 1975, the Group of Seven (G7) had acted indirectly to control carbon emissions and thus climate change through the general principle and instrument of energy conservation. It did so at its inaugural summit at Rambouillet, France, in November 1975 and again at its second summit in San Juan, Puerto Rico, in the United States in 1976 (Kirton 2002a). This was reinforced by its calls for energy efficiency and energy alternatives at the G7 summits in London in 1977 and Bonn in 1978. At the latter, which Germany's Social Democrat chancellor Helmut Schmidt chaired, G7 leaders first noted that it was vital to pay attention to environmental protection in the process of energy development. The proximate cause of this explicit conditionality was the recent shock in March 1978 of the *Amoco Cadiz* oil tanker, which ran aground off the coast of France and disgorged 1.6 million barrels of crude oil into the sea. It was the largest crude oil spill thus far.

From this initial, ever strengthening energy-environment nest, climate change emerged directly as an issue for G7 governance at the Tokyo Summit on June 28–29, 1979. There, amidst its focus on energy, the G7 invented global climate governance for the world. Its leaders instituted the first, most ambitious, and most effective global carbon control regime the world has ever seen. On June 29, 1979, at the conclusion of their summit, the leaders boldly declared: "we need to expand alternative sources of energy, especially those which will help to prevent further pollution, particularly increases of carbon dioxide and sulphur oxides in the atmosphere" (G7 1979). They thus acknowledged the need to halt immediately, at 1979 levels, the concentration of carbon dioxide and sulphur oxides in the world's atmosphere. In doing so they established the issue of climate change on the G7 and global policy agenda, acceptance of the scientific fact that it was occurring, and the core elements of its cause and its cure. They also identified, at least implicitly, the appropriate targets, timetables, baselines, and participation procedures, and did so in ways more ambitious than those that arose since (Kirton 2008–2009). In the following five years, they and their partners in the Organisation for Economic Co-operation and Development (OECD) moved in the desired direction, as their carbon emissions into the atmosphere regularly declined (Sustainable Energy Development Center 2006; Ikenberry 1988). G7 governance did all this before the United Nations system at any political or official policy level had recognized the fact, the causes, the consequences, and the need for control of climate change. The G7 thus created the global governance of climate change.

The Debate

The G7's historic contribution to global climate governance in 1979 has gone unnoticed by mainstream accounts of global climate governance and by scholars of the G7 as well. The general literature on global environmental and climate governance, focused firmly on the UN system, devotes its attention to the 1972 Stockholm conference, from which climate change was absent as an issue, and to scientific studies fostered by the World Meteorological Organization (WMO) (Hoffmann 2005; Bernstein 2001). The literature on the G7 focuses on Tokyo's great achievements in countering the second oil shock, now from revolutionary Iran, while leaving the environmental implications or advances untouched (Putnam and Bayne 1984, 1987). The puzzle is why, four years after Rambouillet, when the G7 leaders had explicitly forged the energy-environment connection in their communiqué, did scholars of G7 summits not probe more deeply into possible advances when hosting passed to the environmentally committed countries of Germany in 1978 and Japan in 1979.

The Argument

This chapter argues that in 1979 the G7 summit created the global governance of climate change, forming a control regime more ambitious and effective than any before or since. It continued to control carbon through its moves on the economy, energy, and the environment, with diminishing force after 1980 until Germany and Britain identified the environment as an issue in its own right at the 1984 London Summit. This cadence was driven by a confluence of severe climate-relevant, cognate energy shocks in 1979, the failure of multilateral organizations to recognize climate change as an issue at a political or policy level, the predominant capability and internal equality of G7 members in climate control capability, and the G7 continuity and personal commitment of Germany's Helmut Schmidt. The post-1980 decline in summit attention and action was caused by the diversionary shocks of the Soviet invasion of Afghanistan and consequent start of the new cold war, the rise of terrorism, and the arrival of a deep recession in G7 economies, the latter of which helped to deliver the G7's 1979 carbon control goals.

The Preparatory Process for the 1979 Tokyo Summit

The Background, 1975–1978

Unlike the UN, the G7 was created with environmental values as an integral part of its founding charter, codified in its informal inaugural summit communiqué (Kirton 2002a). The G7 steadily strengthened its environmental dedication, still with a focus on energy, at its summits in 1976, 1977, and 1978. The *Amoco Cadiz* oil spill in March 1978 provided shocking visual evidence of the environmental damage caused by oil. Acid rain, which stripped the forests in Europe and North America, showed in a more chronic but cumulative, visually compelling fashion that coal also caused environmental damage up in the atmosphere and then damage back down on earth. The next step was to identify and act on the invisible processes much further up in the atmosphere by directly addressing the emissions that caused climate change. This the G7 leaders did at Tokyo in 1979.

Helmut Schmidt's Leadership

The G7 leaders' bold pronouncement on carbon dioxide and sulphur oxides at Tokyo was the product of a German and American initiative, coming together at the summit with Canadian support. In Germany, by the first half of 1979, Social Democrat chancellor Helmut Schmidt had become so concerned about rising carbon dioxide emissions that he raised the issue in his television addresses to his citizens at home. He was driven in part by the visible damage from acid rain in Germany's Black Forest and by the appearance of a Green Party that threatened to bleed electoral support from his governing Social Democrats. As a former finance minister, Schmidt had long disliked Germany's expensive coal subsidies. He perhaps saw a campaign against sulphur oxides–creating acid rain and carbon dioxide, another coal-based atmospheric pollutant, as a popular way to reduce subsidies, over the opposition of the coal miners who formed a core part of his political base.

As the summit approached, Schmidt spoke out publicly about climate change during his pre-summit tour of the United States. On June 6, 1979, he journeyed across the US on his way to meet President Jimmy Carter in the White House to discuss the energy crisis and the G7 summit to be held in Tokyo on June 28–29. In an interview in the US edition of *Time*, published on June 11, Schmidt spoke directly about the problem of climate change and the need for its control:

> Foreseeably, we will within the next one or two decades get into a worldwide debate about the irrevocable consequences of burning hydrocarbons—whether oil or coal or lignite or wood or natural gas—because the carbon dioxide fallout, as science more or less equivocally tells us, results in a heating up of the globe as a whole. This leads to the third point, namely the necessity to put up rather large sums of money in order to develop scientifically, and from the engineering side, sources of energy like nuclear, geothermal, solar energy, all of which enable us to avoid the CO2 consequences (Schmidt 1979a).

It was clear from Schmidt's reference to "heating up of the global as a whole" that he saw global warming, rather than particular parts afflicted by acid rain, as the ultimate threat.

Jimmy Carter's Support

In the United States, the administration of Democratic president Jimmy Carter had shown steadily growing interest at home in the science and associated implications of climate change. It sponsored the first report of the United States National Academy of Sciences on the subject in 1979. In the immediate lead-up to the Tokyo Summit, the director of the Office of Science and Technology asked the National Academy of Sciences "to summarize in concise and objective terms our best present understanding of the carbon dioxide/climate issues for the benefit of policy-makers" (US National Academy of Sciences 1979, ix). The report was delivered after a gathering of scientists at Woods Hole, Massachusetts, on July 23–27, one month after the G7 summit had taken place. It starkly concluded:

> We now have incontrovertible evidence that the atmosphere is indeed changing and that we ourselves contribute to that change. Atmospheric concentrations of carbon dioxide

are steadily increasing, and these changes are linked with man's use of fossil fuels and exploitation of the land. Since carbon dioxide plays a significant role in the heat budget of the atmosphere, it is reasonable to suppose that continued increases would affect climate (US National Academy of Sciences 1979, vii).

The report thus concluded that climate change was an important problem that governments must take seriously and act upon (Speth 2004).

During the summit preparatory process, on April 9, the Americans prepared a paper on "Carbon Dioxide Build-Up and World Climate Change." It was distributed to the personal representatives ("sherpas") of their G7 partners. Its content was consistent with that of the later US National Academy of Science report. The paper noted that the best scientific evidence suggested that the combustion of fossil fuels was creating an accumulation of atmospheric carbon dioxide that would lead to climate change and difficult decisions on energy, industrial, and agricultural matters within the next decade or two. It predicted that, if current trends continued, there would be serious, adverse global impacts flowing from changing temperature and rainfall. Developing countries with dense populations in particular would suffer from a decline in their ability to produce food.

By May 10, America's summit partners were aware, at the highest level, of the growing concern over atmospheric pollution and climate change and the prospective place of this issue as one of the summit's top five priorities. Their sherpas discussed whether to have leaders deal with the issue as part of their political discussions over dinner or during a plenary session that would be attended also by foreign and finance ministers. The Germans, in a coalition government with the Free Democratic Party, whose leader Hans-Dietrich Genscher was foreign minister, favoured discussing it in the plenary.

Canadian and Japanese Support

Support for a leaders' discussion of climate change came from Canada. Its ministry of the environment, Environment Canada, had participated in the pooling of scientific evidence on which the Americans' conclusions were based. Canadian prime minister Joe Clark was advised that the best available science indicated that under present trends there would be serious adverse global effects due to carbon build-up, which would bring drought in Western Canada, weather vulnerability in Eastern Canada, and severe reductions in food production in the densely populated developing world. He was further told that the related problem of sulphur dioxide emissions leading to acid rain was of rapidly mounting concern to Canada. Thus the summit might like to underline the global significance of the climate change problem, the need for scientific cooperation, and the need for energy and development policies to act in a supportive way. As the summit drew near, Clark indicated his support for greater attention and action on environmental issues there.

The Japanese found it easy to accept this German, American, and Canadian thrust. Given the close friendship between Schmidt and French president Valéry Giscard d'Estaing, it is likely that Schmidt's advocacy for the issue helped bring France on board, especially as France was well equipped with nuclear power rather than having to rely on carbon-generating coal to produce its energy at home. Moreover, 1979 was the first summit that would be attended by Schmidt's fellow European leader, British prime minister Margaret Thatcher. As a leader prepared to confront her country's powerful coal unions, and one familiar with chemistry, she was well positioned to provide support (Thatcher 1993).

The sherpa for the leader of the European Community was a British national named Crispin Tickell. Two years earlier Tickell (1977) had spent time at Harvard University producing a study on what he concluded was the real and serious problem of climate change. With this study, the European Community was well equipped to provide strong support to its major member, Germany.

The broader pattern of alignment at the summit fostered this consensus. Germany and Japan cooperated on North-South relations and nuclear safety. Carter's visit to Japan from June 24 to 27 fostered agreement on reducing oil consumption through setting targets and developing alternative energy.

One week before the summit started, on June 22, US sherpa Henry Owen advised Carter that:

> you might remind Schmidt of this *Time* magazine statement that he wants to see a lot more money go into developing new energy technologies. He is particularly interested in investing more money in solar energy, since he fears that increased use of coal in any form will add dangerously to the carbon dioxide levels in the atmosphere. We don't yet know enough to judge in what degree his fears are well founded. He believes that after the environmentalists weary of nuclear energy, they will turn to CO-2 (Owen 1979).

At the 1979 Tokyo Summit

At the opening working session of the summit, on the morning of June 28 at the Akasaka Palace, climate change issues were raised in terms of the energy crisis facing the leaders. As host, Japanese prime minister Masayoshi Ohira made the opening statement, followed by remarks by Carter, Italian prime minister Giulio Andreotti, and then Giscard, who said:

> In the short term the only alternative sources of energy were coal and nuclear power, and on these the meeting should express its determination to increase production. Everybody was concerned about the safety of nuclear power, but this should not be an absolute pre-condition of nuclear development (Minutes 1979).

Then Schmidt spoke. His intervention was recorded as follows:

> The environmentalists sought to prevent the greater use of coal, and they were also fighting, with the help of some court decisions, his Government's plans to increase the production of nuclear energy … Governments would have to spend far more than hitherto on pure and applied research on alternative energy sources so that by the turn of the century we were using not only nuclear energy but solar energy and, possibly, geothermal energy. He personally foresaw that in the early part of the next century there would be pressure not to use hydro-carbons any more because of the dangers of over-heating the outer atmosphere (Minutes 1979).

Thatcher added that "70 per cent of Great Britain's electricity was produced from coal and only 15 per cent from oil. In the longer term we must make much more use of nuclear energy" (Minutes 1979).

After the European Community's Roy Jenkins spoke, Joe Clark intervened. His remarks were recorded as follows:

An expression of serious concern about the energy situation would help Governments like his own who had to introduce unpopular conservation measures. Alternative sources of energy often raised environmental problems. There was, for example, evidence in acidic rain resulting from the use of coal ... The Canadian Government was actively developing an energy policy for the 1990s embracing substitution and conservation (Minutes 1979).

The available evidence thus suggests that Schmidt was in the lead on the issue of climate change. While he and Carter had a difficult personal relationship, reducing the probabilities of collaboration at the summit, at Tokyo Schmidt held bilateral meetings with Clark and Ohira (Schmidt 1989, 266; Carter 1982, 110–20). Schmidt was also part of the extensive European Community effort to coordinate its members' policy on energy security (Putnam and Bayne 1984, 1987).

Moreover, Schmidt's report on the summit to the Bundestag on July 4 noted his bilateral meetings at Tokyo with Clark and with Ohira, whom he had known for many years. Schmidt (1979b) stated that these "in both cases reflected the excellent state of our relations" (see also Deutscher Bundestag 1979). He also declared: "Furthermore, the combustion of coal, oil and gas has led to a greater overall burden on the environment. During the last three decades the world-wide emission of carbon dioxide has increased three fold. The possible consequences for the world's climate—for example in the Sahel zone but not only that region—cannot yet be estimated reliably, but they will have to be taken into consideration in long-term energy policy decisions." Schmidt acknowledged that the short-term need, given the energy crisis, was to rely more heavily on Germany's own abundant coal. But he also declared that in "two or three decades" the world should stop using coal for energy and start to develop immediately, as the summit had agreed, "inexhaustible sources of energy which are less harmful to the environment."

Importantly, Schmidt noted the implications of the dual shocks of the oil crisis that prompted the search for energy alternatives, and the nuclear reactor explosion at Three Mile Island near Harrisburg, Pennsylvania, in March that threatened to end nuclear energy as an attractive alternative. But he did so in the following very particular way:

As for the accident risks inherent in the peaceful use of nuclear energy, we have again been made fully aware of these by what happened at Harrisburg. But the accident risks as regards coal, for example, are also very high ... a figure which I find very disturbing, especially since it is not based on theoretical calculation but reflects very sad, actual losses of human life ... When speaking of coal, we must bear in mind that coal miners have a hard and dangerous job. During the 30 years between 1949 and 1978, the number of miners who lost their lives in German coal mines ... amounted to no less than 15,500. Let me repeat that: during those 30 years, 15,500 miners lost their lives! Anyone who claims that coal is a source of energy which can be used without harm to mankind is clearly speaking without thinking" (Schmidt 1979b; see also Deutscher Bundestag 1979).

For America's Jimmy Carter (1979), the equivalent post-summit national address was his fifth speech to the nation on July 15, 1979. Carter referred directly to the Tokyo Summit, declaring that America would reduce oil imports even below the quotas agreed there. But this determination to comply was to be met by switching to coal as "our most abundant energy source," and to oil shale. There was no reference to any environmental restraints on this switch. Indeed, the environment was mentioned only once, as follows: "We will protect our environment. But when this Nation critically needs a refinery or a pipeline, we will build it." There was no reference to carbon or climate change at all.

In what others soon labelled his "malaise" speech, Carter (1979) referred at length to the "shocks and tragedy" that drove these moves. He stated:

> We were sure that ours was a nation of the ballot, not the bullet, until the murders of John Kennedy and Robert Kennedy and Martin Luther King, Jr. We were taught that our armies were always invincible and our causes were always just, only to suffer the agony of Vietnam. We respected the Presidency as a place of honor until the shock of Watergate. We remember when the phrase "sound as a dollar" was an expression of absolute dependability, until 10 years of inflation began to shrink our dollar and our savings. We believed that our Nation's resources were limitless until 1973, when we had to face a growing dependence on foreign oil. These wounds are still very deep. They have never healed (Carter 1979).

A repeatedly and multiply shocked America under Carter thus turned to coal with no carbon or environmental constraints at all.

Dimensions of Performance

In its climate change performance, the 1979 Tokyo Summit was a striking, historic success, given the state of the physical problem, consensus knowledge about it, and the context of global governance at the time.

Deliberation and Direction Settings

As a result of German leadership, the summit, with little difficulty, made its first, far-reaching, if very brief, statement on climate change. It did so in a paragraph of 28 words, taking 1.3 percent of the communiqué (see Appendices A-1 and C-1). It did so in the context of energy, atmospheric pollution, and sulphur dioxide in particular. It took the form technically, in the specific language used in the communiqué, of a direction-setting consensus rather than a decisional commitment with substantial precision and obligation (Abbott et al. 2000). But its content was clear, and more ambitious than anything before or since. The baseline was zero—the current year. The timetable was zero—action now. The target was zero—stabilization of concentrations of carbon in the atmosphere and not merely of the emissions into it. The components were comprehensive in the sense that stabilization was the focus rather than emissions, although the energy nest and concern with only carbon dioxide placed a heavy emphasis on sources rather than sinks. The temporal focus was entirely on mitigation rather than adaptation, as prevention was the only principle affirmed. Participation was entirely restricted as only "we" the G7 powers were asked to act, with no recognition that anyone outside could or should assist. This was G7 governance alone, for the global good and for that of the developing countries that—the Germans, Americans, and Canadians had agreed—would be harmed.

Delivery

Even more strikingly, the consensus was delivered, fully, by all members of the G7, insofar as they physically could for the following five years. In part this was because leaders returned home to aggressively implement their energy alternatives and efficiency commitments as well as their national targets on containing their imports of oil, and thus the carbon that its combustion contained. In the case of Jimmy Carter's America, a key instrument used was oil-price decontrol and a resulting price rise. Here the summit consensus was critical to inducing a previously unwilling Congress to finally make the move (Ikenberry 1988). Reinforcement came from several more targeted measures, related to transportation and buildings, highlighted at the Venice Summit the following year (Kirton 2009a). Compliant action on these helped the effective fulfilment of the commitments. Also important was conscious G7 macroeconomic policy action to have the G7 and global economy defeat inflation and the power of an oil- and resource-rich South that was asserting its commodity power over the North. In the five years following the G7 consensus at Tokyo, the emission of carbon dioxide by recession-ridden G7 and OECD countries into the atmosphere declined (Sustainable Energy Development Center 2006, 48; Barnes 1994, 42).

Development of Global Governance

Helping ensure such effective, multi-year delivery was the Tokyo Summit's institutional development of global climate governance within the G7 (see Appendix F-2). Here energy was the nest. The 1979 summit created three official-level bodies devoted to energy in environmentally enhancing and climate-controlling ways: the High Level Group on Energy Conservation and Alternative Energy, the International Energy Technology Group, and the compliance-focused High Level Group to Review Oil Import Reduction Progress. The following year the 1980 Venice Summit added two more: the International Team to Promote Collaboration on Specific Projects on Energy Technology and the High Level Group to Review Results on Energy. G8 institutionalization thus embraced the demand and supply side of the energy source component, and the instruments of alternatives, technology, and trade. Also contributing was a meeting of G7 energy ministers held in Paris in 1979.

The 1980–1984 Demise

Following the G7's pioneering move at Tokyo in 1979, climate change in its own right disappeared as an issue in the summit's communiqué for the following five years. However, there was a continuing concern with energy and with the broader environmental issues that embraced and affected climate change. Indeed, environmental concern expanded from energy to broader resources and the overall economy. The sustained appearance of the environment in its own right began in 1984, covering all ambient media (air, water, land) apart from living things. This helped expand G8 climate governance in several ways.

Venice, 1980

At Venice in 1980, the introductory preambular "chapeau" of the G7 communiqué identified energy as central to the solution of all other problems. G7 leaders expressed confidence that these problems would be met by "our democratic societies, based on individual freedom and social solidarity" (G7 1980). They endorsed energy conservation, alternatives, reduced consumption, oil-consumption ceilings (as well as production increases), and a wide range of measures in the sectors of power generation, building, transportation, nuclear, solar, and coal to give this effect. While committing to double coal production, they promised, as

Schmidt had wished in 1979, to "do everything in our power to ensure that increased use of fossil fuels, especially coal, does not damage the environment." The 1981 Montebello-Ottawa communiqué repeated this approach, while adding geothermal and biomass to the specified list of alternatives endorsed. The 1982 Versailles communiqué, in a brief paragraph, endorsed this approach more generally, but with a strong emphasis on nuclear, coal, and technology and no direct attention to the environment at all.

During this time, however, the G7 continued to create and use official-level, limited-duration working groups that addressed energy, the environment, and thus climate change. In 1980, it continued (eventually to 1985) the work of the High Level Group on Energy Conservation and Alternative Energy that it had created in 1979. The second 1979 creation—the International Energy Technology Group—ended its work in 1980. But that year the G7 created the International Team to Promote Collaboration on Specific Projects on Energy Technology. Thus, at the official level, the G7 provided continuous climate-related governance, in the form of energy conservation and technology during the "dark ages" when climate itself disappeared from the leaders' own communiqués.

Montebello, 1981

In 1981, responsibility for hosting the summit passed to Canada, now led by Pierre Trudeau, who had returned to power with his third majority mandate after his electoral defeat of Joe Clark. In the lead-up to the summit, a background paper was prepared for the meeting in Washington of the G7 leaders' personal representatives on October 14, 1980. Entitled "Global Population, Resource and Environmental Problems: The United States' Response," it dealt with the just-released Global 2000 report that had been commissioned by the Carter administration. The 11-page paper stated:

> Internationally, the U.S. has just signed an agreement with Canada to study and assess transboundary air pollution ... The problem of inadvertent climate modification is high on the U.S. environmental agenda. Operating within the framework of a new National Climate program, the Department of Energy is carrying out a major research effort on the climate effects of carbon dioxide accumulation in the atmosphere. U.S. experts are contributing to a variety of bilateral and multilateral efforts to address this same problem within the new World Climate Program. Stratospheric ozone depletion is another matter of concern.

Within the G7, leadership on climate change now passed to the US and Canada, which had been supportive from the start. Yet the Montebello Summit, preoccupied with East-West relations, the economy, and development, did little on the environment and nothing directly on climate change.

Versailles, 1982

The shock-driven G7 attention to energy in its climate-related components, which had diminished sharply in 1981, disappeared in 1982. In both cases, the immediate economic recession took centre stage. Moreover, the dominant energy issue at Versailles, over the participation of the energy-dependent Europeans in a Soviet gas pipeline project, arose in the context of East-West relations during the rising new cold war. The bitter divisions between the socialist host French president François Mitterrand willing to engage with the

Soviets and the Republican US president Ronald Reagan, who resisted, preoccupied the summit, which ended in failure as a result.

Williamsburg, 1983

At Williamsburg in 1983, with Reagan as G7 summit host, energy took one paragraph in the communiqué, in addition to a brief reference to directing official development assistance to reduce energy poverty in the developing world. The energy paragraph endorsed conservation and alternatives despite the falling price of oil. The more general issue of the environment reappeared, with a health link, and took a paragraph of its own. The leaders said: "We have agreed to strengthen cooperation in protection of the environment, in better use of natural resources, and in health research" (G7 1983). The latter reference may have had atmospheric change, notably stratospheric ozone depletion, as a possible cause of the increasing incidence of melanoma.

London, 1984

The descent into the dark ages began to reverse at Margaret Thatcher's London Summit in 1984. Here Germany, supported by Britain, took the lead. Indeed, Germany's leadership was explicitly acknowledged in the compliment to it in the communiqué. In its public deliberation, London for the first time gave the environment a full, stand-alone multi-sentence paragraph in the communiqué, released on June 9, 1984. It identified the three ambient environmental media—the atmosphere, water, and land—that pollution harmed.

In developing G7 environmental governance, London also made a major advance. It instructed ministers responsible for environmental policies to identify areas for continuing cooperation. It invited the existing Working Group on Technology, Growth, and Employment to identify specific areas for environmental research to limit pollution of air, water, and land, and to report on progress by the end of the year. Importantly, the leaders concluded, in paragraph 14, that "in the meantime we welcome the invitation from the Government of the Federal Republic of Germany to certain Summit countries to an international conference on the environment in Munich on 24–27 June 1984" (G7 1984). That "Multilateral Conference on the Causes and Prevention of Damage to Forests and Water by Air Pollution in Europe" did not directly address climate. But its focus on atmospheric sources of pollution from coal-fired generation continued the acid rain driver, through the fusion of sulphur oxides and carbon dioxide, that the G7 had established in 1979. Together, these three advances in the development of G8 environmental governance showed that the G7 was still determined to take the lead, rather than defer to a UN still not in the climate change control game.

Causes of Performance

The G7's 1979 creation of global climate governance and the highly ambitious regime it produced were the product of changes in the six causes highlighted by the concert equality model of G8 governance (Kirton 2013a). The first cause was a series of sharp, severe shocks that highlighted G7 vulnerabilities in the similar or related fields of energy and food, activating the visible vulnerability that carbon-laden coal was bringing in the form of chronic acid rain to the forests, lakes, and coal miners of Germany, the United States, and Canada. The second cause was the failure of the poorly equipped UN multilateral system, devoid of organizations dedicated to energy or the environment as a whole, to respond in any policy way. The third was the predominant equalizing, specialized capability of carbon and clean energy that the G7 countries controlled. These forces were reinforced by the fourth cause of the common, democratic, indeed multicultural commitment of the G7 members. They were further reinforced by the fifth cause of the domestic political control, capital, continuity, commitment, and political support of G7 co-founder Schmidt, and less so Carter or Clark, and by the sixth cause of constricted participation in a compact, leaders-driven, four-year-old G7 club. Of central importance was the unique personal commitment of the leader of the G7's third most capable country, Germany, who made a scientific judgement about the future and acted upon it, before an epistemic community or consensus had formed (Haas 1992; Bernstein 2001).

The post-1979 demise arose as the substance of the shocks to G7 members strongly shifted, diverting leaders' attention to classic political-security issues amid the new cold war and away from newer nonstate threats such as climate change. Changing leadership in G7 countries also contributed, with the electoral defeat of Joe Clark, and then Jimmy Carter, and the replacement of the latter by Republican Ronald Reagan who was a strong skeptic of human-induced acid rain and thus climate change. Only when a climate-competent Margaret Thatcher, a veteran of the 1979 start, hosted her first summit in London in 1984, and a new climate-committed German chancellor arrived in the person of Helmut Kohl did a revival of G7 climate governance begin.

Shock-Activated Vulnerability

The primary cause of the G7's ambitious initiation of global climate governance was the particular configuration of severe shocks to G7 members, as they were interpreted and interrelated by the leaders at the time. Shocks are severe, surprising threats either to the national interests or, secondarily, to the shared and distinctive national values that every country has.

The central thrust came from the second oil shock of 1979, inspired by the Iranian revolution and reaching its peak impact on oil prices as the Tokyo Summit was taking place. That year, world oil prices rose from $49.28 a barrel in January to $59.88 in June to $94.69 in January 1980 (see Appendix G-3). This spurred the G7 leaders to focus their Tokyo Summit fully on energy, arrive flexibly at bolder solutions than those assembled in the preparatory process, and take unprecedentedly ambitious, domestically intrusive action that they complied with, even at personal political cost, for the following years (Ikenberry 1988). The first oil shock, in 1973, had led oil-dependent Germany and Japan to adjust, in a way that the US had not, giving them first-mover experience when the second shock hit in 1979.

The impact of the 1979 oil shock was intensified by its severity, and by the fact that it was the second such oil shock from the Middle East within six years. However, it was now driven by and associated with a new, nonstate-directed terrorist revolutionary shock. This was generated by Islamic fundamentalism and aimed directly at the United States, by virtue of "the great Satan's" very existence more than merely its foreign policy preferences such as support for Israel, as in 1973. With the invasion of the American embassy and seizure of American diplomats in Tehran, the Iranian revolutionaries assaulted the basic norms of the interstate system—the protection of diplomats that had prevailed for centuries before. They also attacked and occupied sovereign America territory and held its diplomats hostage, a clear assault on America's core national interests. This was thus an unprecedented terrorist attack on America from abroad. The failure of America's attempt all alone to rescue its hostage through military force showed just how vulnerable an instinctively unilateralist America had become in its once mighty military sphere, in response to a second such shock coming just four years after its final defeat in the war in Vietnam. America was thus ready to adjust.

This second energy supply and price shock, coming a short six years after the first, inflicted the most economic damage on overall powerful but oil-short America, Japan, and Germany rather than on the energy-rich smaller G7 members of Britain and Canada. It was Canada, the G7's least capable country both overall and in the military realm, that played the decisive role in getting some of the American diplomatic hostages out of oil-rich Iran, now suddenly a revolutionary state. It was the many measures unleashed by the G7, so that it would never again be afflicted by a Middle East energy shock that drove the complete compliance of G7 members with the energy-embedded climate consensus they forged.

These oil and associated terrorist and military shocks caused the singular focus of the Tokyo Summit on oil and energy, necessary to drive the G7 leaders' implicit climate consensus on energy conservation, efficiency, and alternatives. It thus entrenched the new norm, created and collectively approved at the 1978 Bonn Summit the year before, that energy solutions must proceed in an environmentally protective way.

The particular alternative source of energy supplies selected, among the broad array identified, was skewed by a second, autonomous energy shock of a new, nonstate, non-directed sort. This was the nuclear reactor explosion at Three Mile Island on March 28, 1979 (see Appendix G-6). Even though no one died from the blast, it made American and other G7 leaders cautious about choosing the nuclear power alternative, leading them to add and intensify nuclear safety and nonproliferation to those conditions that they had imposed on nuclear power at summits before. This propelled the choice of an energy solution toward carbon-intensive, climate-destroying coal, as the G7's, and above all America's, immediately available, abundant, affordable, domestically secure energy source.

How then did the G7's historic climate control consensus break through such a strong material confluence of shocks? One possibility is that the nuclear shock was not sufficiently strong or shared, as it was only a first and not a second shock and no one died from it in the US or anywhere else. Another possibility is that G7 members selected alternatives according to the particular specialized capability in energy that they most possessed at home. But coal-rich Germany led the climate consensus, supported by coal-rich America, as well as oil- and electricity-rich Canada. And all members showed multi-year compliance, whether or not they were endowed with coal or other alternatives to oil. Yet the G7's collective predominance in energy technology—the preferred solution—was consistent with the choice, as noted below.

Behind these shocks lay three related vulnerabilities that had become visible. The first was a spike in the number of accidents involving oil tankers, rigs, and related things, from 23 in 1978 to 34 in 1979 (see Appendix G-4). The second, as the Tokyo communiqué passage on climate also noted, was sulphur oxides, then known to be a key cause of acid rain. It was physically and thus politically afflicting Germany, Canada, and America with visible damage deadly to wildlife, flora, and fauna, if not humans themselves. It was Germany that led with the US and Canada in strong support of acid rain and carbon control. Indeed, Canada's summit strategy importantly sought to use the occasion, bilaterally, to get Jimmy Carter from immediately neighbouring America to control its emissions that were creating the acid rain that was killing Canadian lakes and trees. It was thus the visible acid rain problem, and its sulphur oxides catalyst, rather than an always invisible and then unrecognized ozone and chlorofluorocarbon problem that gave birth to the G7's global climate governance regime (cf. Hoffmann 2005).

A third relevant visible shock-activated vulnerability was the food crisis afflicting developing countries rather than G7 ones in any damaging economic and deadly human way. Although this shock or subject did not make a prime-time appearance in the summit communiqué, the initial American preparatory paper had highlighted the food and agricultural pathway from uncontrolled climate change to avoidable drought, and agricultural destruction and death. The Canadians also noted the impact of climate change on drought and food, especially in Western Canada, where Clark was from and where his party had its electoral base. And Schmidt's (1979b) reference in his post-summit report to the food shortages of a drought-stricken Sahel confirmed that a visible shock near Europe had causal force.

This balance of visible vulnerabilities at the initial stage was important in setting the path by which the G7 and global community governed climate change in the ensuing years. The G7 alone took ownership of the problem and asked only itself to act. The emphasis was overwhelmingly on emissions, particularly of carbon dioxide and sulphur oxides, and specifically those from energy sources, rather than the agricultural sinks that remained buried in the US concept paper (see Appendix G-7). And both Schmidt himself and the concept paper from an America with a Democratic president noted that difficult decisions—hard commitments rather than a soft consensus—could be deferred for a decade or two. Indeed, America and its G7 partners would wait to take decisive action until 1989.

Beneath these present vulnerabilities lay a profound embedded shock and resulting sense of vulnerability in America itself. Carter's post-summit "malaise" address showed how poignant the shocks accumulated since 1971, including Vietnam and Watergate, still were in the most powerful country in the G7 and in the world.

After Tokyo, the G7's concern was increasingly diverted by more intense, classic, state-to-state, non-energy-related shocks. On Christmas Day 1979, the Soviet Union invaded Afghanistan, to start a new cold war that dominated the following five years. On September 17, 1980, Iraq invaded Iran and proclaimed sovereignty over the Shatt al-Arab waterway. As the 1980 Venice communiqué said, these shocks evoked the common democratic purpose of the G7. The 1980 shock was closely related to the central problem of energy. But the larger geopolitical shock from the Soviet Union and the resulting new cold war would take precedence over energy shocks from the Middle East. It flourished over the Soviet gas pipeline dispute at Versailles, for although Soviet gas was more climate friendly than Middle East oil or German coal, the acute new cold war crowded out such considerations, in the view of the US and others.

Only in 1984 did a classic Middle East energy shock return, to support Germany's leadership on the environment at the London Summit. March 1984 saw the start of a nine-month "tanker war" in which 44 ships, including those from Iran, Iraq, Saudi Arabia, and Kuwait, were attacked by Iraqi or Iranian planes or hit by mines. The escalation of terrorism from oil-rich Libya, directed against the UK, was also relevant in spurring G7 action on the environment at London in 1984.

Multilateral Organizational Failure

The second cause of the G7's 1979 creation of ambitious climate governance was the failure of the multilateral system to address these new visible climate-related vulnerabilities in any way. The UN system from the 1940s was centred on a UN charter from which the environment was absent. It contained no functional organizations dedicated to the environment or energy as a whole (Biermann and Bauer 2005). It offered only the WMO, which was just beginning its climate research program as stand-alone science, closely connected to the weather, and unrelated to any organized multilateral action in the political or policy world. It also offered the UN Environment Programme (UNEP), a program rather than a specialized agency that had been founded only in 1972 and that had also not discovered climate change. In the cognate field of energy, there was only the Atlantic-centric plurilateral International Energy Agency (IEA), created in 1974, which too had not discovered climate change (Keohane 1978). Elsewhere in the UN system, neither the Food and Agriculture Organization nor UNEP made any connection from its core concerns to climate change.

The WMO, the most functionally relevant core organization from the established UN multilateral system, was silent on the subject of climate change at any policy level. Its major contribution was to sponsor the first World Climate Conference, in Geneva on February 12–23, 1979 (Bernstein 2001). Scientists participated from several disciplines at plenary sessions and in four working groups on climate data, topics, impacts, and research. They concluded that it was plausible that burning fossil fuels and deforestation could produce warming, from which many consequences could come (Hoffmann 2005). But this led not to policy attention or action, only to the creation of the World Climate Programme, sponsored by the WMO, UNEP, the UN Educational, Scientific, and Cultural Organization (UNESCO), and the International Committee of Scientific Unions (now the International Council for Science). It also led to another scientific conference a full six years later, in 1985, in Villach, Austria.

This full failure of multilateral organizations in global climate governance in the policy domain, apart from the WMO's scientific contribution, helps account for the G7's invention of it in 1979 and the revival of G7 environmental interest in 1984. But as a constant throughout the period, it does not explain the demise and disappearance of G7 climate and environmental governance from 1980 to 1983.

Predominant Equalizing Capability

The third cause of the G7's 1979 invention of climate governance was the globally predominant and internally equalizing overall and specialized relevant capability of G7 members. In overall capability, as measured by gross domestic product at market exchange rates, in 1979 the G7 had clear if declining global predominance and internal equalization (Kirton 1999b). That equalization was driven generally by the relative rise since 1975 of second-ranked Japan (save for the 1979 oil shock interruption) and third-ranked Germany, and that of relatively small but energy-abundant Canada and the UK. The reversal of these dynamics after 1980, with the soaring US "superdollar" through to the Plaza Accord created by France, Germany, Japan, the US, and the UK in September 1985, America's gathering growth surge, and the decline in the energy premium for Canada and the UK, is consistent with the demise of G7 climate and energy governance during this time.

In the most climate-relevant specialized capability, the configurations and causal connections were more complex. G7 members were globally predominant in energy consumption, and thus in the specialized capability of energy conservation and efficiency as pathways to a cure. They were also globally significant in their high-capacity carbon sinks, thanks to Canada and the US, and predominant in their scientific capability to identify the problem, its causes, and consequences, with the US in the lead. However, in energy production, those in the Soviet sphere, the Organization of Petroleum Exporting Countries (OPEC), and the developing South had a more substantial share. It was thus consistent that in 1979 G7 members chose to cut back oil imports as their dominant carbon emissions control response and took the global lead.

Moreover, in this array of specialized climate control capability, there was considerable internal equality. The smallest member, Canada, had the highest net carbon sinks, while the largest, the US, had the lowest. However, US scientific capability continued to show strength from 1979 on. The US paper in the lead-up to the 1981 summit linked energy to the climate effects of carbon dioxide accumulation in the atmosphere. The pathway was acid rain, although stratospheric ozone depletion made its first appearance then. Still, leading US scientific capability alone was not enough to make Ronald Reagan's America the leader of an epistemic community or of a G7 or global control regime on climate change.

Common Democratic Purpose

The fourth cause was the common democratic purpose of the G7 members. In general, this was a constant, as all members remained democratic throughout the period of initial invention, subsequent demise, and the environmental mini-revival in 1984 (see Appendix J). But more specific changes in democratization help explain this particular path of G7 climate change performance.

First, in 1979 the G7 was only four years from the codification of its distinctive foundational mission, in its constitutional-equivalent charter at the start of its inaugural 1975 communiqué. This mission was to protect within its members and globally promote the values of "open democracy" and "individual liberty" (G7 1975). Schmidt along with Giscard had been a founding father of the summit and thus co-author of that statement (Putnam and Bayne 1984). And while Japan was the initiator of that democratic clause, Germany and Japan—both defeated World War II powers—particularly cherished their postwar democracy and acceptance as democratic powers in the G7 club (Dobson 2004). After 1979, the freshness of that foundational 1975 moment faded and Giscard soon left, leaving Schmidt alone as a founder.

Second, US democracy was on the rise from its Watergate depth in the first half of the 1970s. Carter's "malaise" speech showed that the Watergate trauma was very much on his mind and the minds of his fellow Americans. This syndrome was suddenly and sharply reduced when sunny Ronald Reagan with his "morning in America" arrived. But the shared democratic tradition was not enough in its application to the environment to transcend the post-1980 Cold War division between the continental European socialist version of democracy led by Schmidt and Mitterrand and the Anglo-American conservative version led by Reagan and Thatcher. It was only in 1984, when the new cold war showed the first signs of thawing and of a return to easier G7-wide consensus on East-West relations that the G7's common democratic devotions regained strength. These soon flourished as dominant principles in the summit statement on the 40th anniversary of the end of World War II at the second Bonn Summit in 1985.

Third, America's defeat in Vietnam, also referred to by Carter, and the expansion of Soviet influence afterward assaulted a democratic world in decline and activated its determination to preserve it in its G7 core. The 1979 Iranian shock added another formidable assault, now from a noncommunist and semi-state source. Another came from the largely undemocratic OPEC countries that lay behind the two oil shocks. G7 leaders thus decided to halt the inflation caused by the sudden, severe spike in world oil prices, and to do so by, or despite, the cost of high interest rates and the severe recession into which the G7 and global economy, its energy use, and its carbon emissions predictably plunged (Thatcher 1993, 65–71). As these threats from multiple sources of non-democratic actors compounded after the Soviet invasion of Afghanistan, compliance with the 1979 climate control consensus continued, even as other dimensions of climate performance declined.

Fourth, another common purpose, related to political openness and democratic inclusiveness in its demographic dimension, was multiculturalism. It was the first to directly link democracy to climate change, mentioned in Schmidt's speech to the Bundestag upon his return from Tokyo. He referred poignantly to the Turkish workers who mined coal in Germany and the many deaths they suffered as a result.

Fifth, underlying these shared democratic principles was an emerging ecological one. The unifying principle and norm for all G7 members from 1975 was the conservation component and, from 1978 onward, the additional belief that energy could and should not harm the environment. In 1979, in the United States, such principles were barely visible in Carter's post-summit speech. But in Germany, the climate leader at Tokyo, they loomed large in Schmidt's post-summit report. They were not sufficiently embedded, however, to endure to spur G7 climate change control during the 1980–1983 Cold War Reagan years.

Political Cohesion

The fifth cause of the invention and subsequent decline of G7 climate governance was the particular configuration of political cohesion, composed of the political control, capital, continuity, competence, commitment, and public support of the G7 leaders at home (see Appendices K-1 and K-2). High levels of these components in Germany, France, and the UK drove the German-led 1979 initiation, while lower levels of continuity, conviction, and public support fuelled the subsequent decline, until a scientifically competent Thatcher, with her second majority government, attended her sixth summit and hosted her first in 1984.

In 1979, Germany's Schmidt, the G7's leader on climate change and veteran co-founder, attended his fifth summit in a row. At home, he led a coalition government that controlled the Bundestag, was popular with the public, and was on track to win its third election, which he did on October 5, 1980. He shared a growing awareness of G7 leaders and their publics about atmospheric pollution, especially acid rain. And, as his public speeches showed, he alone among G7 leaders was publicly and personally committed to the climate/carbon control cause. He also faced mounting pressure from Germany's nascent Green Party. After 1979 these forces strengthened, but were overwhelmed by others to keep Schmidt's G7 climate leadership from appearing again. In 1984, however, as the G7 explicitly noted in its communiqué compliment to Germany, its environmental advance was a German initiative, with its new chancellor, Christian Democrat Helmut Kohl, sharing Thatcher and Reagan's market-friendly approach.

In the most powerful US, the sharp shift from the Carter to Reagan years accounts well for the 1979 rise, post-1980 demise, and 1984 mini-revival of G7 climate and environmental governance. In 1979 Jimmy Carter was attending his third summit. As a nuclear engineer, he had some professional scientific competence related to nuclear energy. But he ranked low in approval in public opinion polls, had poor control of his Congress, and was facing re-election in November 1979. This very weak domestic political position induced him to look to the G7 consensus to move his Congress on energy at home (Ikenberry 1988). It also gave him an incentive to respond to a public that, in its elite strata at least, was becoming mobilized about climate change (Leiserowitz 2007). In 1979 an editorial in the leading science journal argued that "the release of carbon dioxide to the atmosphere by the burning of fossil fuels is, conceivably, the most important environmental issue in the world today" (*Nature* 1979; see also Briffa et al. 1990). Leading researchers told the Council on Environmental Quality that "man is setting in motion a series of events that seem certain to cause a significant warming of world climates unless mitigating steps are taken immediately" (Pomerance 1989, 260). From civil society, an epistemic consensus and pressure were starting to emerge. This combination helped stimulate Carter to support Schmidt, despite the poor personal relationship between the two.

Carter's 1981 replacement by Reagan brought a sharp reversal on most of these dimensions. Reagan, attending his first G7 summit at Montebello, had no personal commitment or domestic political incentive to advance climate change control. However, by 1983, at the third summit he attended and the first he hosted, Reagan saw the G7 (1983) agree "to strengthen cooperation in protection of the environment, in better use of natural resources, and in health research." And in 1984, at his fourth summit, facing an election in the fall, he had the experience and willingness to follow the lead of his trusted friend Thatcher as summit host on environmental and other concerns.

In Japan, the G7's second most powerful and highly energy-dependent member, Masayoshi Ohira, host of the 1979 summit, faced an imminent election, which came on October 7, 1979. He had strong incentives to adjust to the German lead and US support to make his summit a success, above all on energy and related environmental concerns (Dobson 2004).

In Canada, the G7's least powerful member and another supporter of Schmidt's 1979 initiative, Joe Clark was attending his first summit but had a minority government and an electorate increasingly concerned about acid rain. His incentive to act boldly on energy and environmental matters at Tokyo continued on his return to Canada. His immediate compliance with Tokyo's energy conservation targets through the introduction of an energy tax led directly to his defeat in the House of Commons, and then in the ensuing election, which brought back Liberal prime minister Pierre Trudeau, now with a majority government. While Trudeau was close to Germany's Schmidt and France's Mitterrand, as a constitutional layer he had no substantial competence in or commitment to atmospheric environmental issues, including climate change.

In France, Giscard—Schmidt's fellow European G7 co-founder and five-year veteran in 1979—was replaced by 1981 with socialist François Mitterrand (1982), whose later interest in the environment seemed more political than personal. The communist party members in Mitterrand's cabinet were similarly preoccupied with non-environmental issues, and fuelled his ideological dispute with Reagan and Thatcher that blacked out environmental concerns.

From the UK in 1979, the climate change–supportive Margaret Thatcher was attending her first summit. But she had the competence to comprehend the science of climate change, having earned a university degree in chemistry (Thatcher 1993). By London in 1984, she was now the host, a veteran of five G7 summits, the only leader that had been at Tokyo in 1979, a recently re-elected prime minister with her second majority government, and a leader whom Ronald Reagan trusted and followed. By then she was well positioned to revive the G7's 1979 interest in the environment in its atmospheric domain.

Constricted Controlled Participation

The sixth cause of this cadence of invention, demise, and mini-revival was the changing constricted, controlled participation in the summit club (see Appendix L).

By 1979, the G7 remained a compact club with the same seven country members for the four years in a row from 1976 to 1979. The one partial addition of the European Community in 1977 had brought to Tokyo as the European sherpa Crispin Tickell, with his strong competence and commitment on climate change. By then, the G7 had become an interpersonal club of leaders and their personal representatives, at its fourth meeting in four stress-ridden years. This gave rise to a process that allowed leaders to be leaders in a major push in the lead-up and, at least passively, in a collective act of spontaneous combustion or acquiescence at the summit itself. On climate change control, none of the preparatory advice prior to the summit was nearly as ambitious as the resulting communiqué consensus was. Nor is there any evidence that this component came as part of a carefully constructed package deal resulting from a two-level game (Putnam and Bayne 1987; Putnam 1988). The long time the leaders spent in Tokyo, and the numerous bilateral meetings there, could have helped fuel this outburst of collective leadership on site. The last-minute French defection from an earlier all–European Community position permitted the bold oil control deal. The second oil shock in six years and the new shock from a terrorist-ridden, revolutionary Iran was strong enough to create an interpersonal bond among the leaders. Thus the far-reaching historic climate consensus appeared to come largely from leaders when they were together alone.

After 1979, the diversionary shock of the new cold war and then the arrival of Reagan and Mitterrand as agents of an acute intra-G7 ideological divide in response, eroded the interpersonal bonds, especially at the highly formal summit held in the splendour of Versailles in 1982. By 1984, a new generation of leaders had met often enough, and the sub-summit network of institutions and participants had become sufficiently dense, to recreate an interpersonal sense of clubbiness on to atmospheric environmental concerns.

Conclusion

At Tokyo in 1979, the G7 invented global climate change governance. It did so by acting ambitiously and effectively at a time when the UN system had not recognized, at any policy level, the problem of climate change. The problem and prescriptions for global environmental governance solutions had been proclaimed by leading authorities in scholarly articles, including those in the prestigious and influential American journal *Foreign Affairs* (Kennan 1970). They were also known through public US government reports, from the civilian National Academy of Sciences (1979), the Department of Transportation's Climatic Impact Assessment Program (1975), and the security community's Central Intelligence Agency (1976). But among the institutions claiming to offer global governance, it was the G7 alone that answered the call and did so in a convincing way. It was led by the personal commitment of German chancellor Helmut Schmidt, supported by America's Jimmy Carter and Canada's Joe Clark, and their colleagues brought together by the force of the many new environment-related shocks that exposed and equalized their vulnerabilities in 1979. After a subsequent four-year demise in G7 climate governance, driven by the diversion of a divisive new cold war, the G7 began to revive its attention in 1984 as it rediscovered that those vulnerabilities remained.

References

Abbott, Kenneth W., Robert Keohane, Andrew Moravcsik, et al. (2000). "The Concept of Legalization." *International Organization* 54(3): 401–20.

Barnes, James (1994). *Promises, Promises! A Review: G7 Economic Summit Declarations on Environment and Development*. Washington DC: Friends of the Earth.

Bernstein, Steven (2001). *The Compromise of Liberal Environmentalism*. New York: Columbia University Press.

Biermann, Frank and Steffen Bauer, eds. (2005). *A World Environmental Organization: Solution or Threat for Effective International Environmental Governance*. Aldershot: Ashgate.

Briffa, Keith R., Thomas S. Bartholin, Dieter Eckstein, et al. (1990). "A 1,400-Year Tree-Ring Record of Summer Temperatures in Fennoscandia." *Nature* 346(6283): 434–9. doi: 10.1038/346434a0.

Carter, Jimmy (1979). "'Crisis of Confidence' Speech." July 15. http://millercenter.org/president/speeches/speech-3402 (January 2015).

Carter, Jimmy (1982). *Keeping Faith: Memoirs of a President*. New York: Bantam Books.

Climatic Impact Assessment Program (1975). "Impacts of Climatic Change on the Biosphere." Monograph 5, DOT-TST-75-55, September, United States Department of Transportation, Washington DC.

Deutscher Bundestag (1979, 4 July). Stenographischer Bericht, Plenarprotokoll Nr. 08/167. Deutscher Bundestag. http://dipbt.bundestag.de/doc/btp/08/08167.pdf (January 2015).

Dobson, Hugo (2004). *Japan and the G7/8, 1975–2002*. London: RoutledgeCurzon.

G7 (1975). "Declaration of Rambouillet." Rambouillet, November 17. http://www.g8.utoronto.ca/summit/1975rambouillet/communique.html (January 2015).

G7 (1979). "Declaration." Tokyo, June 29. http://www.g8.utoronto.ca/summit/1979tokyo/communique.html (January 2015).

G7 (1980). "Declaration." Venice, June 23. http://www.g8.utoronto.ca/summit/1980venice/communique (January 2015).

G7 (1983). "Williamsburg Declaration on Economic Recovery." Williamsburg, May 30. http://www.g8.utoronto.ca/summit/1983williamsburg/communique.html (January 2015).

G7 (1984). "London Economic Declaration." London, June 9. http://www.g8.utoronto.ca/summit/1984london/communique.html (January 2015).

Haas, Peter M. (1992). "Introduction: Epistemic Communities and International Policy Coordination." *International Organization* 46(1): 35.

Hoffmann, Matthew (2005). *Ozone Depletion and Climate Change: Constructing a Global Response*. Albany: SUNY Press.

Ikenberry, G. John (1988). "Market Solutions for State Problems: The International and Domestic Politics of American Oil Decontrol." *International Organization* 42(1): 151–77.

Kennan, George (1970). "To Prevent a World Wasteland: A Proposal." *Foreign Affairs* 48 (April): 401–13. http://www.foreignaffairs.com/articles/24149/george-f-kennan/to-prevent-a-world-wasteland (January 2015).

Keohane, Robert O. (1978). "The International Energy Agency: State Influence and Transgovernmental Politics." *International Organization* 32(4): 929–51.

Kirton, John J. (1999). "Explaining G8 Effectiveness." In *The G8's Role in the New Millennium*, Michael R. Hodges, John J. Kirton, and Joseph P. Daniels, eds. Aldershot: Ashgate, pp. 45–68.

Kirton, John J. (2002). "Embedded Ecologism and Institutional Inequality: Linking Trade, Environment, and Social Cohesion in the G8." In *Linking Trade, Environment, and Social Cohesion: NAFTA Experiences, Global Challenges*, John J. Kirton and Virginia W. Maclaren, eds. Aldershot: Ashgate, pp. 45–72.

Kirton, John J. (2008–09). "Consequences of the 2008 US Elections for America's Climate Change Policy, Canada and the World." *International Journal* 64(1): 153–62.

Kirton, John J. (2009). "Governing Global Trucks and Buses: The G20 and G8 Roles." Keynote address, Truck and Bus Forum, May 12, Lyon. http://www.g8.utoronto.ca/scholar/kirton-lyon-090511.pdf (January 2015).

Kirton, John J. (2013). *G20 Governance for a Globalized World*. Farnham: Ashgate.

Leiserowitz, Anthony (2007). "American Opinions on Global Warming." Yale University/Gallup/ClearVision Institute Poll, Yale School of Forestry and Environmental Studies, New Haven CT.

Minutes (1979, 28 June). G7 Tokyo Summit (Session 1). Margaret Thatcher Foundation. http://www.margaretthatcher.org/document/112029 (January 2015).

Mitterrand, François (1982). *The Wheat and the Chaff*. New York: Seaver Books.

Nature (1979, 3 May). "Costs and Benefits of Carbon Dioxide." 279(5708): 1. Editorial. doi: 10.1038/279001a0.

Owen, Henry (1979, 22 June). G7 Tokyo Summit: Henry Owen Brief for President Carter. Margaret Thatcher Foundation (original source: Carter Library). http://www.margaretthatcher.org/document/111681 (January 2015).

Pomerance, Rafe (1989). "The Dangers from Climate Warming: A Public Awakening." In *The Challenge of Global Warming*, Dean Abrahamson, ed. Washington DC: Island Press, pp. 259–69.

Putnam, Robert (1988). "Diplomacy and Domestic Politics: The Logic of Two-Level Games." *International Organization* 42(3): 427–60.

Putnam, Robert and Nicholas Bayne (1984). *Hanging Together: Co-operation and Conflict in the Seven-Power Summit*. 1st ed. Cambridge MA: Harvard University Press.

Putnam, Robert and Nicholas Bayne (1987). *Hanging Together: Co-operation and Conflict in the Seven-Power Summit*. 2nd ed. London: Sage Publications.

Schmidt, Helmut (1979a). "An Interview with Helmut Schmidt." *Time*, June 11.

Schmidt, Helmut (1979b, 4 July). Policy Statement Made by the Chancellor of the Federal Republic of Germany Herr Helmut Schmidt before the German Bundestag. G7/G8 Research Collection in the John W. Graham Library, Trinity College, University of Toronto. Locator ID: 28.1062.

Schmidt, Helmut (1989). *Men and Powers: A Political Retrospective*. Translated by Ruth Hein. New York: Random House.

Speth, Gustav (2004). "Red Sky at Morning: America and the Crisis of the Global Environment." Interview with Joanne Myers, April 22. http://www.carnegiecouncil.org/studio/multimedia/20040422/index.html (January 2015).

Sustainable Energy Development Center (2006). *Global Energy Security: Summary of Russia's G8 Presidency*. Moscow: Sustainable Energy Development Center.

Thatcher, Margaret (1993). *The Downing Street Years*. New York: HarperCollins.

Tickell, Crispin (1977). *Climatic Change and World Affairs*. Cambridge MA: Center for International Affairs, Harvard University.

United States Central Intelligence Agency (1976). "USSR: The Impact of Recent Climate Change on Grain Production." Report ER 76-10577 U, October, Washington DC. http://

www.foia.cia.gov/sites/default/files/document_conversions/89801/DOC_0000499885.pdf (January 2015).

United States National Academy of Sciences (1979). "Carbon Dioxide and Climate: A Scientific Assessment." Report of an Ad Hoc Study Group on Carbon Dioxide and Climate, Woods Hole, Massachusetts, July 23-27, to the Climate Research Board, Assembly of Mathematical and Physical Sciences, National Research Council, Washington DC. http://web.atmos.ucla.edu/~brianpm/download/charney_report.pdf (January 2015).

Chapter 3
Rediscovery, 1985–1986

The Bonn Breakthrough, 1985

The second German-hosted Bonn Summit on May 2–4, 1985, brought a rediscovery of Group of Seven (G7) governance on climate change. Above all, it brought direction-setting achievements, defining "climate change" for the first time as an issue in its own right. Moreover, G7 leaders declared the environment and the economy to be equally valuable and mutually supportive, gave these principles prominence in their communiqué, and affirmed the principle of "polluter pays."

These important advances came from a summit widely regarded as a great failure, due to bitter American-French divisions over America's Strategic Defense Initiative, France's desire for a new monetary arrangement, and a date to launch new multilateral trade negotiations—culminating in François Mitterrand's threat to walk out of the G7 and summit institution itself over its treatment of those trade negotiations (Putnam and Bayne 1987). But where economics and security divided, the environment and its extension to African drought and desertification brought G7 leaders together to rediscover and more expansively govern the new issue of climate change.

The concert equality model provides a sound explanation of this strong performance on climate change, with a premium again placed on personal leadership in the absence of severe, climate-specific shocks.

Preparations

The climate change advances at Bonn were largely the work of the German hosts, working in tandem with Canada's new prime minister, Progressive Conservative Brian Mulroney. In sharp contrast to the Tokyo Summit in 1979, Bonn's supportive climate coalition included another fellow conservative, America's Ronald Reagan, with France's socialist Mitterrand the most active opponent on the core environmental principles involved.

Canada's summit team began its journey to Bonn by listing its five objectives: economic growth, unemployment, trade, the developing world, and an enhanced emphasis on the need to protect the environment. As Mulroney recorded immediately after the summit:

> I was concerned that the Bonn summit give appropriate emphasis to the problems of the environment, and that there should be specific reference to acid rain, air pollution, protection of the ozone layers, and the management of toxic chemicals and hazardous wastes. It is, in my view, especially significant that a consensus emerged on the idea that governments and private industry have a joint responsibility in preserving the environment and on the proposal that the 'polluter pays' principle should be developed and applied more widely (Mulroney 2007, 373–4).

In the preparatory process, Canada worked hard to secure the language it wanted on the environment in the summit's thematic paper and the final communiqué. At the start of

May, the draft thematic paper was very satisfactory to the Canadians as the basis for the communiqué. Yet they still looked for opportunities to advance their key issues of acid rain, the crisis in sub-Saharan Africa (which they saw arising partly due to environmental mismanagement), and international chemicals management. On Africa the communiqué might refer to improved early warning systems and efforts against desertification.

At a late-stage sherpa meeting, France proposed an initiative on desertification, including an extension of early warning systems through financing two satellite reception stations on the African continent, one in Commonwealth Nairobi and the other in francophone Ouagadougou. There was not much specific reaction from the other sherpas. The Americans were guarded, preferring to focus on long-term strategies and agricultural research. Canada supported the French initiative. But, mindful of its own budget deficit, Canada did not endorse the specific French suggestions to establish a new fund for desertification.

At the Summit

At the summit table, Mitterrand was on the offensive, pushing the fight against desertification. Both Mulroney and Joe Clark, now foreign minister, supported the objectives of the French proposal and the importance of the summit showing its concern about Africa. Despite some reluctance from Reagan, host Helmut Kohl secured agreement for Germany's proposal to create an expert group on Africa, chaired by Germany, to report back to G7 foreign ministers by September.

Mulroney continued to push on the environment and African desertification. At the summit table and in private conversations, Mulroney overcame Mitterrand's reservations on the environment. In the leaders' discussions, Mitterrand objected to what he viewed as an assertion in the proposed text that market forces provided a basis for solutions to all environmental problems. He thought this view flowed from an obsession with the ideology of markets, which he did not share. Others defended the text and the tension level rose. Then Mulroney intervened with bridging language, based on his recent "Shamrock Summit" with Reagan in Quebec City. Thus the Bonn declaration stated: "we shall harness both the mechanisms of governmental vigilance and the disciplines of the market to solve environmental problems" (G7 1985). Kohl congratulated Mulroney on his "Irish language." A little later, during the most divisive moment for the summit, when its very existence was in doubt, Kohl, responding to Mitterrand's threat to leave the summit over the issue of multilateral trade negotiations, noted the leaders' success in producing a communiqué on the environment to which both the US and France had agreed. Mulroney noted at his subsequent news conference that this result on the environment had been tough to get. But by May 13, just over a week after the leaders had returned home, memories had mutated to emphasize the strong like-mindedness shown by all the leaders on the environmental front.

Results

At the end of the summit, Canada listed among the top three accomplishments the consensus on dealing urgently and sympathetically with the problems of Africa and strengthened environmental cooperation. It particularly valued the discussion of environmental issues, the communiqué reference to acid rain, and the recognition of the link between economic development and environmental protection. Canadian officials felt they had positively influenced the agreement on an excellent and important environment/acid rain formulation. In an interview on national German television on May 4, Kohl himself referred to the close cooperation between the "Canadians and Germans on environment issues—a subject of great importance for both countries." Privately, German officials also shared this assessment. They regarded the polluter-pays principle, accepted for the first time at the summit, as an especially valuable achievement.

Dimensions of Performance

The Bonn Summit was a strong success for climate change, above all in the deliberative, direction-setting, and decision-making dimensions (see Appendix A-1).

Deliberation

In Bonn's public deliberation, the environment appeared in seven paragraphs and climate change in one paragraph of the communiqué (see Appendix C-1). Bonn dedicated, for the first time, a full section in the communiqué to the environment. Entitled "Environmental Policies," it contained four paragraphs (G7 1985). They began with the following statement:

> New approaches and strengthened international cooperation are essential to anticipate and prevent damage to the environment, which knows no national frontiers. We shall cooperate in order to solve pressing environmental problems such as acid deposition and air pollution from motor vehicles and all other significant sources. We shall also address other concerns such as climatic change, the protection of the ozone layer and the management of toxic chemicals and hazardous wastes (G7 1985).

With these words, the issue of climate change—absent since 1979—returned to the leaders' declaration. For the first time it was now named "climate change" rather than the narrower category of carbon dioxide or sulphur oxides in the atmosphere as components or causes of pollution. It thus created the concept and category that endure to this day. Bonn also added, if not with an explicit causal connection, an accompanying concern with climate change adaptation, through action on Africa and the acute drought, desertification, and famine afflicting many there.

Direction Setting

In the realm of principled direction setting, Bonn's results on the environment and climate change were striking (see Appendix C-1). For the first time, the environment in general entered the communiqué chapeau, signalling that it was a priority fact and value for the G7 and the world. The leaders began their communiqué with the ringing words "conscious of the responsibility which we bear, together with other governments, for the future of the world economy and the preservation of natural resources" (G7 1985). The communiqué thus opened by proclaiming that the economy and the environment had equal value as goals in their own right. The chapeau ended by proclaiming that "economic progress and the preservation of the natural environment are necessary and mutually supportive goals. Effective environmental protection is a central element in our national and international policies." Necessity, integration, and centrality for the environment were thus added to equality as statements of causation and rectitude for global environmental governance. Elsewhere in the communiqué the polluter-pays principle was affirmed for the first time.

Decision Making

These principles came with a decision-making achievement on climate change. For the first time, G7 leaders produced a clear, specific, future-oriented commitment on the environment, with the words: "we shall also address other concerns such as climatic change" (see Appendix D-1). They prioritized climate change as an urgent issue equal to ozone, toxic chemicals, and hazardous waste, although less urgent than the "pressing" problems of acid deposition and automotive emissions. In this summit in Germany, acid rain came first. To achieve these ends, G7 leaders chose governmental vigilance, market discipline, the polluter-pays principle, and science and technology. They also identified "improved and internationally harmonized techniques of environmental measurement" as a necessity, laying the foundation for the Intergovernmental Panel on Climate Change (IPCC).

Delivery

In the delivery of this decision, the performance of G7 members was strong. The level of compliance was positive, with an overall average of +0.50, on a scale where +1.00 is full compliance, 0 is partial compliance or a work in progress, and −1.00 is no compliance or action antithetical to the commitment (see Appendix E-1). Among the eight members, the United States, Germany, France, and the European Union fully complied, while the United Kingdom, Japan, Canada, and Italy partially complied.

Development of Global Governance

To act on their desire for improved environmental measurement, the leaders also developed global governance institutionally inside and outside the G7. They invited the experts in their environment working group to consult with appropriate international bodies about the most efficient ways to achieve progress in techniques of environmental measurement. They also welcomed the contribution made by the environment ministers to closer international cooperation beyond the G7. They decided to "focus our cooperation within existing international bodies, especially the OECD [Organisation for Economic Co-operation and Development" (G7 1985). The leaders thus remained determined to develop global environmental governance on a G7-controlled plurilateral platform, rather than defer to a United Nations system not yet seriously in the climate change game.

Causes of Performance

Shock-Activated Vulnerability

These achievements at Bonn were the result in the first instance of the highly visible vulnerability of G7 members to assaults arising directly in the environmental domain. In host Germany, acid rain from its neighbours was denuding the forests so beloved by its citizens. Publics within both compact, crowded Germany and a relatively empty, transcontinental Canada were concerned with the ecological and accompanying economic destruction being caused in their countries by the acid rain created by their coal-burning neighbours—Britain and the United States respectively. This led both Germany and Canada to place a premium on issues dealing with atmospheric chemistry, on controlling coal-fired power generation in the source "aggressor" states, and on using the G7 summit to establish the prominence and the principles to reinforce their bilateral priorities. This concern on the part of both countries' governments was well known. In preparing for bilateral talks at the summit, Canada included acid rain among the issues it wished to raise with the US. And on May 1, on the eve of the summit, Japan's *Mainichi* newspaper reported that Canada was also interested in environmental issues because air pollutants from major industrial areas in the US were being carried across the border, causing atmospheric pollution and acid rain in Canada.

A second vulnerability came from a shock from the cognate food and agriculture field. The food crisis in Africa was highlighted by a famine in Ethiopia that had attracted widespread media and public attention and ministerial action in many G7 countries, including Canada, in the autumn of 1984. In the summit process Canada considered environmental mismanagement, if not climate change specifically, as a cause.

Multilateral Organizational Failure

The second cause of Bonn's environmental and climate change achievements was the belated but still slender attention that the UN system was finally giving to global issues of atmospheric chemistry. As the summit text suggests, climate change was pushed onto the G7 agenda because of the now known problem of ozone depletion. Six weeks before the Bonn Summit, the UN system concluded its convention to combat ozone depletion, in German-speaking Austria next door.

The signing of the Vienna Convention on the Protection of the Ozone Layer by 20 countries took place on March 22, 1985. Discussions on the depletion of the ozone layer had begun in the 1970s with the first scientific publication in 1974 linking the chemicals produced by humans to the potential harm they could cause to the stratospheric ozone layer. The discussions continued into the 1980s. During the Vienna Convention negotiations, countries discussed a possible protocol that would provide specific targets for certain chemicals, but no consensus was reached (Benedick 1991).

The UN system also started to act more directly on climate change, holding its second scientific conference on the subject in Villach, Austria, in 1985. But this conference, which was organized by the United Nations Environment Programme, the World Meteorological Organization, and the International Committee of Scientific Unions, took place on October 9–15, long after the G7 had met in Bonn. And it was once again a gathering of scientists, now from 29 developed and developing countries, in search of scientific knowledge and consensus, rather than policy change.

Moreover, in both Canada and Germany, as the climate change leaders, the concern was still predominantly with acid rain and other forms of atmospheric pollution, rather than with ozone or climate change itself. The UN-ozone precedent much emphasized by some scholars in global climate change governance was not the dominant driver of G7 action in Germany in 1985 (Hoffmann 2005). And while it was on Mulroney's list of concerns, he only took ownership of it two years later, when the Montreal Protocol on Substances that Deplete the Ozone Layer was concluded in Canada in 1987. In the field of acid rain—a primary concern of both Canada and Germany—even with the discovery of long-range transboundary pollution, the UN system played only a minor role.

Of far more impact were informal plurilateral processes, with Germany and its G7 partners of Canada and France in the lead. Just prior to the G7's London Summit, in March 1984 the environment ministers from Canada, Austria, Denmark, Germany, Finland, France, the Netherlands, Norway, Sweden, and Switzerland met and agreed to a 30 percent reduction of national levels of sulphur dioxide by 1993. At the meeting, Canadian environment minister Charles Caccia, stated "we will proceed independently from the United States in developing a Canadian solution on the matter of acid rain and we hope that the U.S. will join us at the earliest possible date" (Franklin et al. 1985, 156).

The London Summit was the first time the environment had received a full, stand-alone multi-sentence paragraph in the communiqué. The leaders had concluded, in paragraph 14, "we welcome the invitation from the Government of the Federal Republic of Germany to certain Summit countries to an international conference on the environment in Munich on 24–27 June 1984" (G7 1984). The Multilateral Conference on the Causes and Prevention of Damage to Forests and Water by Air Pollution in Europe had subsequently been held. It concluded with East Germany, Bulgaria, and the USSR agreeing to a reduction of 30 percent levels of sulphur dioxide by 1990 (Franklin et al. 1985). Furthermore, it requested the executive body for the Convention on Long-Range Transboundary Air Pollution "adopt a proposal for a specific agreement on the reduction of annual national sulphur emissions or their transboundary fluxes by 1993 at the latest" (UN 1985).

The Munich conference contributed to the G7 summit process in several ways. It generated momentum for the Convention on Long-Range Transboundary Air Pollution and determined the general outline of an international agreement concerning sulphur dioxide emissions. It also demonstrated that not all participating members were prepared to accept an agreement. And it deemed it necessary to reduce sulphur dioxide by 30 percent by 1993 (Gehring 1992, ch. 4).

Although the Munich conference did not address climate directly, its focus on atmospheric sources of pollution from coal-fired generation continued the acid rain driver of G7 climate governance—through the fusion of sulphur dioxide and carbon dioxide—that the G7 had established in 1979.

Predominant Equalizing Capability
The third cause of Bonn's climate achievements was the continuing predominance, if not internal equality, of the G7 members in overall and climate-relevant relative capability (see Appendix I). Scientifically and economically, they remained even more strongly in the global lead, as plummeting oil and commodity prices drove the Soviet Union—the primary rival long bogged down in Afghanistan—into precipitous decline. Yet with America still enjoying its Reagan restoration and soaring "superdollar," Reagan felt no need to defer to the superior climate or overall capability of his G7 partners (Kirton 1999b). He did, however, appreciate the increased defence efforts that Mulroney's Canada and Kohl's Germany had just made. Even after America's overall relative capability, based on its suddenly strong dollar, plummeted after the Plaza Accord signed by the finance ministers of France, Germany, Japan, the US, and the UK in September 1985, Reagan's America remained slow to accommodate its G7 partners on the environment and climate change.

Common Principles
The fourth cause of Bonn's climate change advances, if one of restricted relevance, was the common democratic purpose of the G8 and its members (see Appendix J). The Cold War context brought to the fore the common democratic devotion of the G7, highlighted by the emphasis of this summit taking place on the 40th anniversary of the end of World War II (Reagan 1990, 376–7). But there was no direct connection made between democracy and the environment or climate, and the unifying democratic spirit failed to diminish the differences on international trade. The ideological distinctions between the pro-climate conservative coalition of Kohl and Mulroney, supported generally by Reagan on the one hand and opposed by a resistant socialist Mitterrand on the other, was consistent with the dominant alignments and divisions on these issues. But it did not prevent the accommodation that produced agreement in the end.

Political Cohesion

The fifth cause of Bonn's success on the environment and climate change was the high domestic political cohesion of G7 leaders from their political capital, control, continuity, competence, commitment, and civil society pressure (see Appendices K-1 and K-2). Political control was high for Kohl's coalition government, whose next election was not until January 25, 1987. Mulroney won his historic massive majority government less than a year before Bonn. Reagan won his second term in November 1984. Mitterrand did not have to go to the polls again until the spring of 1988. Civil society pressure also counted. Kohl as host, under pressure from a growing Green Party, and Mulroney were both pushed by their publics to do more for the environment and especially to control acid rain.

Constricted Club Participation

The sixth cause of Bonn's environmental success was its constricted participation as a personal club (see Appendix L-1). Together with the leaders' relative domestic freedom, it allowed them to act freely and informally to come together to some degree. Reagan's affection for his "best" friends from Canada, Germany, Britain, and Japan helped to broaden the pro-climate coalition. But spontaneous combustion at the summit was evident only on the issue of drugs, and noticeably absent on trade, the environment, and climate. It took active mediatory diplomacy by Mulroney to get the environmental agreement that was reached.

The Tokyo Twilight, 1986

As with the Tokyo invention in 1979, the G7's Bonn rediscovery of climate change in 1985 was not sustained. At the subsequent Tokyo Summit (Tokyo II) on May 4-6, 1986, the environment dwindled to receive only a single paragraph at the end of the communiqué. Climate change as an explicit subject disappeared. But behind the scenes, the legacy of Bonn remained alive. Once again, the concert equality model adequately, if not entirely, explains both the Tokyo II disappearance and the momentum that remained.

Preparations

By April 15, 1986, Germany's Helmut Kohl was, in public interviews, noting that the decline in oil prices now underway had not been expected. It was having a beneficial economic effect on many countries including Germany, but hitting the least developed countries and their debt burden hard. He warned of the danger from the view of many that low oil prices were here to stay because they would thus let down their guard. In Kohl's prescient view, low prices might last for a few years, but prices would then rise once again. G7 members should thus not stop their conservation efforts now.

Canada, Germany's traditionally reliable environmental colleague, identified its summit priorities as debt, official development assistance, economic growth and policies, and agriculture. The environment was no longer on the list. Sherpa Sylvia Ostry calculated that energy would be discussed at the summit, that the current line taken by the International Energy Agency (IEA) could not be sustained beyond the autumn of 1986, but that it would be a mistake to change it now. Her officials agreed. They advised the summit team not to change longstanding policies designed to promote energy conservation, diversification of energy types and sources, and more efficient energy use in their economies. This was because IEA members calculated that the effects of low oil prices on energy security would not be evident for some time. Canada's environmental support thus remained intact.

At his pre-summit news conference on April 30, 1986, Mulroney stated that acid rain would be at the top of his agenda at the various meetings he would have with Reagan. In preparing for the Mulroney-Reagan bilateral meeting on May 7, the Canadians identified their priorities as softwood lumber, the US farm bill, Arctic sovereignty, and acid rain, specifically a recent statement by US energy secretary John Herrington that raised doubts about the US commitment to the report of the special joint envoys on acid rain. Mulroney was to suggest that they first meet before July 1. He was also to express concern about the comments, but instead expressed satisfaction that John Negroponte of the US and Don Campbell of Canada had been named as special envoys.

As of May 1, the draft communiqué for the Tokyo Summit contained a draft passage on the environment. By May 3 it contained the environmental paragraph that would appear unchanged in the final communiqué released on May 6. By then the shock of the Chernobyl nuclear accident in the USSR on April 26 had come onto the summit preparatory agenda in a major way. Some feared that the Americans, and perhaps the British and French, would be tempted to use the secrecy of the Soviet reaction to show that little had really changed since Mikhail Gorbachev had arrived as Soviet leader. The Canadians had a different view, concluding that Gorbachev was indeed different from Leonid Brezhnev and even thinking that the US strong reaction to the Chernobyl accident may have made the Soviets less forthcoming. The main message of the summit in Canada's view should be the inadequacies in the international regimes governing nuclear safety and international environmental catastrophes more generally. Castigating the Soviets for their sins would harm the central task of improving international mechanisms for enhancing nuclear safety and environmental programs.

On Sunday morning, May 4, in a bilateral meeting with a summit colleague, Japanese prime minister Yasuhiro Nakasone noted that his government had detected small amounts of iodine in the rain that had fallen on Japan in the past few days. While not dangerous, they came from Chernobyl. He proposed that the summit consider whether the International Atomic Energy Agency (IAEA) should be strengthened to deal with major nuclear accidents. He noted that the IAEA had not been established with a major nuclear accident in mind, that its safeguard procedures were not adequate, and that its mandate and procedures equipped it only to deal with problems of an internal or national nature, not a large-scale disaster affecting many countries at once. Under existing agreements, notification of accidents to the IAEA was voluntary and any country choosing to do so had 40 days in which to act. Nakasone said he was not interested in criticizing the Soviets but in rectifying defects in the IAEA regulatory arrangements to better protect humankind.

Mulroney shared this view. He thought G7 leaders should consider whether an international conference could be useful to discuss the expansion of nuclear safeguards and nuclear disaster management. He offered Canada's facilities for such an event.

At the Summit

At the summit table, the leaders discussed nuclear safety, including on the morning of May 5, when the leaders reviewed the draft communiqué text. Thatcher and Reagan thought the draft on Chernobyl was too weak, while Italy's Bettino Craxi cautioned against causing greater alarm. His soft line was supported by Germany's Kohl and France's Mitterrand. Mulroney tried to find a middle way but soon yielded to support Mitterrand. In the end,

Reagan's proposed language won out, but it began on a dovish note. It started "we note with satisfaction the Soviet Union's willingness to undertake discussions this week with the Director-General of the International Atomic Energy Agency" (G7 1986b).

Leaders also discussed how to respond to the new reality of cheaper oil. The question was whether to let the lower prices flow through to consumers, leading them to consume more. Leaders had no inclination either to intervene to keep prices high or to make any changes to their existing energy policy designed to reduce their dependence on imported oil. The EU leaders and Mitterrand were preoccupied with the risks of the falling price. They were opposed by Thatcher, her chancellor of the exchequer Nigel Lawson, and America's secretary of state George Shultz and treasury secretary James Baker. Kohl proposed a compromise. His foreign minister Hans-Dietrich Genscher argued that the decline in oil prices was not so clearly beneficial as the existing text proposed. Nakasone concluded the argument, coming down on the Anglo-American side by accepting the text as it was.

Acid rain was left to bilateral discussions. Mulroney's bilateral with Reagan was cancelled because the president was tired. But in an informal discussion during the main summit, Mulroney raised with him the issues of acid rain and Canada's free trade initiative, but not the other two priorities of softwood lumber and Arctic sovereignty that Canada had identified for the bilateral.

Results

On nuclear safety, the leaders issued a separate, four-paragraph statement that began with sympathy for the victims of Chernobyl, an offer of assistance, an affirmation of properly managed nuclear power as an increasing energy source, and a stress on the transboundary consequences of nuclear accidents (G7 1986a). It called for strengthening the IAEA and the rapid creation of an international convention on nuclear safety. The once national, highly political, national security issue of nuclear power had now gone global, and become a human security issue of health and environmental safety in the physical world. It did so in the G7 leaders' shared perception, principles, and call for the development of global governance through a reformed UN body and new international law.

Dimensions of Performance

The 1986 Tokyo Summit was a very small success on climate change. It did nothing on the subject specifically, but did not make its governance worse and added general support through its performance on more general and related environmental issues.

Domestic Political Management

The summit was a domestic political management success for Kohl on the Chernobyl front. German media reaction to the summit focused on terrorism, Chernobyl, and exchange rates. Chernobyl was reported as a personal victory for the chancellor, who was described as having seized the initiative for the summit discussion. Genscher was reported as saying that Chernobyl showed the need for better East-West cooperation on new technology. However, Kohl's summit partners felt his views were aimed at a domestic audience, with a campaign trail quality.

The Chernobyl shock was experienced perceptually and politically as well as physically in Italy on the summit's eve. Italian media coverage of the summit on May 3–4 was overshadowed by the Chernobyl explosion.

Deliberation
On oil, the communiqué paragraph noted: "bearing in mind that the recent oil price decline owes much to the cooperative energy policies which we have pursued during the past decade, we recognize the need for continuity of policies for achieving longterm energy market stability and security of supply" (G7 1986b; see Appendix A-1). The same successful spirit of cooperation and compliance that had created the great climate control advances in 1979 was still alive.

Direction Setting
The leaders' communiqué paragraph on the environment endorsed the principles of international cooperation, pollution prevention and control, and management of natural resources, while adding cooperation with developing countries to the list (see Appendix C-1). It read: "we reaffirm our responsibility, shared with other governments, to preserve the natural environment, and continue to attach importance to international cooperation in the effective prevention and control of pollution and natural resources management" (G7 1986b).

Development of Global Governance
The communiqué also noted: "In this regard, we take note of the work of the environmental experts on the improvement and harmonization of the techniques and practices of environmental measurement, and ask them to report as soon as possible. We also recognize the need to strengthen cooperation with developing countries in the area of the environment" (G7 1986b). Yet it made no reference to the work of the environment ministers. Measurement had become key and G7-centred environmental governance was the exclusive way to get this done.

Overall, G7 climate governance was kept alive on several flanks. One was the continuing conviction on the part of Canada and Germany that the principles of energy conservation and efficiency should be affirmed despite plummeting oil prices in the first half of 1986. The second was the G7's effort, in response to the Chernobyl nuclear explosion on the summit's eve, to strengthen the international nuclear safety regime and affirm the value of nuclear power as an energy source. The third was Canada's bilateral intervention with the US at the summit on acid rain. While none of these three advances was explicitly linked to climate change, each offered support on the flanks until the time arrived when all could be joined.

Causes of Performance

Shock-Activated Vulnerabilities
The first cause of this climate-supportive governance was shock-activated vulnerability. Rising oil price shocks had long been a cause of summit success, as the first Tokyo Summit had clearly showed in 1979. This time, at Tokyo II, they worked in reverse. As recently as January 1986, oil had been trading at $47 a barrel. But by the summit's eve it had dropped to about $26 per barrel, for a plunge of almost 50 percent (see Appendix G-3). The price of grain had fallen dramatically too. The shock of plunging oil prices was discounted despite the political pain it caused oil-producing countries such as Canada. The memories of the shock sequence of 1973 and 1979 were too strong to be forgotten. Indeed, the plummeting oil price was interpreted as confirmation of the success of the G7's cooperative, conservationist oil policies of the past.

The disappearance of climate change directly from the agenda at Tokyo II and its diffusion to indirect, climate-enhancing action on oil, nuclear energy, and acid rain was also driven by the eruption of a new shock that activated G7 members' vulnerability to old and new threats. The shock of Chernobyl on the eve of the summit in the cognate field of nuclear energy produced an outburst of G7 cooperation. It was a second, now very deadly, nuclear safety shock, following the Three Mile Island accident in America in 1979 in which no one had died. It hit on April 26, just a few days before the summit opened on May 4. It was delivered directly to G7 leaders as they stood in Tokyo in the rains that carried radioactive waste from the explosion half a world away. This time the G7 response was not to abandon climate-friendly nuclear energy as too unsafe to rely on. Rather, it was to defend nuclear power in all places save the Soviet Union and to push for a stronger international nuclear safety regime.

The Canadians saw the Chernobyl disaster not merely as a nuclear power one, but as a shock that signalled environmental catastrophes more generally and as a spur to improve international mechanisms for cooperative environmental programs across the East-West divide. Chernobyl catalyzed in the Canadians—in sharp contrast to the Americans—a desire for global environmental cooperation rather than a return to the old Cold War. The new nonstate existential global human security threat trumped the old state-to-state national security one. The new, unintended global human security vulnerability had arrived to trump the old national security logic of an earlier age.

Multilateral Organizational Failure
In the face of the Chernobyl shock, multilateral organizational failure also spurred Tokyo II's climate-supportive advance (see Appendix G-6). The IAEA team, led by Hans Blix, was on site at Chernobyl from May 5 to 9, when military helicopters dumped 5,000 tons of sand, clay, dolomite, boron carbide, and lead on the burning reactor core (Park 1989). However, it is now known that it did not cover the core, but only the "red glow," which was a minor portion of it (Sich 1996). To investigate the causes of the nuclear accident, the IAEA created the International Nuclear Safety Advisory Group. In August 1986, the IAEA held a review meeting. The board of governors conducted a post-accident expert analysis and gave directions on several concrete points to deal with weaknesses in international collaboration on nuclear energy safety standards. They agreed on early warning systems and on the provision of machinery for emergency assistance. They further identified the need for a global network to transmit figures on levels of radiation. In the end, the IAEA concluded that accident management and strengthened collaboration were required (Blix 1986). But even with the IAEA's immediate response, there were criticisms of the level of transparency and the failure to adequately transmit information to the general public and the international community at large.

Common Principles
The G7's devotion to open democracy produced a geographic bifurcation (see Appendix J). In the West, it added to the remembered energy success of the past to produce a safe, clean nuclear energy result. Across the East-West divide, it was reinforced by the advent of Gorbachev's new thinking in favour of political openness in the Soviet Union. This offered a glimmer of hope that the G7 would be successful in its push for a new international nuclear safety regime.

Conclusion

The 1985 Bonn Summit marked a rediscovery of G7 and thus global climate change governance. G7 leaders produced their first climate commitment and complied with it well. The environment had priority placement in the chapeau of the communiqué. The language illustrated the G7's preparedness to deal with environmental issues formally within the G7 platform. The leaders declared the environment and the economy to be equally valuable and mutually supportive.

Acid rain was a major focus. An impetus for this concern largely came from Germany and Canada, which were both feeling the effects of coal-burning neighbours, respectively Britain and the United States. The famine in Ethiopia, brought on by drought, attracted widespread public attention. Even though there were contradictory ideological stances between the conservative pro-climate coalition of Kohl and Mulroney, on the one hand, and a reluctant Mitterrand and Reagan, on the other, ultimately this did not prevent a successful outcome and they managed to reach an agreement on environmental issues. Mulroney successfully mediated the negotiations to produce a communiqué passage. A group on Africa, proposed by Germany, was also born.

At the 1986 Tokyo Summit, however, the communiqué on the environment was reduced to one paragraph. Nonetheless, G7 climate-supportive governance was still pursued indirectly but with rigour, again mainly due to the efforts of Canada and Germany. Acid rain remained at the top of Mulroney's list of issues to discuss with Reagan. With the Chernobyl nuclear accident and plummeting oil prices on the eve of the summit, G7 leaders turned to other shock-driven issues but still governed energy in climate-protective ways.

References

Benedick, Richard E. (1991). *Ozone Diplomacy: New Directions in Safeguarding the Planet*. Cambridge MA: Harvard University Press.

Blix, Hans (1986). "The Post-Chernobyl Outlook for Nuclear Power: A View on Responses to the Accident from an International Perspective." *IAEA Bulletin*, Autumn, pp. 9–12. http://www.iaea.org/Publications/Magazines/Bulletin/Bull283/28304780912.pdf (September 2014).

Franklin, Claire A., Richard T. Burnett, Richard J.P. Paolini, et al. (1985). "Health Risks from Acid Rain: A Canadian Perspective." *Environmental Health Perspectives* 63: 155–68. http://www.ncbi.nlm.nih.gov/pmc/articles/PMC1568495/pdf/envhper00446-0153.pdf (January 2015).

G7 (1984). "London Economic Declaration." London, June 9. http://www.g8.utoronto.ca/summit/1984london/communique.html (January 2015).

G7 (1985). "The Bonn Economic Declaration: Towards Sustained Growth and Higher Employment." Bonn, May 4. http://www.g8.utoronto.ca/summit/1985bonn/communique.html (September 2014).

G7 (1986a). "Statement on the Implications of the Chernobyl Nuclear Accident." Tokyo, May 5. http://www.g8.utoronto.ca/summit/1986tokyo/chernobyl.html (January 2015).

G7 (1986b). "Tokyo Economic Declaration." Tokyo, May 6. http://www.g8.utoronto.ca/summit/1986tokyo/communique.html (January 2015).

Gehring, Thomas (1992). *Dynamic International Regimes: Institutions for International Environmental Governance*. Frankfurt: Peter Lang.

Hoffmann, Matthew (2005). *Ozone Depletion and Climate Change: Constructing a Global Response*. Albany: SUNY Press.

Kirton, John J. (1999). "Explaining G8 Effectiveness." In *The G8's Role in the New Millennium*, Michael R. Hodges, John J. Kirton, and Joseph P. Daniels, eds. Aldershot: Ashgate, pp. 45–68.

Mulroney, Brian (2007). *Memoirs*. Toronto: McClelland and Stewart.

Park, Chris C. (1989). *Chernobyl: The Long Shadow*. Abdingdon: Routledge.

Putnam, Robert and Nicholas Bayne (1987). *Hanging Together: Co-operation and Conflict in the Seven-Power Summit*. 2nd ed. London: Sage Publications.

Reagan, Ronald (1990). *An American Life*. New York: Pocket Books.

Sich, Alexander R. (1996). "The Denial Syndrome." *Bulletin of the Atomic Scientists* 52(3): 38–9.

United Nations (1985). "Protocol to the 1979 Convention on Long-Range Transboundary Air Pollution on the Reduction of Sulphur Emissions or Their Transboundary Fluxes by at Least 30 Per Cent." Entered into force 2 September 1987. http://www.unece.org/fileadmin/DAM/env/documents/2012/EB/1985.Sulphur.e.pdf (January 2015).

Chapter 4
Revival, 1987–1988

Climate change, which first appeared on the summit agenda of the Group of Seven (G7) in its own right in 1985, disappeared the following year. The 1986 Tokyo Summit's references to pollution, natural resource management, and environmental measurement were the closest it had come to addressing climate change. However, unlike the long sabbatical from 1980 to 1984, climate change returned the year after Tokyo. The 1987 G7 Venice Summit (the second held in Venice) dealt with the issue as part of a stand-alone list of environmental issues that included environmental measurement, technological innovation, and stratospheric ozone depletion. Venice also marked the second time the leaders issued a concrete climate-related commitment in their communiqué.

Henceforth, environmental and climate change issues became a regular and often robust part of the G7 agenda. Under the impulse of Canadian prime minister Brian Mulroney, in 1988 climate change started to be defined and expanded. The Toronto Summit called for the creation of an intergovernmental panel on climate change, under the auspices of the United Nations Environment Programme (UNEP) and the World Meteorological Organization (WMO). It further welcomed the international conference on "the Changing Atmosphere" that would be held immediately soon after the summit. As of then, global environmental and climate change issues found a prominent place on the G7 summit agenda, often generating ambitious commitments that the leaders of the most powerful democracies implemented to a substantial degree.

Venice, 1987

The Venice Summit, taking place from June 8 to 10, 1987, was hosted by Italian prime minister Amintore Fanfani, with help from Giulio Andreotti, minister for foreign affairs, Giovanni Goria, minister of the treasury, and, above all, given the electoral distraction of the leaders and ministers, Renato Ruggiero, the Italian prime minister's personal representative, or sherpa. Venice brought back climate change in its own right as a concern, even if ozone depletion now leapt ahead in the list of global environmental problems the summit addressed. The communiqué contained three full paragraphs on the environment, triple the number from the previous year, even if the environment and climate change were still absent from the communiqué's introductory preamble.

Venice marked a turning point in G7 climate governance in other ways. Most notably, ozone replaced the still present acid rain as the dominant environmental focal point. Policies related to environmental measurement now represented the key thrust. And consistent with these shifts, the G7 explicitly handed the leadership for such issues to the established, still largely scientific multilateral institutions of the United Nations, most notably UNEP.

Preparations

Canada was admitted to the G7 meeting of finance ministers and central bank governors at Tokyo in 1986, which expanded the country's role in economic policy coordination at the summits, although Canada and Italy were excluded from the short opening session of the finance ministers and central bank governors of France, Germany, Japan, the United Kingdom, and the United States in Paris in February 1987. Italy responded by refusing to take part in the rest of the meetings. In contrast, Canadian finance minister Michael Wilson patiently accepted his exclusion from the first meeting, participated in the rest of the discussions, and was ultimately a signatory to the resulting Louvre Accord. Afterward, Canada would no longer be excluded from the finance meetings, which had begun to play an increasingly prominent role in setting the agenda for the global economy. But in 1987, it was left to Canada's prime minister to advance the environmental cause.

In contrast to Mulroney's high-profile behaviour at previous summits, he arrived at Venice quietly and immediately went into deep retreat. There were two particular issues he wished to advance: agricultural subsidies and apartheid in South Africa. He wanted the leaders to commit to including agricultural subsidies in the upcoming Uruguay Round of negotiations of the General Agreement on Tariffs and Trade. On South Africa, he felt strongly about apartheid and wanted it addressed by the leaders.

In preparation for Venice, the European Commission had, for the first time, drawn up a detailed working paper outlining the problems of sub-Saharan African debt, analyzed the specific circumstances of the most heavily indebted countries in this region, and proposed special measures to remedy their situation.

Thus, as Venice approached, it became increasingly clear that debt management, North-South relations, macroeconomics, terrorism, and now HIV/AIDS would feature high on the summit's agenda. There was little expectation that the environment would be a prominent part.

At the Summit

At the summit there was a debate about adding non-economic issues such as East-West relations, arms control, freedom of navigation in the Persian Gulf, South African apartheid, HIV/AIDS, and the global narcotics trade to the communiqué. Some argued these issues should be included due to their mention in past communiqués. Others, including the US, argued that repetition merely dulled the communiqué, cheapened its currency, and weakened its impact. Moreover, the communiqué, negotiated by officials, reflected delicate compromises that were often stated in language unintelligible to those who read the final text. The Americans lost this debate. Subsequently, prior to the 1988 Toronto Summit, US president Ronald Reagan sent a letter to host Brian Mulroney saying that the communiqué should contain only things that the leaders themselves had actually discussed. The Americans, with Reagan approaching his last full year in office, had lost not only their leadership at the summit table, but their veto as well. Yet even with this agenda expansion, the environment received no document of its own.

The principal economic themes emerging at the summit included the Uruguay Round of multilateral trade negotiations, agricultural subsidies, and the plight of the developing world. The Canadians continued to pursue the North-South debt issue, proposing—along with France and the United Kingdom—a structural adjustment facility for the International Monetary Fund. Indeed, during the summit, Mulroney telephoned Kenneth Kaunda, president of Zambia, to inform him that his country was being used as an example of the need for greater assistance to the developing world, setting the stage for what would emerge as a more prominent North-South dialogue at future summits (Heeney 1988). The integration of the environment and the economy, under the concept of sustainable development, had not yet taken hold.

Results

Venice produced a record number of non-economic statements on natural and human issues, including East-West relations, arms control, freedom of navigation in the Persian Gulf, South African apartheid, AIDS, and the global narcotics trade (Hajnal 1989; G7 1987). All found their way into the communiqué, despite their relative absence in the draft text prepared by the sherpas.

Amid the general expansion of the agenda, Venice marked the start of continuing increases in the amount of time and attention leaders dedicated to climate change. The final communiqué dedicated a three-paragraph section to environmental issues. The first noted the importance of harmonizing techniques and practices related to environmental measurement. It encouraged UNEP "to institute a forum for information exchange and consultation" in cooperation with other countries and interested international organizations (G7 1987). The second paragraph noted the leaders' intent to further examine innovation on clean, cost-effective, and low-resource technologies, including the "promotion of international trade in low-pollution products, low-polluting industrial plants and other environmental protection technologies" (G7 1987).

The second paragraph also directly addressed climate change. It now placed climate change primarily within the context of ozone as well as acid rain. In doing so, the summit generated the G7's second climate commitment. It underlined the leaders' own responsibility to encourage efforts to tackle "stratospheric ozone depletion, climate change, acid rain, endangered species, hazardous substances, air and water pollution and the destruction of tropical forests" (G7 1987).

A year later, the seven members had complied with this commitment at an average of +0.27. This was a little lower than the average of +0.31 for the economic and energy commitments from 1975 to 1989 combined, and much lower than G7 members' compliance with their energy commitments (Daniels 1993). Still, it was in the positive range.

Moreover, the environment in general began to infuse other areas of the communiqué, beyond the traditional energy domain. The market-friendly passage on agriculture added the need to give consideration to social and other concerns, such as food security, environmental protection, and overall employment. The environment-agriculture connection thus first appeared.

Dimensions of Performance

The Venice Summit was a solid success for climate change, particularly in deliberation, decision making, and delivery (see Appendix A-1).

Deliberation

In its deliberation, the summit restored the attention to climate change specifically in its communiqué. Climate change appeared in one paragraph, accounting for 1.5 percent of the total words in the communiqué (see Appendix C-1). In this paragraph, G7 leaders underlined their responsibility to deal effectively with global environmental problems "such as stratospheric ozone depletion, climate change, acid rains … and the destruction of tropical forests" (G7 1987). They also stated their intent to examine other issues such as environmental standards and the development of clean, cost-effective, and low-resource technologies.

Decision Making

In their decision making, G7 leaders at Venice produced their second commitment on climate change. It appeared in the paragraph noted above (see Appendix D-1).

Delivery

On their delivery of this decision, the G7 leaders' performance was small. The level of compliance was positive, but only at an overall average of +0.29 (see Appendix E-1). The United Kingdom and Canada fully complied, while the United States, Japan, Germany, France, and Italy partially complied.

Development of Global Governance

The Venice Summit developed global climate governance not within the G7 itself, but by handing it over to the UN. To be sure, the G7-generated experts were to provide the blueprint. But the work was to be done by UNEP, with the independent International Organization for Standardization (ISO) and International Council of Scientific Unions (later the International Council for Science). While two of the three specified bodies were scientific in nature, UNEP brought in a political dimension and the inevitable UN politics of the old North-South divide.

Causes of Performance

Venice's small performance on climate change flowed from the limited new climate-relevant shocks, the distraction of other issues, and low direct political control by the leaders at home. With the Soviet Union under the leadership of Mikhail Gorbachev undergoing *glasnost* and *perestroika*, G7 leaders focused on how far these proposed changes would shift the USSR from its traditional path. Consequently, the leaders were attuned to non-environmental issues and discouraged policy change, preferring to reinforce previous policies and those made within other international forums (Heeney 1988).

A significant factor was the presence of electoral concerns in some members, primarily Italy and the UK. A preoccupation with presidential elections in the US and France emerged. Reagan was also under severe criticism at home for his involvement in the Iran-Contra scandal, in which senior officials in his administration were accused of secretly facilitating the sale of arms to Iran, the subject of an arms embargo, in an effort to use the proceeds to fund the anticommunist Nicaraguan Contras.

Shock-Activated Vulnerabilities

The small achievements at Venice were due in part to the shock-activated vulnerabilities to G7 members that year. The first shock was the 10 accidents involving oil tankers and rigs, which each spilled more than 700 tonnes of oil (see Appendix G-4).

Predominant Equalizing Capability

The second cause of Venice's small success was the increase in G7 members' overall capability (see Appendix I). Each member experienced more growth in gross domestic product than in the previous year.

Common Principles

The third cause was the common democratic purpose of the G7 and its members (see Appendix J). On the Polity score index, each member scored a +10, representing a consolidated democracy, with the exception of France, which scored a still high +9.

Political Cohesion

The fourth cause was domestic political cohesion. The lack of political control held by Italian host Fanfani, who was not re-elected in the subsequent election, may partly explain why Venice's success was small (see Appendices K-1 and K-2).

Constricted Club Participation

The fifth cause of Venice's success was its constricted participation (see Appendix L-1). The summit was as usual attended by the leaders of the seven member states plus the two leaders of the European Communities, with no external participants, allowing for a free exchange of ideas and the forging of consensus.

Assessment

To a considerable extent, Venice reaffirmed existing economic policies (Hainsworth 1990). The global environment had not yet taken centre stage, although the leaders were beginning to realize the benefits of a healthy and sustainable environment.

Toronto, 1988

The Canadian-hosted Toronto Summit from June 19–21, 1988, saw a major advance in G8 climate governance. The 34-paragraph communiqué contained three paragraphs devoted to the environment, although it remained absent from the chapeau. The first paragraph endorsed the Brundtland Commission report and the concept of sustainable development (World Commission on Environment and Development 1987). The second proclaimed the Montreal Protocol on the ozone to be a milestone that all countries should sign and ratify. Flowing from this context, the third paragraph "encouraged as well as the establishment of an inter-governmental panel on global climate change under the auspices of UNEP and the World Meteorological Organization," with the leaders "welcoming" the Conference on the Changing Atmosphere to be held in Toronto the following week (G7 1988).

The signature achievements of Toronto were thus to help create the Intergovernmental Panel on Climate Change (IPCC) and endorse the international conference on the environment and thus, implicitly, the conclusions both were set to reach. Toronto also developed the environmental protection agenda and sustainable development principle, which would both dominate the summit at Paris in 1989 and form a core element of G7 summitry in the years beyond. More broadly, Toronto reinvigorated the summit process, as the Canadian host had intended, by adding an initial half-day discussion on economic issues, inserting a mini-retreat to focus on the future, and injecting issues of environmentally sustainable development, education and literacy, and aging populations. In the end, the G7 led global governance on the environment, for which no specialized UN agency or charter principles existed at all.

Preparations

In shaping the Toronto Summit, Canada sought to advance five domestic and international priorities—debt of the poorest, trade, agricultural subsidies, structural adjustment, and the environment (Kirton 2007b). Yet although the environment was now included, economic objectives took pride of place. Canadian sherpa Sylvia Ostry insisted that the event be called the "Toronto Economic Summit" and that its economic declaration alone be called the "Summit Communiqué." In this spirit, others recommended not issuing separate G7 statements on issues such as AIDS and drugs (as had been done the previous year at Venice), the environment, or minor regional political concerns.

The leaders themselves were eager to get a communiqué of manageable size. Mulroney as host said he would not read a long text. The drafters thus struggled to incorporate all the suggestions at each sherpa meeting and keep the overall length to a minimum. Sherpa teams became obsessed with the question of length. The leaders felt the sherpas were out of control when they saw draft communiqués that were nine, ten, or eleven pages long. Drafting was a debilitating exercise, particularly in regard to the introduction, conclusion, and the three long paragraphs on the environment. But in the end, those three paragraphs—the most vulnerable item in the quest for compression—survived.

Indeed, on the summit's still relatively new agenda of global or transnational issues, the environment took the prime spot. Although Mulroney was still considering what issues to discuss at his 90-minute mini-retreat at Hart House on the campus of the University of Toronto, by the middle of May, the environment, along with East-West economic relations, appeared possible as the main item on the summit's broader agenda. The environment had arisen during Mulroney's bilateral visit with Reagan in Washington in April.

By mid May, the environment was becoming a more salient topic in domestic Canadian politics. Indeed, by the end of 1988, the Canadian public had come to view the government as a regulator and guarantor of quality-of-life issues, and see the environment as "Canada's single most important issue" (Kopvillem 1989). This cadence was important, as Mulroney was considering calling a general election that autumn, on the fourth anniversary of his first victory. Mulroney also recognized that the environment and the Brundtland Commission were key issues with which he wanted to be associated at both his G7 summit and the subsequent Conference on the Changing Atmosphere.

By the time of the sherpa meeting on June 3, it was clear that the environment would indeed be discussed by the leaders. The focus would be on climate concerns, specifically, the ozone problem and the greenhouse effect. Both Germany and Canada were substantially stronger than the others in pushing this issue, which was strongly opposed by the US and Britain. Italy, for domestic political reasons, would now join Canada and Germany's side. The new Italian prime minister, Ciriaco De Mita, now felt pressure to steal the environmental initiative from the opposition party he faced.

It was thus agreed that environmental issues would be addressed in the plenary meetings (where leaders were joined by their foreign and finance ministers) and at the end of the summit. Germany also felt it important to include the environment in the communiqué. There was a consensus that the 90-minute Hart House session would discuss two issues—demography in industrialized countries and education, along with the environment. None of the Hart House items, save the environment, would be carried over into the deliberately relatively short Toronto communiqué.

On the summit's eve, its agenda still focused on macroeconomics and trade, but also offered an attractive opportunity for Canada to show its strong commitment to the environment. It was now slated to be dealt with during the separate session on the afternoon of Monday, June 20. Here, Canada sought greater international cooperation and broad adherence to the ratification and implementation of existing conventions such as the Montreal Protocol. It also sought greater cooperation on the integration of environmental considerations into economic decision making, and G7 leadership on understanding and planning to mitigate the effects of climate change, through such instruments as the Conference on the Changing Atmosphere. Germany was expected to be very supportive of a discussion of global environmental issues, including affirming the principle of an environment-economy link. Italy by then had stated that it too wanted an environmental discussion. France, the European Commission, and Japan supported a general discussion but not a coordinated summit conclusion or action. The most resistant member was the UK, especially in regard to air pollution. The US appeared pleased to discuss ozone and climate change, but not be confronted on acid rain. There being no coordinated initiative proposed, the Canadians foresaw an occasion for a good exchange of views.

At the Summit

Mulroney opened the Toronto Summit on Sunday, June 19 (Burney 2005; Mulroney 2007; Kirton 2007b). He called for freewheeling discussions and noted that Canada's priorities were debt, agriculture, and trade, especially the free trade agreement (FTA) with the US. Germany's Chancellor Helmut Kohl stressed the need to discuss the environment, which affected the least developed countries, and to refer to the environment in the communiqué. This led Mulroney to note that environmental issues would be discussed, along with education and training, at the Monday night dinner and during informal sessions. Japan's Prime Minister Noboru Takeshita noted his country's desire to become a major global aid donor, and agreed that the summit should provide an impetus on agriculture (Dobson 2004). Italy's De Mita resisted singling out agriculture, and called for attention to the environment as an integral issue transcending all global problems. The EU's Jacques Delors also defended European agricultural policy, but added a welcome for the FTA.

At their dinner on Monday, the foreign ministers discussed the Iran-Iraq war. At Japan's initiative, they also dealt with North Korea and the forthcoming Seoul Olympics. Germany's foreign minister Hans-Dietrich Genscher then raised environmental issues, describing the ozone treaty as the absolute minimum needed and noting the need for a new generation of domestic policies with international effects.

Results

The release of the final communiqué on June 21 yielded 34 paragraphs. They included economic policy coordination plans, the multilateral trading system, development, developing country debt, and issues related to new, industrializing economies. The environment received a stand-alone section, with the leaders dedicating three paragraphs there to global environmental concerns. In the first paragraph of that section, the leaders agreed that protecting and enhancing the environment was "essential." For the first time, they endorsed the concept of "sustainable development," stressing that "environmental considerations must be integrated into all areas of economic policymaking if the globe is to continue to support humankind" (G7 1988).

The second paragraph acknowledged that "threats to the environment recognize no boundaries" and that "their urgent nature requires strengthened international cooperation among all countries" (G7 1988). They remarked on the "significant progress" made on several issues, particularly the "milestone" of the Montreal Protocol on Substances that Deplete the Ozone Layer that "all countries [were] encouraged to sign and ratify."

Climate change specifically was referred to in the third paragraph. It noted that "priority attention" and "further action" were needed through the establishment of an intergovernmental panel on global climate change to be led by UNEP and the WMO (G7 1988). The leaders also welcomed the Conference on the Changing Atmosphere.

At the conclusion of the summit, in the estimate of the Canadian government, Canada had achieved its main objectives of restoring market confidence, enhancing the credibility of the summit forum through useful discussions and good organization, and advancing debt of the poorest, trade and agriculture, international economic policy coordination, structural adjustment, and the environment.

Dimensions of Performance

The Toronto Summit was a significant success for climate change, marking the first time the issue had been dealt with two years in a row, making it a permanent part of the G7 agenda, and, above all, for the developing global governance outside the G7 by advising the UN to act (see Appendix A-1).

Domestic Political Management
In its domestic political management, Toronto was a strong success, above all for its host (see Appendix A-1). Mulroney used the momentum of a summit broadly viewed as successful at home to help win the election he called for November. While the key issue was the FTA that the G7 leaders had endorsed, the environment was linked to the new trade agreement at a late stage of the campaign. Mulroney's G7 summit thus enhanced his credentials as an environmentalist, helping him become re-elected with a second majority mandate.

Deliberation
In its deliberation, Toronto's performance was significant. It devoted 140 words to climate change in the final communiqué. This was the highest amount to date and significantly more than the 85 words on climate change at Venice in 1987, the 88 at Bonn in 1985, and the 28 at Tokyo in 1979 (see Appendix C-1).

Direction Setting
In its direction setting, the Toronto Summit made important advances in giving prominence, backed by new principles, to environmental issues. For the first time, it endorsed the concept of sustainable development and the integration of the environment into all economic decision-making elements. It declared environmental protection and enhancement to be "essential" and climate change to be addressed "urgently" (G7 1988).

Decision Making
In its decision making, Toronto produced no specific commitments on climate change. However, the Canadian government accurately saw the summit as forwarding movement

on several relevant issues: debt of the poorest in particular, as well as trade and agriculture, international economic policy coordination (through the leaders' endorsement of two commodity price indicators including gold), and environmental and structural reform.

Development of Global Governance

In its development of global governance on the environment and climate change, Toronto helped launch an ongoing process for the changing global atmosphere. It did so almost exclusively beyond the G7, initially through the Montreal Protocol, then more robustly through the Conference on the Changing Atmosphere, and, ultimately, through the IPCC and the United Nations Framework Convention on Climate Change, with specific references to UNEP and the WMO (see Appendix F-5). Hence, there emerged a defined and agreed track of UN-centred international meetings on environmental issues, particularly as they related to changing atmospheric conditions.

Causes of Performance

The causes of the climate change advances at Toronto came in the first instance from the enthusiasm of Canada's Brian Mulroney as host, with the support of Germany's Kohl, Italy's De Mita, and the openness of the others, save for Margaret Thatcher alone. Yet behind this large consensus in favour of general principles and processes lay deeper drivers.

Shock-Activated Vulnerabilities

The first was the shock on the same subject as global warming from the unusually hot summer, especially in North America, and the drought and deaths that the heat wave caused. In both the US and Canada, this heat wave in the US received heavy attention, especially on television network news. In turn, this heavy coverage made global warming in particular, and the environment in general, a highly salient issue and plausible phenomenon to mass publics as seen in the opinion polls.

Multilateral Organizational Failure

A second cause was multilateral organizational failure. The UN system still did little at any policy level to confront this visible climate shock that was harming and killing Americans at home. Emphasis was thus placed on new, plurilateral institutions, notably the G7, the ad hoc Toronto Conference on the Changing Atmosphere, and the UN's scientific IPCC.

Predominant Equalizing Capability

A third cause was changing relative capability (see Appendix I). Overall, the G7's predominance increased as the Soviet Union stagnated. Internally, equality among G7 members increased as America's 1980–1985 surge brought by the Reagan restoration and the soaring superdollar had by 1988 all but disappeared (Kirton 1999b).

Common Principles

The fourth cause was the shared common principles of democracy among G7 members (see Appendix J). All members scored high on the Polity IV ranking of democracies and thus were committed to the G7's fundamental mission of promoting democracy.

Political Cohesion

A fifth cause was political cohesion—the solid political control, capital, continuity, commitment, competence, and civil society responsiveness of G7 leaders (see Appendices K-1 and K-2). Most were long-time veterans of G7 summits. Looming large was the proximity of elections for relatively unpopular leaders, above all Mulroney as host, who would call his election soon after the summit and win a second majority mandate in November. In such a situation, Mulroney sought to make the summit a domestically compelling event. The environment, reinforced by the Conference on the Changing Atmosphere, was an attractive candidate here in view of the polls and the domestic criticism he faced over acid rain and his central bilateral FTA with the United States. According to the 1988 *Maclean's* Decima Poll published on January 2, 1989, when Canadians were asked about the most important problem facing Canada today and the one they were most concerned about "pollution/ environment, followed by unemployment, and then free trade (*Maclean's* 1989). While Reagan's America was unyielding on acid rain, it was open to a summit discussion on climate change. And while a lame-duck Reagan would soon leave office, his vice-president George H. Bush sought to succeed him in November by appealing to American voters who increasingly had environmental concerns. Italy's De Mita, with a fragile coalition that could fall at any time, similarly had the domestic electoral attractiveness of environmental issues much on his mind.

Constricted Club Participation

The sixth cause of the climate change advance at Toronto was the constricted, club-like participation at the summit (see Appendix L-1). The interpersonal solidarity bred by this continuity and constricted participation was enhanced by the innovative format of the Hart House "focus on the future" session. It allowed leaders to be leaders in a free-flowing, unscripted, relaxed exchange among themselves alone. This led directly to the important innovative deliberative and direction-setting achievements of adding education and literacy to the G7 summit agenda, sharpening the principles for its evolving environmental agenda, and providing the foundation for the many major breakthroughs on the environment that François Mitterrand's Paris Summit would make the following year. In particular, this format allowed the leaders to deal with climate change without having to make many hard, precise, public commitments in the final communiqué.

After the Summit

The week following the summit, from June 27 to 30, Canada hosted the Conference on the Changing Atmosphere in Toronto as a means of finding ways to improve the world's capacity for forecasting environmental change. Its key objectives were to increase international awareness of the changing atmosphere, develop strategies and actions to stabilize and reduce the adverse human influences, and increase global cooperation in programs to forecast change and reduce harmful emissions (WMO et al. 1988). Conference delegates included senior government officials from all G7 members, ambassadors, legal and policy advisors, environmental and social scientists, energy industry representatives, and nongovernmental organizations. Issues discussed over the four days included food security, urbanization, conservation, and settlement, as well as coastal and marine issues, forecasting, trade, and investment. The energy working group set ambitious targets to reduce annual carbon dioxide emissions by 20 percent by 2005 through improved energy efficiency, alternative energy supply, and conservation (p. 300). Although that represented a progressive target, the working group did not emphasize the economic implications of reducing emissions by such levels—particularly in the developing world—resulting in an overall lack of support from fellow delegates for such ambitious targets and timetables.

Recommendations did emerge, however, on the creation of a legal regime through an umbrella or framework convention that would lend itself to the further development of specific agreements or protocols laying down international standards for the protection of the earth's atmosphere. This recommendation would ultimately form the foundation for the UN Framework Convention on Climate Change (UNFCCC). Such a convention would include a recognition of the means available to states for dealing with atmospheric problems, encouraging compliance through incentive mechanisms, and providing a forum for the coordination of scientific activities and technological research (WMO et al. 1988, 346).

However, given the diversity of representatives, sectors, interests, opinions, and knowledge, the working groups in general failed to produce more acute and practical suggestions for either halting undesirable changes in the atmosphere or adapting in any meaningful way to these changes (Rosenberg et al. 1989). In the end, the conference succeeded in making climate change and the changing atmosphere globally recognized. It was well staged, well organized, and well orchestrated and garnered a significant amount of international media attention. The conference's success was seen in the knowledge it produced on all aspects of changing atmospheric conditions. Organizers stressed in their closing remarks that the meeting had accelerated the momentum in dealing with environmental issues that were having detrimental consequences not only on humanity, but on the planet more broadly (WMO et al. 1988, viii). The most notable success of the conference, however, lay in establishing the foundation for the UNFCCC and the promotion of a legal climate change regime that would set the stage for future global climate governance.

Conclusion

The 1987 Venice and 1988 Toronto summits revived the issue of climate change and returned it to the leaders' agenda, after its absence from Tokyo in 1986. The release of the Brundtland Commission report in March 1987, as the core of a new epistemic community, set the stage for the sustainable development approach to climate change that would unfold in Venice that year and eventually at Rio de Janeiro at the UN Convention on Environment and Development in 1992. There was no major mention of outside international institutions related to climate change in the G7 summit communiqués until the 1988 Toronto Summit, where the UN was noted in the form of UNEP and the WMO. Toronto G7 leaders urged the two organizations to collaborate to create a UN-based intergovernmental panel on climate change. The communiqué also approved the Conference on the Changing Atmosphere, to be held in Toronto soon after summit's end.

Propelling the resurgence of G7 attention and action on climate change at Toronto was the arrival of the first, plausibly direct climate or "global warming" shock to the US, in the form of the deadly and destructive heat wave that afflicted the G7's most powerful member in the summer of 1988. Yet that shock did not determine the content of the G7's response, which took the form of its historic choice of UN institutions to assume the lead. That preference flowed from the multilateral instincts of summit host Brian Mulroney and the fact that action focused on measurement and science, where the UN's functional expertise and the consensus codified in the Brundtland Commission stood out.

Following the 1987 and 1988 summits, global environmental and climate issues became a regular part of the G7's summit agenda. The G7 generated a steady stream of ambitious and time-bound agreements that advanced environmental principles that the world's most powerful countries were prepared to accept. Thus, the G7 summit in 1987 and 1988 revived climate change for good, even as the UN system still struggled to discover and develop it for the first time.

References

Burney, Derek (2005). *Getting It Done: A Memoir*. Montreal: McGill-Queen's University Press.

Daniels, Joseph P. (1993). *The Meaning and Reliability of Economic Summit Undertakings* New York: Garland Publishing.

Dobson, Hugo (2004). *Japan and the G7/8, 1975–2002*. London: RoutledgeCurzon.

G7 (1987). "Venezia Economic Declaration." Venice, June 10. http://www.g8.utoronto.ca/summit/1987venice (January 2015).

G7 (1988). "Toronto Economic Summit Economic Declaration." Toronto, June 21. http://www.g8.utoronto.ca/summit/1988toronto/communique.html (January 2015).

Hainsworth, Susan (1990). "Coming of Age: The European Community and the Economic Summit." Country Study No. 7, G7 Research Group. http://www.g8.utoronto.ca/scholar/hainsworth1990/index.html (January 2015).

Hajnal, Peter I. (1989). *The Seven-Power Summit: Documents from the Summits of Industrialized Countries, 1975–1989*. Millwood NY: Kraus International Publishers.

Heeney, Timothy (1988). "Canadian Foreign Policy and the Seven Power Summits." Country Study No. 1, Centre for International Studies, University of Toronto. http://www.g8.utoronto.ca/scholar/heeney1988/heenl.htm (January 2015).

Kirton, John J. (1999). "Explaining G8 Effectiveness." In *The G8's Role in the New Millennium*, Michael R. Hodges, John J. Kirton, and Joseph P. Daniels, eds. Aldershot: Ashgate, pp. 45–68.

Kirton, John J. (2007). "Mulroney's 1988 Toronto Summit." Unpublished manuscript.

Kopvillem, Peeter (1989). "A Keener Earth Watch." *Macleans*, January 2, p. 18.

Maclean's (1989). "Looking Ahead." *Macleans*, January 2.

Mulroney, Brian (2007). *Memoirs*. Toronto: McClelland and Stewart.

Rosenberg, Norman, Joel Darmstadter, and Pierre Crosson (1989). "Overview: Climate Change." *Environment* 31(3): 2.

World Commission on Environment and Development (1987). *Our Common Future* (Brundtland Report). Oxford: Oxford University Press.

World Meteorological Organization, United Nations Environment Programme, and Government of Canada (1988). "The Changing Atmosphere: Implications for Global Society." June 27–30, Toronto. http://www.cmos.ca/ChangingAtmosphere1988e.pdf (September 2014).

PART III
Shaping the Divided UN Regime

Chapter 5
Reinvention, 1989–1992

The years 1989 to 1992 marked a period of reinvention in global climate governance, as the Group of Seven (G7) retreated in favour of the new divided, development-first regime of the United Nations. G7 summit attention and action on climate change mutually spiked to new highs that would not be surpassed on a sustained basis until the Gleneagles Summit in 2005. The 1989 Paris Summit in particular saw a notable rise in both the volume and scope of climate deliberations, devoting one third of its final declaration to global environmental concerns. This trend continued, albeit to a lesser extent, at the US-hosted 1990 Houston Summit, where the G7 recognized, identified, prioritized, and defined the parameters for UN-led international action on climate change and arrived at interim agreements that would ultimately bring the new UN-embedded framework to life. In 1991 at London, the G7 summit focused specifically on the forthcoming UN Earth Summit scheduled for June 1992 in Rio de Janeiro, galvanizing collective leadership and support for ensuring the successful completion of a climate change convention by that time. In 1992, just three weeks after the Earth Summit, G7 leaders in Munich took firm responsibility for the effects of global warming and recognized that specific, concrete actions were needed to reduce the irreversible effects of greenhouse gas emissions in the earth's atmosphere. From 1989 to 1992, summit leaders made 23 climate-related commitments, compared to only two in the years before.

This reinvention was propelled to new heights by summit leaders who collectively recognized the key role they played in filling the international institutional void during this time of global climate change. It also marked a major retreat from the G7's bold beginning in 1979, when the leaders sought and successfully secured effective climate change governance all by itself. Now the G7 assigned global leadership to the full UN. Here the values of natural resource development trumped those of ecological sustainability, and the major carbon polluters of the future were given a political and legal right, and thus incentive, to behave in carbon-unconstrained ways. What might be termed "developmental environmentalism" became the defining principle of this new approach.

Paris, 1989

The 1989 "Summit of the Arch," in Paris from July 14–16, marked both the beginning of the third cycle of summitry and the dawn of a new era in the global political landscape. The profound political changes that took root that year, beginning with the collapse of the Soviet Union and Mikhail Gorbachev's quest for inclusion in the democratic G7 club, enabled the summit to address the rapidly increasing number of non-traditional security matters, including the proliferation of HIV/AIDS, the global flow of narcotics, assistance to the developing world, debt strategies for highly indebted countries, protectionism and free trade, human rights violations, and transfers of cross-border pollutants.

As environmental issues increasingly became identified as important challenges facing western, eastern, and southern governments, the protection of the environment evoked a determined and concerted international response, through the adoption of global policies based on the principles of sustainable development. This heightened concern was prompted by a torrent of scientific research in 1987 that included studies on the processes of acid rain, the harmful effects of chlorofluorocarbons (CFCs) on the ozone layer, and the warming trends resulting from the burning of fossil fuels.

As political leaders began recognizing that environmental abuse was driven largely by economic phenomena, the need to emphasize environmental issues at global economic meetings intensified. The Paris Summit responded to these challenges by making the largest, full-scale treatment of global environmental governance to date. More than one third of the 25-page final communiqué was devoted exclusively to environmental issues, including the depletion of the ozone layer to global warming, deforestation, ocean dumping, hazardous wastes, energy conservation, alternatives to CFCs, limits to toxic emissions, and the establishment of a global climatological network to detect climatic changes. With the significant rise in both the volume and scope of components and causes related to climate change, the Paris Summit marked the reinvention of G7 governance on the issue.

Preparations

The year between the 1988 Toronto and 1989 Paris summits saw a significant volume of policy coordination on global environmental issues, as the environment increasingly became a preoccupation for both political leaders and the general public. Several climate-related activities and conferences took place: the new *Clean Air Act* was proposed by the Bush administration in the US; the Acid Rain Accord was reached between Canada and the US; the elimination of CFCs dominated the agenda at the annual meeting of the Organisation for Economic Co-operation and Development (OECD); discussions on stratospheric ozone depletion took place in London and Helsinki; and numerous other meetings on greenhouse gas emissions were held in London, Davos, and Geneva.

Immediately following the 1988 Toronto Summit, Canada hosted a separate but G7-approved international meeting to consider ways to improve the world's capacity for forecasting environmental change. Taking place on June 27–30, 1988, the Conference on the Changing Atmosphere sought to increase international awareness of, and responses to, the consequences of a changing atmosphere, and to develop strategies and actions to stabilize and reduce the adverse human influences on the global atmosphere. The conference further sought to promote global cooperation in programs that attempted to forecast change and reduce harmful emissions (World Meteorological Organization [WMO] et al. 1988). Regarded as a very successful event—well staged, well organized, and well orchestrated—it drew global attention to climate change and the changing atmosphere. The fundamental achievement of the conference, however, was the knowledge it provided on all aspects of the changing atmosphere (Rosenberg et al. 1989).

The most notable success of the conference came in its recommendation for an umbrella framework convention that would lend itself to the development of specific agreements or protocols laying down international standards for the protection of the atmosphere. This recommendation, which followed the convention-protocol path set by the UN Convention on the Law of the Sea (UNCLOS) and the Montreal Protocol on Substances that Deplete the Ozone Layer, would ultimately lay the foundation for the UN's Framework Convention on Climate Change in 1992. The most notable contribution to the climate conference was made by J. Alan Beesley, Canada's former ambassador to the UN in Geneva at the UNCLOS negotiations and chair of the Legal Dimensions Working Group at the Toronto conference.

His working group urged the recognition of different means available to states for dealing with atmospheric problems and encouraging compliance through incentive mechanisms (which would later become the flexibility mechanisms). It argued that the umbrella convention should provide the coordination of scientific activities and technological research—which would later be done by the Intergovernmental Panel on Climate Change (IPCC) (Rosenberg et al. 1989). Beesley's work and dedication to law and the environment were key drivers behind recommendations that would emphasize the urgent need for an international legal framework for protecting the atmosphere based on scientific evidence (Rosenberg et al. 1989).

Momentum from the Toronto conference continued through the spring of 1989, as a series of international conferences on preserving the global atmosphere took place. On March 5–7, 1989, British prime minister Margaret Thatcher hosted the London Conference on Saving the Ozone Layer, attracting members from more than 120 countries. Endorsing new and alarming scientific data on ozone depletion, Thatcher urged delegates to "go further and act faster, to accept higher targets and shorter deadlines" (Anderson and Sarma 2002, 305). At London, 20 countries committed to sign the 1987 Montreal Protocol, through a 50 percent reduction in the production of ozone-destroying CFCs by the end of the century.

Less than one week later, on March 11, the Hague Declaration on the Protection of the Atmosphere was signed by 24 countries, including all G7 countries with the exception of the US. The declaration called for the development, within the UN framework, of a new institutional authority to combat global warming, supported the negotiation of international conventions on the global atmosphere, and urged all countries to sign and ratify conventions to protect the atmosphere and counter climate change (Grubb 1990).

Momentum continued into May when 80 contracting parties to the Montreal Protocol agreed in Helsinki to support the ban of eight industrial chemicals that depleted the ozone layer, including CFCs, as soon as possible, but not later than 2000 (Whitney 1989).

Also in May, at its annual ministerial conference, the 24-member OECD made the world's deteriorating environment a top economic policy concern, along with developing country debt and trade. In the lead-up to the ministerial, a senior US government official from George H. Bush's administration noted that an international approach was essential to bring green issues to the centre of economic decision making. In addition to being tasked with providing firm analytical data, OECD officials considered several approaches to managing the environment, including developing country grants to curb environmental abuse and tougher environmental criteria for loans by the world's top lenders (Kilbourn 1989).

Through these global events and initiatives, the political impetus arose for a very rich and active global environmental agenda and dialogue at the G7's Paris Summit. The French hosts, supported by the British, were eager to make climate change the centrepiece of their summit, even though debt relief for middle income countries was the primary concern at the first two sherpa meetings, held in the Périgord region near Lascaux, and then in St. Martin, in the Caribbean.

The third sherpa meeting, on Lake Geneva at Evian, was consumed by the new shock of China's massacre of unarmed students at Tiananmen Square. Yet as British sous-sherpa Nicholas Bayne recalls:

French sherpa Jacques Attali wanted the summit to concentrate on the global environment, which had never been a lead subject before. We readily supported this, as Margaret Thatcher had been converted to the cause by the persuasive Crispin Tickell, now her environment adviser. The focus was on principles for domestic policy action ... This approach proved uncontroversial and the sherpas prepared a long text that made up nearly half the economic declaration (Bayne 2010, 147–8).

In the immediate lead-up to Paris, William Nitze, a senior environmental official at the US State Department, noted that the environment "is now an issue of consequence that has risen to the top of the international agenda" (Kilbourn 1989). One official said that it was "very important to the President." US officials stated that the US would seek to "take the lead on environmental issues" and press for "international cooperation on combating pollution and preserving natural resources at the summit meeting next week in Paris" (Shabecoff 1989). At a White House press briefing just before the summit, William Reilly, head of the US Environmental Protection Agency (EPA), noted that there was "something of a race on by the summit leaders to see who can be the greenest" (Shabecoff 1989). And on the eve of the summit, White House chief of staff Governor John Sununu noted a strong sense of agreement on the environment. "I don't sense any disagreement in the development of the drafts of the statements that they would be considering," noted Sununu. "It sounds like, from all the preliminary work done by the Sherpas, that the leaders—the seven leaders, the heads of state, are on the same wavelength. And I don't think that's going to be a difficult issue for them to come to an agreement on at all" (White House 1989).

At the Summit

The leaders themselves met in Paris on July 14–16, 1989. Given the strong push by Canada to advance the global environmental agenda following the Toronto climate conference, Canada was particularly pleased with the environment's strong place at Paris. In the end, in the opening passages of their final declaration G7 leaders acknowledged three global priorities, the third of which was "the urgent need to safeguard the environment for future generations" (G7 1989). For the second time at the summit thus far, the global environment was referred to in the introductory chapeau of the communiqué, firmly establishing the environment as a priority. The leaders noted that protection of the environment required a "determined and concerted international response" and called "for the early adoption, worldwide, of policies based on sustainable development." This was the second time the G7 promoted the idea of sustainable development in their final communiqué and in relation to climate change, two years after it had been envisioned by the UN in the Brundtland Report produced by the UN's World Commission on Environment and Development (1987).

The environment thus became the principal theme discussed by leaders over their three-day summit, with two full sessions devoted to it. Canada urged others to do something about the environment, specifically climate change. Fearing that a global warming convention would not be accomplished, Canadian prime minister Brian Mulroney intervened personally, urging his summit partners to act on atmospheric concerns. He wanted the OECD to add environmental data to its repertoire and develop an "environmental 'report card'" to assess members' progress in cleaning up their acts (*Toronto Star* 1989). Eventually, Mulroney argued, these standards would be applied globally through the UN. The outcome for Mulroney was positive. He noted during his final summit press conference that the G7 "shared a common view of the importance of a convention on climate change and the need to make progress on marine pollution, resource depletion, and on deforestation" (Mulroney 1989). Noting that the "environment was a major preoccupation by all leaders," Mulroney pushed for climate indicators, arguing that "new environmental indicators are needed that will allow governments, business and private citizens to measure the state of the environment and the relationship of environmental factors to economic development."

However, American opposition to a framework convention with binding targets and timetables was becoming increasingly apparent. The Bush administration expressed deep concern about the impact of a climate convention on the business community. The US—the world's largest emitter of carbon emissions, with 25 percent of the global total—found itself both pressured and isolated at Paris, as its G7 partners attempted to push it closer to a commitment to limit emissions. Bush's summit partners were determined to not give the US a free ride on sacrifices borne by other countries to reduce their emissions. But the EPA's Reilly, present at Paris, resisted pressure for even the most modest reductions, arguing that a "summit of industrial countries was the wrong place to negotiate international commitments that could have serious implications for some important developing countries not represented at the meeting" (Paarlberg 1992, 216). The Bush administration argued that a "scientific basis for policy action did not yet exist and to advise caution, pending the conduct of more research." Other G7 members reacted quite strongly to the US position, including the Germans, who remarked that "gaps in knowledge must not be used as an excuse for world-wide inaction." According to some, this "early failure of the Bush administration to endorse some minimally inconvenient actions on global warming drew significant domestic and international criticism." Bush, however, seemed unphased by the criticism, as he continued to take a stand as the "defender of American business and jobs against what he termed 'environmental extremism'" (Hill 1994).

Results

Although criticized by several nongovernmental organizations (NGOs) as lacking substance, fully one third of the final communiqué was devoted to a broad range of global environmental issues, making the 1989 summit the largest, full-scale, sustained global environmental summit of the G7 by far. Climate change was particularly prominent here.

Regardless of the US position, summit leaders recognized that if the Paris commitments were to be implemented, high-level input and broad-based cooperation around the summit table would be required. Continued opposition from the US, however, resulted in yet another failed agreement on specific qualitative targets on greenhouse gas emissions. The Bush administration continued to argue that a truly committed US position on emissions reductions was not likely given the "rather large anticipated economic costs of achieving a rapid reduction in greenhouse gas emissions and given the uncertain or difficult-to-measure short-term environmental benefits" (Paarlberg 1992, 216). Arguing that the American population was simply too dispersed, Bush felt that the US would have serious difficulties meeting the same carbon reductions that existed in the more densely populated regions of Western Europe. Bush officials argued that "per capita emissions in the U.S. were more than twice as high as in France and could not be reduced quickly to the French level without high economic costs, greater reliance on nuclear power, or both" (Paarlberg 1992, 211).

By accepting the language of the Paris communiqué, however, Bush agreed to several commitments that would initiate a process aimed at establishing "specific protocols" and "concrete commitments" for a framework convention on climate change (G7 1989). The push by the summit leaders in Paris thus set the process firmly on track for what would soon result in the mobilization of global support for the principles, guidelines, and obligations underlying the framework convention on climate change that would emerge at Rio in 1992.

Dimensions of Performance

The Paris Summit's performance on climate change was very strong, across most of the six dimensions of summit performance (see Appendix A-1). Nicholas Bayne awarded the overall summit a score of B+, highlighting its environmental achievements along with those in helping Central Europe and debt (Bayne 2000b, p. 195).

Deliberation

In terms of deliberation, climate change soared from 140 words at Toronto to 422 at Paris, thus more than doubling its share from 2.7 percent to 6 percent of the total communiqué (see Appendices A-1 and C-1). Paris introduced seven new climate- relevant issues.

Direction Setting

The leaders noted that the "depletion of the stratospheric ozone layer is alarming and calls for prompt action" (G7 1989).

Decision Making

Paris was the first time the summit had made more than one commitment on climate change (see Appendix D-1). Four commitments were made, including the advocacy of common efforts to limit greenhouse gas emissions, the need to strengthen the global network of greenhouse gas observatories, the adoption of sustainable forest management practices, and the urgent requirement for an international framework convention on climate change containing protocols and commitments (G7 1989). The latter was a key pledge in setting the foundation for the framework convention on climate change.

Delivery

However, compliance with these four commitments in the following year was a failure, at only −0.07 (see Appendix E-1). A respectable performance on greenhouse gases (+0.43) was offset by an umbrella climate change convention (0), adoption of sustainable forest practices (−0.29), and, above all, the establishment of a global climatological reference network within the WMO (−0.43). Regardless, Bayne (2010, 148) notes that "the G7 decisions led to greater attention to environmental policy in all industrial countries, so that it was integrated into overall governmental strategy."

Development of Global Governance

In its development of global governance, Paris saw the US environment administrator attend the summit—the first such minister to do so. Several multilateral organizations were referred to within the text of the final communiqué, including a commitment that supported the WMO (see Appendix F-5).

Causes of Performance

Shock-Activated Vulnerability

The strong climate change performance at Paris was driven in the first instance by shock-activated vulnerability. Several mounting ecological concerns affecting G7 members directly drove the summit leaders to respond to the growing threat of environmental degradation in 1989. In the most severe, visible, classic, energy-related shock, the US *Exxon Valdez* oil tanker disaster off the coast of Alaska in March spilled 11 million gallons of crude oil into Prince William Sound, devastating the environment of "one of the most bountiful marine ecosystems in the world" (Miller 1999). Oil tanker traffic had long produced a clear chronic threat arising from energy interdependence. About 95 percent of oil spills into the oceans came from routine tanker operations from their stowage and off-loading, and the wash-over, ballast, and bilge water thrown into the sea. But the *Exxon Valdez* shock was concentrated in a single time and space, large in volume and highly visible to locals and G7 citizens around the globe, watching the disaster unfold on television. Moreover, it was the first serious oil spill to strike America's shores since G7 summitry had begun. It attacked the once promising lifeline to America's Prudhoe Bay, that, when discovered in 1968, had promised to restore America's autarkic energy independence once again. The ecological costs and consequences of the *Exxon Valdez* disaster were thus simply too high to ignore.

Similar smaller shocks arose elsewhere. In Italy, all political parties were "hearing screams from Italians" who hit the beaches only to discover brown sludge contaminating the waters of the Adriatic (Simpson 1989). The West Germans were giving East Germans "a pile of money" to control the air pollution that was "drifting across the barbed wire separating the two countries," prompting the German government to task its central statistical agency to devise a yearly environmental equivalent of the country's economic profile (Simpson 1989). The Paris communiqué was the first to refer to climate change in the context of a "threat" (see Appendix G-1). There were three sentences in which G7 leaders acknowledged the existence of serious threats to the environment that could lead to future climate changes.

Predominant Equalizing Capability
The predominant equalizing capability of G7 members was another cause of summit performance (see Appendix I). Most G7 members continued to experience increases in their gross domestic product (GDP).

Common Principles
Another contributing factor was the democratic common principles of G7 members, which were all committed and consolidated democracies (see Appendix J).

Political Cohesion
Political control, capital, continuity, commitment, and civil society pressure also played a role in having summit leaders take important steps in protecting the environment and tackling climate change (see Appendices K-1 and K-2). Margaret Thatcher, addressing the UN General Assembly in November 1989, announced a greater British commitment to scientific research by establishing a centre for the prediction of climate change in the UK, along with a pledge of $158 million of bilateral assistance for the preservation of tropical forests over the next three years (Riddell 1989). Arguing that "the prospect of irretrievable damage to the atmosphere, to the oceans, to earth itself" was now as menacing as the threat of war, and that its consequences were "even more far-reaching" than the discovery of how to split the atom, Thatcher (1989) urged the UN to create a new global convention to conserve plant and animal life, particularly in tropical forests, where many species were in danger. Wearing both hats as a scientist and politician, Thatcher proposed solutions based on "sound science and sound economics," calling on industry to conduct the research needed to produce environmentally safe products. She argued for "sound scientific analysis" for greenhouse gas reductions that would allow economic growth to continue. She rejected the call for new institutions, arguing instead that the framework convention on

climate change should be ready by the time of the 1992 UN Conference on Environment and Development (Riddell 1989).

In North America, the United States and Canada had set their sights on negotiating an acid rain agreement that would benefit both countries (Stewart 1989). Bush had told Mulroney in February 1989 that he intended to propose legislation to Congress on limiting emissions that cause acid rain. That discussion, which began in earnest in 1989, would culminate in the adoption of the Canada–United States Air Quality Agreement in 1991, better known as the Acid Rain Accord, to address transboundary air pollution. Just one month prior to the Paris Summit, on June 12, Bush proposed a $19 billion/year clean-up package "that would saddle industry with dozens of costly new clean-air requirements" (Rosewicz and McQueen 1989). Bush's proposal, which aimed to cut power plant emissions that caused acid rain by 50 percent by 2000, would mean an estimated expenditure of $33 billion by industry on annual air pollution controls. Although Bush's proposal would let polluting plants select their own means of reducing emissions, industry would be responsible for most of the costs associated with these new proposals.

These actions and the 1989 summit came at a time when public opinion across the G7 acknowledged that voters wanted something done to protect the environment. As Jeffrey Simpson (1989) noted following Paris, "the leaders did not decide to suddenly go green because they liked the colour. Powerful public sentiment in their respective countries drove them towards the fine words in the communiqué." Although perceived as "incongruous for an economic summit to be preoccupied with the environment," a wave of international support by this time was growing for the idea that economies would flounder if forests were destroyed, the ozone depleted, and oceans filled with toxins and garbage (*Toronto Star* 1989).

The G7's strong climate performance in 1989 came because several G7 leaders made climate change a priority as part of their wider domestic environmental policies. The overall placement of climate change as a priority for member governments helps explain the G7's high performance in climate deliberation and decision making during this time.

Constricted Club Participation
The final factor contributing to this strong summit performance was its constricted club participation. Paris was attended by only the members themselves without any external invitees, even if many other guests of the French president held a parallel dinner (see Appendix L).

After the Summit

As the year progressed and climate concern continued to mount, several other international conferences were held to address the issue. At the May 1990 Bergen Conference on Sustainable Development in Norway organized by the UN's Economic Commission for Europe, environment ministers from 34 leading economic powers plus the European Union's commissioner for the environment agreed that stabilizing carbon dioxide emissions at 1990 levels by 2000 was "essential" (Information Unit on Climate Change 1993). In issuing their "Joint Agenda for Action," the ministers noted that concerted measures in the transport sector, energy efficiency and conservation, renewable energy sources, and regulatory tools were all key to combating climate change.

Houston, 1990

The 15th annual G7 summit was held under the presidency of the United States in Houston, Texas, on July 9–11, 1990. It represented a substantial if uneven step forward in the process of securing an international agreement on the principles and practices of sustainable development. It came amid the distinctive geopolitical challenges facing the leaders as they prepared for their summit. The Cold War had effectively ended with the fall of the Berlin Wall in November 1989, an event that had significant consequences for summit planning and preparations (Bayne 1997). One result of this positive diversionary shock was that the issue of the environment received somewhat less attention at Houston than it had at Paris the year before. Houston's environmental achievements failed to meet the expectations of either the nongovernmental environmental community or such activist governments as West Germany. However, despite strong resistance from the host United States, Houston succeeded in setting international environmental priorities, making conceptual advances on the key issue of climate change, and taking action in several other important environmental areas (Kirton 1990).

Preparations

The summit preparatory process saw the leaders' personal representatives take up the key components of climate change and search, however agonizingly, for areas of consensus. It enabled them to lock in early areas of relatively easy but still substantial agreement. At their first meeting, held in early January 1990 in Key West, Florida, the sherpas identified the environment as one of the three big summit issues, on the grounds that domestic public opinion was still heavily engaged in the issue, and that their leaders' performance in Paris had raised expectations. They were aware that after 1989's relatively easy advances, which had largely tasked international organizations to move the climate debate forward, Houston would have to be more concrete to satisfy public demands. At the same time, it was clear that those worried about the economic impact of climate change action held the upper hand in the US administration, and that progress here would be slow. There were complex and interdependent issues to be dealt with, and varying demands from country to country.

The less developed countries were not interested in environmental clean-up. Until someone paid them, especially on carbon emissions, it was hard to persuade them to act. Yet some of the pollution in their larger cities would—and should—be an indigenous concern for them. On the environment, Houston would be the conclusion to Bergen and Paris. There was not much consensus among the summit members. Canada advised very strongly that Houston establish a priority focus and create a synthesis around the table. Climate change was very important but there were divergent views, with the US standing alone and differences between Canada and some Europeans. Canadian environment minister Lucien Bouchard said publicly that Canada agreed on the need for setting timetables and targets, but would come to specifics at the end of a process of domestic consultation (Lewington and Howard 1990). This position went beyond the Americans but not as far as the Europeans.

At the second sherpa meeting, held in San Francisco, environmental issues took a back seat to East-West relations. However at the third sherpa meeting, in Paris in May, the impasse on climate change came to the fore. Although all agreed it was a major issue, they agreed on little else. America's colleagues left knowing the subject would be a major focus at the subsequent sherpa meeting and a major challenge for the US as summit host. Realizing that they represented a minority of one, the Americans understood that the summit would have to achieve a consensus, which would require all to compromise—with them needing to adjust the most. The Paris meeting did make rapid progress on an agreement to take action on tropical rainforests, an issue of particular concern to the Germans and Japanese. Subsequent sherpa meetings codified this progress, while leaving the central issue of climate change unresolved. The challenge was how to package a message that set priorities and get away from the laundry-list approach at Paris of each country adding the particular environmental issues of concern to it domestically.

Managing competing expectations seemed a daunting process for the sherpas. But two major achievements in the lead-up to Houston effectively forced them to arrive at positions that would be negotiable when the leaders met face to face. The first was the reversal of the US refusal to contribute to a fund for developing countries to find substitutes for CFCs. US National Security Council officials managing the summit preparations were confident that such a reversal should take place just prior to the Houston Summit, both to exempt Bush from foreign criticism and to give him an easy success as an environmentalist. The US sherpa team saw much consensus on the Montreal Protocol, which was a test case in its procedures and structures. Developing countries had been very strong in declaring they needed help, but no one had been able to fix a price. Therefore, the US was concerned about open-ended commitments. Canada felt this would be the key to resolving the matter.

The second major action-inspiring achievement was the Canada-US agreement to begin negotiating an acid rain accord. Announced by Bush and Mulroney following their bilateral meeting on the day before the Houston Summit opened, it represented a change in the previous US position that negotiations would not start until Congress had passed the agreed-upon revisions to the Clean Air Act. US senator Max Baucus, responsible for guiding the legislation through Congress, said the bilateral agreement would give added incentive to pass the act. The otherwise uninvolved William Reilly, head of the EPA, was thus able to fly to Ottawa the following week to begin negotiations with the Canadians as soon as possible. Such negotiations might have taken place in any event, and the agreement could be considered Bush's strategy to achieve an early summit environmental success and deflect more important issues. However, for Mulroney, who had spent half a decade inching the US forward on the acid rain issue, any increase in the pace of negotiations represented a meaningful accomplishment.

The US sherpa team insisted that it did not want to duplicate the work of the IPCC, so it focused on the role of economics and science and their contributions to environmental issues. Bush, in agreement with his key advisor on the environment, agreed that the issue should be prominent at Houston—the public was interested in it and it clearly had a role in all G7 members. But there was less consensus on what exactly the leaders should deal with at Houston. Mulroney and German president Helmut Kohl insisted on the importance of the environment. Canada was particularly concerned with the issue of the marine environment because of its huge coastline that needed protection from overfishing and costal pollution. These were relevant in a domestic context but had international implications. The lengthy communiqué passages on the environment at Paris had given rise to expectations, but the sherpa team did not want environmental references to dominate the agenda at Houston. Factoring the IPCC report into their decision making, they identified the most pertinent environmental issues and the ones to which the leaders could significantly contribute.

At the Summit

At Houston itself, despite the competing claims for attention from pressing issues such as agricultural subsidies, aid to the Soviet Union, and post-Tiananmen aid to China, environmental issues received extensive and serious discussion. They commanded much of the leaders' attention in their afternoon session on Tuesday, July 10, and took much of the sherpas' time that night when they struggled to draft a communiqué their leaders could endorse the next morning. As the leaders took up the subject on Wednesday morning, all the items that remained unresolved—in square brackets—dealt with the environment.

Results

With the release of the final communiqué, the Houston Summit succeeded in securing significant advances on environmental issues. It did so by introducing new items onto the global environmental agenda, setting priorities among them, defining processes to deal with them, forwarding consensus on key issues including global warming, and agreeing to action on important questions such as the world's forests.

On climate-related issues, the communiqué applauded recent progress on the protection of the ozone layer, reflecting the perspectives of Japan's Science and Technology Agency even more than those of the United States. It emphasized the need for international cooperation on science, new technologies, and economic research. The communiqué recognized that the loss of temperate and tropical forests as well as the destruction of other ecologically sensitive areas was continuing at an "alarming pace" (G7 1990). The leaders welcomed Brazil's commitment to trying to halt this process and said they were "ready for a new dialogue with developing countries on ways and means to support their efforts." They asked the World Bank and the Commission of European Communities to prepare a proposal for a pilot project with Brazil to preserve tropical rainforests to be delivered to the next summit. They called for reform of the Tropical Forestry Action Plan and declared themselves ready to begin negotiations on a global forest convention or agreement to be completed by 1992. In deference to French and Japanese preferences in particular, this section stated that nuclear energy could "play a significant role in reducing the growth of greenhouse gas emissions."

Environmental values also received a major boost in the summit's main political declaration, in the consensus on how to deal with the divisive issue of loans and other normal relations with China, one year after the Tiananmen event. Apart from giving Japan the green light to resume bilateral lending if it wished, the summit agreed that World Bank lending would be resumed only for projects that met basic human needs (an old criterion), or that enhanced the environment (a new one). G7 leaders thus placed environmental protection on the same high level as basic human needs as a priority international value. The practical effect of this change, however, remained in some doubt, particularly as World Bank president Barber Conable subsequently vowed that summit decisions would have no effect on its decisions, which remained the prerogative of all its member states.

The most notable success at Houston came as a result of Mulroney's plea to Bush on global warming, when Bush accepted a summit endorsement of the 1992 framework negotiation on climate change being organized by the United Nations Environmental Programme (UNEP). For the summit overall, this was an achievement of process rather than substance: an agreement on a deadline and a forum rather than on targets, programs, and costs. Yet it proved to be a defining and fateful choice, for it assigned the highest level of political leadership on climate change to the UN—dominated by developing countries.

Dimensions of Performance

Houston's final communiqué showed that the leaders had paid significant attention to environmental issues, even if the summit could not maintain the exceptionally high profile for sustainable development that the Paris Summit had produced the year before (see Appendix A-1). From the start, those preparing the Houston agenda knew that environmental groups, the media, and the general public would expect Houston to maintain the cadence established at Paris. Yet they were slow to arrive at concrete ways to meet this objective, given the reluctance of the US host to move on the central issues of the global atmosphere.

Domestic Political Management

In its domestic political management, the results were mixed for the US host. Internationally, the media consensus was that Houston was an environmental disappointment. The *Wall Street Journal* (1990) singled out the National Wildlife Federation's judgement that "Bush's efforts at balance, compromise and consensus-building are killing the world," and declared it to be "the familiar voice of political extremism." This view suggested that the communiqué did not look particularly substantive in dealing with the major disagreement on the environment because it did not commit to reduce the emission of greenhouse gases by a specific date.

Deliberation

Houston was a limited success in terms of deliberation. At 491 words it exceeded the 422 at Paris, with 5.9 percent dedicated to climate change, almost as much as 1989's 6 percent (see Appendix C-1).

More broadly, the three and a half pages and 13 paragraphs devoted to environmental issues in the declaration made it the single largest subject in the communiqué. The environment came ahead of such traditional economic subjects as developing countries and trade.

Direction Setting

For the second year in a row, the environment also appeared in the summit's introductory section, which set the defining priorities and principles for the leaders' work. The third paragraph recognized that "sustainable economic prosperity" depended, among other things, "on an environment safeguarded for future generations" (G7 1990). The core principles of sustainable development—the integration of environmental and economic considerations, and the acceptance of custodianship for future generations—were explicitly affirmed as fundamental to the summit's direction.

Perhaps the most important feature of the Houston communiqué was the spread of explicit references from the environmental section into the economic sections of the document—a recognition that environmental considerations were relevant to previously self-contained economic subjects. Whereas none of the economic sections of the Paris communiqué had such environmental recognitions or reminders, Houston had four.

Of the four sections, two dealt with Eastern and Central Europe. The communiqué recognized that the countries in those regions faced "major problems in cleaning their environment. It will be important to assist the countries of Central and Eastern Europe to develop the necessary policies and infrastructure to confront those environmental policies" (G7 1990). It stated: "we also welcome the recent initiatives in regional cooperation, e.g., in transport and environment."

The other two sections dealt with developing countries. The leaders noted that "the International Development Association replenishment ... agreed to last December ... marks the incorporation of environmental concerns into development lending" (G7 1990). They somewhat cautiously added that "the recent U.S. Enterprise for the Americas initiative to support investment reform and the environment in Latin America needs to be given careful consideration by Finance Ministers."

A further advance was Houston's move from a laundry-list approach to the global environmental agenda to setting priorities. At Paris, the leaders had been late to realize the burgeoning public concern in their countries for environmental matters. They had thus looked to Canada, with the most summit experience in environmental matters, to draft the environment section of a communiqué that would elaborate the issues most prominent in each summit member, without indicating any priorities. Canada and others had pushed strongly for a communiqué that was not only designed for domestic political consumption but would also define priorities for international action. They were largely successful. The communiqué's environmental section singled out "climate change, ozone depletion, deforestation, marine pollution, and loss of biological diversity" as the key issues requiring "closer and more effective international cooperation and concrete action" (G7 1990).

Most importantly, the section defined a key principle or decision rule to guide such cooperation and action. It read: "we agree that, in the face of threats of irreversible environmental damage, lack of full scientific certainty is no excuse to postpone actions which are justified in their own right" (G7 1990). Although the final phrase represented a substantial American-generated qualifier, the acceptance of the overwhelming magnitude of the threat and of the uncertainty principle as a basis for further decision making and action was a major American concession and summit advance.

Decision Making
The Houston Summit made seven commitments on climate change, more than ever before and almost double the four made at the Paris Summit a year earlier (see Appendix D-1).

Delivery
However, the delivery of Houston's decisions was very low, as the four assessed commitments had an average compliance score of only −0.11 (see Appendix E-1).

Development of Global Governance
Similarly low was the development of global governance inside the G7, as no new G7-centric bodies were created at Houston to deal with climate change (see Appendix F-2). However, within the communiqué, G7 members reiterated their support for UNEP and the WMO (see Appendix F-5). Houston thus did much to develop global climate governance outside the G7. It declared that climate change was "of key importance" and committed to "common efforts to limit emissions of greenhouse gases, such as carbon dioxide" (G7 1990). Although it did not specify targets and timetables for actual reductions, as Germany had wished, it did speak of "strategies and measures," of "stabilizing" as well as "limiting" greenhouse gas emissions, and of three deadlines for decision and action. The first deadline was the "opportunity" provided by the Second World Climate Conference. The second was the 1992 date for the completion of a framework convention on climate change to be

negotiated under the auspices of the UNEP and WMO. And the third was a declaration that "work on appropriate implementing protocols should be undertaken as expeditiously as possible and should consider all sources and sinks." While this phrasing was a concession to Bush's skeptical chief of staff, John Sununu, the presence of a demand that such work should comprehensively consider all such sources and sinks, and the call for immediate action, represented a major environmental advance.

The communiqué dealt with the 1992 UN Conference on Environment and Development (UNCED). It stated clearly that "cooperation between developed and developing countries is essential to the resolution of global environmental problems" (G7 1990). Whereas the Paris communiqué had expected the UNCED only to "give additional momentum to the protection of the global environment," the Houston Declaration identified it more strongly as "an important opportunity to develop widespread agreement on common action and coordinated plans" (G7 1989, 1990). It further suggested the conference consider the conclusions of the Siena Forum on the International Law of the Environment that had been held in April 1990. Together with the other references to 1992 as a deadline, this section did much to legitimize UNCED as the central forum and action-forcing event for global environmental decision making. Houston thus saw the G7 shift global leadership on climate change even more to the UN, its developing country majority and its legalized, top-down, convention-protocol approach.

Also discussed was environmental assistance to developing countries. This section called for the strengthening of multilateral development banks in the field of environmental protection. It declared that "debt-for-nature swaps can play a useful role" (G7 1990). It concluded with: "We will examine how the World Bank can provide a coordinating role for measures to promote environmental protection."

Climate finance from advanced to developing countries was thus entrenched. The communiqué referred directly to environmental decision making and information. It endorsed a series of pet national projects, notably an international network for satellite data on earth and the atmosphere, the importance of the private sector, the OECD work on environmental indicators, an international conference on environmental information, and voluntary product labelling. It ended with an endorsement of the Human Frontier Science Program of funding for research in the life sciences, a project much beloved by Japan.

Causes of Performance

Shock-Activated Vulnerability
The increase in attention to and action on climate change at Houston is first explained by shock-activated vulnerability. The G7 leaders acknowledged that climate change posed a threat to world's forests and that world leaders thus had an obligation to address the threat of irreversible environmental damage (see Appendix G-1). There were 13 oil tanker spills and related accidents in 1989 and 14 in 1990, although none as dramatic as the *Exxon Valdez* (see Appendix G-4). Furthermore, in 1990 the G7 members combined produced 15,412 megatonnes of carbon dioxide emissions (see Appendix G-7).

Predominant Equalizing Capability
The second cause of performance is the predominant equalizing capability of G7 members. They all experienced growth in GDP between 1989 and 1990 (see Appendix I).

Common Principles

G7 performance was also caused by the democratic principles shared by its members. In 1990, G7 members continued to show their commitment to democracy. In the same year future G8 member Russia moved from a −4 to a 0 on the Polity IV regime authority spectrum (see Appendix J).

Political Cohesion

Political cohesion also helped. In the US, from the very start of the summit preparatory cycle, the EPA's attempts to gain control of the US government's summit environmental agenda was turned back by the State Department. And although Bush had taken EPA administrator William Reilly to Paris the year before, where he was able to brief the world's media on American environmental aspirations, there were no signs that Reilly's attendance at the hometown Houston Summit was wanted. Ultimately, the president took two cabinet-level officials—trade representative Carla Hills and agriculture secretary Clayton Yeutter—to Houston, in addition to the foreign and finance ministers who routinely attended the summit. The task of explaining American positions, summit progress, and even scientific "facts" on the environment to the world was thus left to John Sununu, Bush's chief of staff. Under Sununu's influence, the US position on atmospheric issues pitted it against all its summit partners. In terms of concrete action, the economically preoccupied United States confronted the environmentally engaged Germany, France, Italy, and the European Community, with Canada, Japan, and the UK in the middle but tilted toward engagement (Kirton 1990).

As the sherpas dealt with this impasse in private, public expectations faded. The United States was trying to focus attention on multilateral trade and agricultural subsidies. And, in the final weeks before the summit, Germany pushed the idea of coordinated western aid to the Soviet Union, and Japan was quietly trying to persuade the summit to approve the resumption of aid to China. Without political pressure, the environment became less of a preoccupation. Because each of the G7's three most powerful members, including Germany, had chosen or acquired more nationally critical causes to champion, the environment had no dedicated, powerful advocate at the summit.

Moreover, the central environmental initiative of the Houston Summit—Germany's proposal for acceptance or endorsement of its far-reaching national plan for carbon dioxide emissions—died an early and decisive death (Guebert et al. 2011). Media reports alleged that even before the summit opened, in a bilateral meeting between Kohl and Bush, the Germans had traded off their push for a strong summit statement on carbon reductions in return for American acquiescence to Germany providing direct economic aid to the Soviet Union.

Nor did the media, unlike Paris the year before, propel environmental issues into the top level of summit-related public consciousness, either before, during, or after summit. English-language newspapers throughout Canada in the two days before, three days during, and three days after the summit did little to highlight the summit's environmental agenda or successes. Canada's elite English-language daily newspaper *The Globe and Mail* devoted 34 articles to the summit between July 7 and 14, but only one highlighted environmental issues and an additional 13 referred to them in passing. Of the 36 summit-related articles during the same period in the *Toronto Star*, Canada's largest circulation English-language daily, six focused on environmental subjects, but all but one of these dealt with the Canada-US bilateral agreement to open negotiations on an acid rain accord.

The leading financial daily, the *Financial Post*, consistent with this pattern, focused only two, and devoted part of another six news items to the environment in its summit coverage.

Constricted Club Participation
The final cause of performance was the constricted participation of the Houston Summit, which was attended by G7 members alone (see Appendix L).

London, 1991

The 1991 London Summit largely helped prepare the upcoming UNCED in Rio de Janeiro in June 1992. London was important for G7 leaders to reach consensus on divisive environmental issues, including climate change and the preservation of global forests (Norman 1991). Geopolitical changes were once again highlighted. The summit's themes of "building world partnership" and "strengthening the international order" were an important tribute marking the end of the Cold War and the collapse of the Soviet Union (G7 1991a, 1991b). The challenges for the Soviet leadership under Mikhail Gorbachev to create an open society and liberalize the economy were pervasive and complex. Moreover, G7 members remained divided over Soviet accession to the General Agreement on Tariffs and Trade (GATT) and negotiating the Soviet-era debt to the Paris Club, while at the same time relieving the debt burdens of the world's poorest countries. Although financial support for Soviet economic reform dominated pre-summit negotiations and the post-summit lunch the leaders held with Gorbachev, G7 leaders were not distracted from discussing global environmental concerns.

Preparations

Prior to the summit, Britain's new Conservative prime minister and host John Major affirmed that the G7 would seek consensus on key environmental issues in preparation for the Earth Summit. There was early agreement in pre-summit meetings to reduce deforestation to cut greenhouse gas emissions, with delegates from both Germany and the European Community supporting the G7's 1990 commitment to fund the pilot program in Brazil (Johnson 1991b). Despite reluctance from some leaders, particularly Bush, to support the project financially, $1.5 billion had been committed and it remained a central topic of debate. The 1991 summit preparations sought support for the program to be in place by the time of the Rio summit in June 1992 (Johnson 1991c).

Canada judged that the UK could not ignore the environment, given the past two G7 summits, but insisted that a "checklist" approach should be used to further negotiations. In the past, the UK had pushed on oceans, sustainable development, sustainable fisheries, and land-based sources of marine pollution, which shared a transregional dimension that required international cooperation. Hence the UK argued for a multilateral approach. Canada felt that such a multilateral approach would provide the needed political spin. It proposed that the G7 recognize the linkages and the need for a world oceans conference to take place after UNCED, like the conference on the Changing Global Atmosphere right after the 1988 Toronto Summit. Canada wanted the London Summit to provide political recognition of the importance of UNCED. Furthermore, it insisted that the summit and the communiqué recognize the interdependence of these issues specifically. The G7 should make it clear that Rio would not be just another routine UN conference and that all leaders needed to make climate commitments in the three cross-sectoral areas of technology transfer, funding, and international institutional reform.

At the Summit

Overall, London marked a historic advance in G7 summitry with the Soviet Union invited to participate, albeit in a limited way. Gorbachev was invited on the final day of the summit to discuss his plans for liberalizing the Soviet economy and creating an open democratic society. His presence signalled that G7 governance was reaching out to embrace Eastern Europe and thus bring its inefficient carbon-intensive and resource-extractive development under the embrace of a G7-centred climate change control regime. Talks with Gorbachev were productive. G7 trade discussions focused on criteria for Soviet accession to the GATT and bridging related differences in the Uruguay Round of trade negotiations, developing country debt, and action against international drug trafficking (Norman 1991).

Results

The London Summit was successful in agreeing on important transnational environmental issues, despite criticism from environmental organizations that considered the United States the main barrier to climate change–related initiatives (Timberlake 1991). Such organizations were critical of the summit process, citing a lack of strong G7 leadership in taking more substantive action (Schoon and Marshall 1991). While bilateral meetings between Kohl and Bush achieved agreement on funding for the Amazonian rainforest pilot project, Bush remained reluctant to contribute financially or to accept any emissions reduction targets for his own country (Johnson 1991b). However, a commitment was reached by the summit's end, with the leaders promising to "financially support the implementation of the preliminary stage" of the Brazil pilot project (G7 1991a). This proved to be a notable summit success (Kokotsis 1999).

Thus, early agreement on the two important issues of tropical forests and greenhouse gas emissions were real advances (Johnson 1991a). Leaders also expressed support for initiatives to incorporate environmental objectives into national economic policies and growth.

Dimensions of Performance

Across the dimensions of performance, the London Summit improved in the area of compliance but otherwise generally declined (see Appendix A-1).

Deliberation
In London's public deliberation, climate change dropped to 2.4 percent of the communiqué, less than half of the 5.9 percent at Houston the year before (see Appendices A-1 and C-1). The leaders discussed the importance of nuclear power generation in diversifying energy sources in order to reduce greenhouse gas emissions.

Decision Making
The London communiqué produced five concrete commitments on climate change, the most notable being the pledge to conclude a convention on climate change by June 1992. Unlike the previous two summits, where language reflected an "urgent need" for a climate

convention, at London the leaders set a definitive timeline for completing the convention. Second, they committed to design and implement "concrete strategies" to limit greenhouse gas emissions with measures to "facilitate adaptation" (G7 1991a). Third, the communiqué stressed the need for more cooperative efforts in environmental science and technology. Fourth, the leaders pledged financial support for the implementation of the preliminary stage of the Brazil tropical forest pilot program. And, fifth, the G7 agreed to mobilize financial resources to assist developing countries tackle their environmental problems, which included, by extension, those problems associated with carbon dioxide sources and sinks. Included in this commitment was the thrust toward supporting the Global Environment Facility (GEF) as the central funding mechanism to assist developing countries meet their convention obligations.

Leaders at London thus expressed not only their concern for environmental issues, but also the necessity for moving forward on concrete negotiations to be completed by the time of the Earth Summit. Negotiations had proved somewhat difficult, with Bush reluctant to accept firm emissions targets. In the end, however, the Americans accepted communiqué language that committed the leaders to provide funding to implement the preliminary stage of the Brazil pilot project, a commitment that Bush eventually signed on to (Kokotsis 1999).

Delivery
Compliance with the two assessed commitments on climate change made at London was substantial, at +0.38 (see Appendix E-1). It surged into the positive range after the negative scores of the previous two years.

Development of Global Governance
In its development of global governance, the G7's performance declined at London. There were no official-level institutions that related to climate change and there was only one reference to an outside organization—UNCED—in the communiqué passages that dealt with the issue (see Appendices F-2 and F-5).

Causes of Performance

This performance was driven by several causes, as described below.

Shock-Activated Vulnerabilities
The Gulf War began with Iraq's invasion of Kuwait on August 2, 1990, and lasted until March 1991, causing a spike in world oil prices. This provided a security and energy shock that spurred G7 climate action, if less intensely than the much larger oil shock of 1979 (see Appendix G-3).

Multilateral Organizational Failure
From the outset, London was geared toward a successful UNCED outcome, with the need to involve all summit members in ensuring a successful framework convention on climate change by the time of the Earth Summit in 1992. However, the US refusal to set targets and timetables for climate change control had already made the UNCED preparatory process a failure by the time the G7's London Summit was held.

Predominant Equalizing Capability

The third cause of London's performance was the continued predominant equalizing capability of G7 members. Between 1990 and 1991, all G7 members experienced growth in their GDP (see Appendix I).

Common Principles

The fourth cause of London's performance was the common democratic purpose of the G7 and its members shown by each member's individual commitment to a democratic regime (see Appendix J).

Political Cohesion

The fifth cause of London's performance was domestic political cohesion. Political control was a concern for host Prime Minister John Major, who was facing an election in April the following year (see Appendix K-1). There were, however, several experienced leaders returning to the summit table (see Appendix K-2).

Constricted Club Participation

The final cause of performance was the constricted participation of the London Summit, which was attended by G7 members alone, although the post-summit meeting with Gorbachev preoccupied leaders during their own meetings (see Appendix L).

After the Summit: UNCED, Rio de Janeiro

By the third session of the UNCED Intergovernmental Negotiating Committee in Nairobi in September 1991, the US position on targets and timetables remained unchanged. Bush remained concerned that any new climate-related programs could result in new taxes, which would directly contradict his longstanding promise heading into the 1992 election campaign not to hike taxes.

The fourth negotiating session took place in February 1992, four months before Rio. By then, the G7 had collectively expressed concern that a climate agreement would not be in place by the time of Rio. Although indications suggested that all G7 leaders would attend the Earth Summit, Bush refused to commit to attend until a "satisfactory climate agreement was achieved" (Hecht and Tirpak 1995, p. 391). The US remained determined to avoid any binding commitments for targets and timetables, but did make a major concession to the developing world by acknowledging, for the first time, "the need to provide financial resources to developing countries to permit their full participation in the convention." Moreover, the US announced a commitment of $50 million to the core fund of the GEF, with an additional $25 million during the next two years to be used by developing countries to inventory their greenhouse gas emissions.

By the final negotiating session in April 1992, the US maintained its position against emissions reduction strategies, largely because "factors such as population growth, world fuel prices and economic growth could seriously push U.S. emissions toward the high end of the projected range" (Hecht and Tirpak 1995, 393). There remained a strong perception in the US administration that a "binding commitment to stabilize could force the U.S. to adopt significant policy actions with unforeseen economic consequences."

British environment minister Michael Howard knew that a climate convention without the US would be weak (Hecht and Tirpak 1995, 392). He proposed text on non-binding emissions goals that captured the interest of both the US and the European Community. US negotiators presented the compromise text to Bush, who called on key leaders, including those of the G7, to accept it. Canada and France acknowledged that the compromise represented would encourage the US president to attend the Earth Summit. Two days later, UNCED secretary general Maurice Strong invited Bush to Rio and the president graciously accepted.

Convening just one month in advance of the G7's 1992 Munich Summit, the Earth Summit was thus attended by all G7 leaders. The foremost achievement of the UNCED process was the creation of five principal documents, including a legally binding treaty that outlined a regime for curbing emissions of carbon dioxide and other greenhouse gases that cause global warming. With the Earth Summit, 1992 marked the launch of a new era in environmental diplomacy (Lanchberry 1996) (see Appendix H-3).

Munich, 1992

The 1992 Munich Summit, with its first ever pre-summit assembly of G7 environment ministers, supported the international focus on global environmental issues that the Earth Summit generated, and began to secure environmental issues on the G7's standing agenda. This tradition would become strongly entrenched in the summit's annual preparatory process after 1992.

Munich marked a clear recognition by G7 leaders that pollution, toxic waste, and the emission of greenhouse gases came at a very high price, and that environmental sustainability would require integration into every aspect of the G7's economic, business, and societal models, as well as its political agenda (Lascelles 1992; Meadows et al. 1992). Contained within this recognition was an acceptance of UNCED's fundamental principle that countries would have common but differentiated responsibilities in their environmental efforts, depending on the extent of their ecological degradation and the amount of money they could ultimately afford to contribute to the cause of climate control.

Preparations

The preparatory process for Munich's environmental agenda was driven largely by preparations for the Earth Summit. In the spring of 1992, German environment minister Klaus Toefler invited his G7 colleagues to a meeting in Germany to discuss, among other international environmental concerns, climate control strategies and proposals to limit greenhouse gas emissions. The meeting marked the emergence of a separate G7 forum for environment ministers, paralleling G7 ministerial bodies for foreign affairs, finance, trade, and employment (Johnson and Kirton 1995).

Just one month prior to Munich, UNCED convened in Rio de Janeiro, attended by all G7 leaders. With US presidential elections only a few months away, Bush realized that it might serve him well to look more climate friendly. This view was intensified by Arkansas governor and probable Democratic presidential nominee Bill Clinton's assertion that the White House had committed grievous errors in failing to commit to global environmental agreements. Clinton argued that Bush failed to understand that "rising global temperatures [could] threaten America's standard of living," and faulted the Bush administration's decision not to sign an agreement to limit carbon dioxide emissions to 1990 levels by 2000 (Ifill 1992).

In the final days at Rio, the G7 leaders took centre stage, making the most ambitious commitments on the environment and climate change. John Major announced that the UK would commit to new and additional resources for developing countries through the GEF. Canada's Brian Mulroney announced the most ambitious financial commitments made by any G7 leader—C$115 million to developing countries for forest management, the elimination of C$145 million in official development assistance (ODA) debt from Latin American countries in exchange for sustainable development projects, replenishment of the GEF, and C$50 million in humanitarian assistance to drought-stricken countries (*Earth Negotiations Bulletin* 1992).

Although more than 140 countries favoured a treaty on climate change with firm targets and timetables, the US prevailed at Rio by encouraging other countries to sign an agreement that contained no obligatory standards or deadlines. It only required government measures to stabilize atmospheric concentrations of carbon dioxide, with the aim of reducing industrial-level emissions to 1990 levels by 2000. Because treaty limits would have no value without the US on board, countries that supported firmer targets and timetables reluctantly accepted the American terms. By securing a treaty with targets weaker than what the Europeans, Canadians, and Japanese had initially hoped for, Bush signed the UN Framework Convention on Climate Change in Rio de Janeiro and called for its speedy implementation.

The last G7 sherpa meeting prior to the Munich Summit continued the debate on targets and timetables, with the US once again blocking progress. US sherpa Bob Zoellick contended that the Americans had conducted the most research on climate issues and their related costs to business. With projected growth rates, it was impossible to meet Rio's aspirational targets and timetables. The US, he argued, preferred a "best efforts" model, with the hope of sufficient future technological advancement to balance growth with emissions reduction goals. This position prevailed, despite a leaked US government report in May 1992 admitting that "cutting emissions even 10 percent below 1990 levels by the turn of the century would create more than 80,000 new jobs" (Oaks 1992).

At the Summit

The main divergences at Munich were between the Europeans and Japanese on one side and the US on the other. The US refused to sign any agreement that committed it to specific carbon-reduction targets that could potentially result in job losses. Europe and Japan felt, as they had at Rio, that environmental protection was "an unavoidable challenge that would strengthen their industry in the long run, not as a sinister threat to their way of life" (Lewis 1992). Germany in particular felt the US was irrationally cautious and needed to be more proactive in this global effort.

Results

In the end, G7 leaders agreed that Rio represented a "landmark in heightening the consciousness of global environmental challenges and giving new impetus to the process of creating a world-wide partnership on development and the environment" (US Department of State 1992). But along with the success of UNCED came the recognition that if Rio was to have any lasting significance, the international community had to act collectively to implement the conventions created there. Thus, at Munich, G7 leaders stressed the importance and urgency of carrying forward the momentum of UNCED. They agreed on a several immediate measures to follow up the conventions and agreements established at Rio including ratifying the Framework Convention on Climate Change and publishing national action plans by the end of 1993, giving additional financial and technical support to developing countries for sustainable development through ODA, and establishing the GEF as a permanent funding mechanism. They called for the establishment of a sustainable development commission under the auspices of the UN, the creation of an international review process on forest principles, and the development and diffusion of energy and environment technologies, including proposals for innovative technology programs. Other mechanisms were discussed, including international peer review mechanisms and data collection from satellite observation programs. The communiqué also emphasized the "enhanced use of voluntary debt conversions, including debt conversions for environmental protection" (G7 1992). Collectively, these mechanisms provided legitimacy to the Rio regime.

The G7 leaders recognized at Munich that because industrial countries were largely responsible for the damage done by greenhouse gas emissions, they would have to take specific and concrete measures to reduce the causes and effects of global warming. Most were prepared to make firm and binding stabilization commitments. However, others—notably the United States—continued to support only general measures aimed at limiting the effects of greenhouse gases, calling instead for more scientific research to support the global warming theory. Nonetheless, the US signed the climate change convention at Rio and became the first industrialized nation (and the fourth country overall) to ratify it, on October 13, 1992.

Dimensions of Performance

Across the dimensions of performance, the Munich Summit followed the trend set in London by improving compliance but making no significant advances in other dimensions (see Appendix A-1).

Deliberation
In its declaration, Munich devoted 137 words and 1.8 percent of its communiqué to climate change, among the lowest in the previous five years (see Appendix C-1). Leaders made note of the fact that rapid action was needed to fulfil their commitments on climate change (G7 1992).

Decision Making
Munich's decisional performance was substantial. Its seven climate commitments tied it with Houston as the highest to date (see Appendix D-1).

Delivery
In delivery, at +0.71 G7 compliance on environmental issues jumped to a new height in 1992 (see Appendix E-1). This was the beginning of a new trend (Kokotsis 1999).

Development of Global Governance
Munich's development of global governance focused on institutions both inside and outside the G7. Inside, the summit created the Nuclear Safety Working Group, which, while not directly related to climate change, was working toward the diversification and safety of alternative energy supplies (see Appendix F-2). Outside, the Munich communiqué pledged

to carry forward the momentum of the Earth Summit and publish the national action plans set out by UNCED (see Appendix F-5).

Causes of Performance

The 1992 summit's substantial climate performance, especially on delivery, flowed primarily not from shock-activated vulnerability but from other causes.

Shock-Activated Vulnerabilities
In 1992, there were 10 accidents that resulted in spills of more than 700 tonnes of oil (see Appendix G-4). Also that year G7 members collectively produced 15,183 megatonnes of carbon dioxide emissions (see Appendix G-7).

Multilateral Organizational Performance
The first key cause of Munich's performance was a multilateral organizational success, rather than failure, although it was a success generated by the G7 itself. Convening just three weeks after the UN's Earth Summit, the G7's Munich Summit endorsed the convention on climate change and brokered deals critical to the Munich delivery of the other Rio conventions on biodiversity and desertification. What followed was a general recognition that if the Rio process was to have a significant and lasting impact, the G7 had to act collectively to implement the Rio conventions, thereby setting the standard for the rest of the international community. The result was an ongoing "Rio effect," whereby sustained compliance with commitments reached at Rio was subsequently endorsed by the G7 summit (Kokotsis 1999).

Common Principles
A second cause of Munich's performance was the common democratic purpose of the G7 and its individual members (see Appendix J). It was reinforced by the start of a democratizing Russia's increasing involvement in the G7 summit.

Political Cohesion
A third cause was domestic political cohesion. Political control was a high for host Helmut Kohl who would be re-elected in 1994 and remain in office until 1998 (see Appendix K-1). There were also six returning leaders at the summit table (see Appendix K-2).

Constructed Club Participation
A fourth cause of Munich's success was the creation of the G7 environmental ministerial process, flowing from the Rio process. It was launched in Germany just prior to the 1992 summit and became an annual event in 1994 (see Appendix F-1). G7 ministers and leaders also operated as a caucus group at the Earth Summit itself. Through the inauguration of this ministerial process came the recognition of the importance of the G7 continuing to implement their Rio commitments. Thus, for every subsequent summit until 1995, the national implementation of the Rio climate commitments was advocated at the environment ministerial meetings and subsequently endorsed by the leaders. The initiation of an environment ministerial forum through which the specifics of the climate change convention could be negotiated reinforced the significance of this issue by the time the leaders met, which led to increased compliance at Munich and beyond (Kokotsis 1999).

Conclusion

As G7 leaders began to realize in 1989 that global environmental protection required a concerted and determined international response, the adoption of global policies based on sustainable development principles and a UN process began to emerge. And as scientific research increasingly showed the harmful effects of toxic emissions on the atmosphere, political leaders began to acknowledge that environmental abuse was driven largely by human economic phenomena. The 1989 Paris Summit thus launched the first full-scale response to these environmental challenges by issuing a final communiqué that dedicated one third of its text to issues directly associated with the environment and climate change. With the significant rise in both the depth and breadth of environmental governance, Paris marked the reinvention of G7 governance on climate change, by shifting the centre of global climate governance firmly from the G7 itself to the UN as a whole.

Houston in 1990 likewise secured significant attention to environmental and climate issues, even if it could not maintain their exceptionally high profile at the Paris Summit. From the early phases of the preparatory process, those involved clearly understood that environmental NGOs, the media, and the public expected Houston to maintain the momentum established at Paris. Yet the Houston Summit was slow to arrive at concrete ways to meet these expectations, given the overall reluctance of the Bush administration to move on the central issues of the global atmosphere. Environmentalists and editorialists, impatient for clear, concrete, far-reaching action, viewed Houston as an environmental failure. Nonetheless Houston played a key role in identifying, prioritizing, and defining the boundaries for international action on climate change and arriving at interim, component agreements to help breathe life into these processes. By these standards, Houston was a substantial, if uneven, climate change success.

London in 1991 was geared toward solidifying a successful Earth Summit, by involving G7 leaders to ensure a framework convention on climate change would be in place in time for Rio. While the Bush administration was determined to avoid any binding commitments on targets and timetables, it did concede to the developing world by acknowledging—for the first time—the importance of financial assistance for developing countries to facilitate their full participation in the UN's prospective climate regime. In early 1992, US committed $50 million to the core fund of the GEF, with an additional $25 million pledged over two years to assist developing countries in producing their country studies.

At Munich in 1992, the G7 recognized that industrial countries were largely responsible for the effects of global warming and would thus have to take specific and concrete measures to diminish its detrimental and irreversible effects. While most members were prepared to make firm and binding stabilization commitments, the US continued to support only general measures, insisting on the need for continued scientific research. However, the US had signed the Framework Convention on Climate Change at Rio a month before and became the first industrialized country to ratify it.

This G7 reinvention of global climate governance from 1989 to 1992 flowed from several key causes contained in the concert equality model. First, shock-activated vulnerability initially drove the summit countries to respond to the growing visible threat of environmental degradation. From the *Exxon Valdez* oil disaster in Alaska to the contaminated Adriatic waters of Italy, the visible ecological costs and consequences of environmental disasters brought by the dependence on oil were rapidly escalating.

Moreover, oil prices and supply shocks coming with Iraq's invasion of Kuwait in 1990–1991, drove G7 climate governance in the area of energy efficiency. The major years for energy efficiency and nuclear energy in the G7's climate agenda during this time correspond to sharp peaks in oil prices. The 1991 summit action on climate came right after the highest spike in world oil prices since the crisis in the early 1980s, due to the Gulf War.

Second, taking up the issue of carbon emissions in 1989, at a time when UNEP and other international institutions were still largely silent on the issue, led the G7 to fill this critical void in global climate governance by taking the necessary steps to create the UN's new climate change control regime at Rio in 1992. High levels of deliberation during this period focused on getting the UN to take concrete action in the form of the IPCC and ultimately creating a global convention on climate change. The G7 thus played a critical role in agreeing to develop, within the framework of the UN, new institutional authority to combat further warming of the earth's atmosphere. The UN multilateral organizations turned from failure to success, due to G7 summit leadership.

Third, domestic political cohesion played a key role as G7 leaders took measures to protect the environment and tackle climate change. The 1989 Paris Summit came at a time of heightened public awareness among G7 voters who wanted action to protect the global environment. Powerful public sentiment throughout the summit countries effectively drove the G7 leadership toward consensus building on several concrete, climate-related commitments. From Margaret Thatcher's call about the "irretrievable damage to the atmosphere" to George H. Bush's pre-election $19 billion proposal to Congress on limiting emissions that cause acid rain, G7 leaders individually and collectively rallied to make climate change a priority within their broader domestic environmental policies.

Fourth, the new G7 environment ministers' institution arose in support. Climate compliance spiked in 1992 following UNCED's Earth Summit and the G7's Munich Summit due in large part to the launch of the new era in environmental diplomacy that Rio generated. With the launch of the convention on climate change and its endorsement at Munich, the G7 began to broker and deliver on deals critical to the full implementation of Rio and its associated conventions on biodiversity and desertification. The inauguration of the G7 environmental ministerial process, flowing on the heels of Rio and launched prior to Munich, set the stage for the institutionalization of an annual forum where environment ministers could meet face to face to negotiate the specifics of the climate convention. The outcomes of this ministerial process would then feed directly into the leaders-level summit meeting, reinforcing the significance of climate issues and accounting for improved delivery on climate commitments over time.

References

Anderson, Stephen O. and K. Madhava Sarma (2002). *Protecting the Ozone Layer: The United Nations History.* London: Earthscan.

Bayne, Nicholas (1997). "Impressions of the Denver Summit." G8 Research Group, Denver. http://www.g8.utoronto.ca/evaluations/1997denver/impression/index.html (January 2015).

Bayne, Nicholas (2000). *Hanging In There: The G7 and G8 Summit in Maturity and Renewal.* Aldershot: Ashgate.

Bayne, Nicholas (2010). *Economic Diplomat.* Durham, UK: Memoir Club.

Earth Negotiations Bulletin (1992, 12 June). "Plenary." 2(11). http://www.iisd.ca/vol02/0211000e.html (January 2015).

G7 (1989). "Economic Declaration." Paris, July 16. http://www.g8.utoronto.ca/ summit/1989paris/communique (January 2015).

G7 (1990). "Houston Economic Declaration." Houston, July 11. http://www.g8.utoronto. ca/summit/1990houston/declaration.html (January 2015).

G7 (1991a). "Economic Declaration: Building a World Partnership." London, July 17. http://www.g8.utoronto.ca/summit/1991london/communique (January 2015).

G7 (1991b). "Political Declaration: Strengthening the International Order." London, July 16. http://www.g8.utoronto.ca/summit/1991london/political.html (January 2015).

G7 (1992). "Economic Declaration: Working Together for Growth and a Safer World." July 8. http://www.g8.utoronto.ca/summit/1992munich/communique (January 2015).

Grubb, Michael (1990). *Energy Policies and the Greenhouse Effect*. Vol. 1: Policy Appraisal. Aldershot: Dartmouth.

Guebert, Jenilee, Zaria Shaw, and Sarah Jane Vassallo (2011). "G8 Conclusions on Climate Change, 1975–2011." Toronto, June 20. http://www.g8.utoronto.ca/conclusions/ climatechange.pdf (January 2015).

Hecht, Alan and Dennis Tirpak (1995). "Framework Agreement on Climate Change: A Scientific and Policy History." *Climatic Change* 29(4): 371–402.

Hill, Dilys M. (1994). "Domestic Policy." In *The Bush Presidency: Triumphs and Adversities*, Dilys M. Hill and Phil Williams, eds. London: Macmillan.

Ifill, Gwen (1992). "Clinton Cites Bush 'Errors'." *New York Times*, June 13.

Information Unit on Climate Change (1993). "The Bergen Conference and Its Proposals for Addressing Climate Change." Châtelaine, Switzerland. http://unfccc.int/resource/ ccsites/senegal/fact/fs220.htm (January 2015).

Johnson, Pierre Marc and John J. Kirton (1995). "Sustainable Development and Canada at the G7 Summit." In *The Halifax Summit, Sustainable Development, and International Institutional Reform*, John J. Kirton and Sarah Richardson, eds. Ottawa: National Round Table on the Environment and the Economy. http://www.g8.utoronto.ca/scholar/ kirton199503/johnson/index.html (January 2015).

Johnson, Rachel (1991a). "G7 Summit in London: Kohl and Bush Focus on Environment." *Financial Times*, July 16.

Johnson, Rachel (1991b). "G7 Summit in London: Summiteers Backpedal on the Environment." *Financial Times*, July 17.

Johnson, Rachel (1991c). "Summiteers Get Ready to Head for the Forest." *Financial Times*, July 12.

Kilbourn, Peter T. (1989). "Environment Is Becoming Priority Issue." *New York Times*, May 14. http://www.nytimes.com/1989/05/15/business/environment-is-becoming-priority-issue.html (January 2015).

Kirton, John J. (1990). "Sustainable Development at the Houston Seven Power Summit." Paper prepared for the Foreign Policy Committee, National Round Table on the Environment and the Economy, September 6. http://www.g8.utoronto.ca/scholar/ kirton199001/index.html (January 2015).

Kokotsis, Eleanore (1999). *Keeping International Commitments: Compliance, Credibility, and the G7, 1988–1995*. New York: Garland.

Lanchberry, John (1996). "The Rio Earth Summit." In *Diplomacy at the Highest Level: The Evolution of International Summitry*, David Dunn, ed. London: Palgrave Macmillan, pp. 220–43.

Lascelles, David (1992). "Getting Down to Earth in Rio." *Financial Times*, June 3.

Lewington, Jenniffer and Ross Howard (1990). "Ottawa Promises Action This Fall on Carbon-Dioxide Emissions." *Globe and Mail*, April 19.

Lewis, Paul (1992). "U.S. at the Earth Summit: Isolated and Challenged." *New York Times*, June 10. http://www.nytimes.com/1992/06/10/world/us-at-the-earth-summit-isolated-and-challenged.html (January 2015).

Meadows, Donella, Dennis Meadows, and Jorgan Randers (1992). *Beyond the Limits: Confronting Global Collapse Envisioning a Sustainable Future*. White River Junction VT: Chelsea Green.

Miller, Pamela A. (1999). "Exxon Valdez Oil Spill: Ten Years Later." Technical background paper for Alaska Will League. *Arctic Connections* 3/99. http://arcticcircle.uconn.edu/SEEJ/Alaska/miller2.htm (January 2015).

Mulroney, Brian (1989, 16 July). Opening Statement by Prime Minister Brian Mulroney at his Final Press Conference, Summit of the Arch. G7/G8 Research Collection in the John W. Graham Library, Trinity College, University of Toronto. Locator ID: 22.665.

Norman, Peter (1991). "A Table Piled High with Problems: The Group of Seven Summit in London Will Expose Both Harmony and Division Between Industrial Nations." *Financial Times*, July 12.

Oaks, John (1992). "An Environmentalist? Bush? Forget It." *New York Times*, May 8. http://www.nytimes.com/1992/05/08/opinion/an-environmentalist-bush-forget-it.html (January 2015).

Paarlberg, Robert L. (1992). "Ecodiplomacy: U.S. Environmental Policy Goes Abroad." In *Eagle in a New World: American Grand Strategy in the Post-Cold War Era*, Kenneth A. Oye, Robert J. Lieber, and Donald S. Rothschild, eds. New York: HarperBusiness.

Riddell, Peter (1989). "Thatcher Warns of Insidious Threat to Planet: The UK Prime Minister's Wide-Ranging Speech to the UN." *Financial Times*, November 9.

Rosenberg, Norman, Joel Darmstadter, and Pierre Crosson (1989). "Overview: Climate Change." *Environment* 31(3): 2.

Rosewicz, Barbara and Michel McQueen (1989). "Clearing the Air: Bush, Resolving Clash In Campaign Promises, Tilts to Environment." *Wall Street Journal*, June 13.

Schoon, Nicholas and Andrew Marshall (1991). "Environmental Issues Given 15 Minutes at London Summit." *Independent*, July 18.

Shabecoff, Philip (1989). "U.S. to Urge Joint Environmental Effort at Summit." *New York Times*, July 6. http://www.nytimes.com/1989/07/06/world/us-to-urge-joint-environmental-effort-at-summit.html (January 2015).

Simpson, Jeffrey (1989). "The Greening of the G7." *Globe and Mail*, July 19.

Stewart, Edison (1989). "PM, Bush Set Sights on Acid Rain Accord." *Toronto Star*, February 11.

Thatcher, Margaret (1989, 8 November). Speech to United Nations General Assembly. Margaret Thatcher Foundation. http://www.margaretthatcher.org/document/107817 (January 2015).

Timberlake, Cotten (1991). "On Environment, Communique Is Long on Words, Short on Specifics." *Associated Press*, July 17.

Toronto Star (1989). "Green Revolution Arrives in Paris." July 14.

United States Department of State (1992). "US Environment Initiatives and the UN Conference on Environment and Development." *US Department of State Dispatch Supplement* 3(4). http://dosfan.lib.uic.edu/ERC/briefing/dispatch/1992/html/Dispatchv3Sup4.html (January 2015).

Wall Street Journal (1990). "Review and Outlook: Environmental Balance." July 13, p. A8.

White House (1989, 14 July). Press Briefing by White House Chief of Staff, Governor John Sununu. G7/G8 Research Collection in the John W. Graham Library, Trinity College, University of Toronto. Locator ID: 22.677.

Whitney, Craig R. (1989). "80 Nations Favor Ban to Help Ozone." *New York Times*, May 3. http://www.nytimes.com/1989/05/03/world/80-nations-favor-ban-to-help-ozone.html (January 2015).

World Commission on Environment and Development (1987). *Our Common Future* (Brundtland Report). Oxford: Oxford University Press.

World Meteorological Organization, United Nations Environment Programme, and Government of Canada (1988). "The Changing Atmosphere: Implications for Global Society." June 27–30, Toronto. http://www.cmos.ca/ChangingAtmosphere1988e.pdf (September 2014).

Chapter 6
Reinforcement, 1993–1997

The 1992 Earth Summit in Rio de Janeiro sparked a subsequent period in which the Group of Seven (G7) increasingly reinforced the United Nations' new divided regime on climate change control. International attention and G7 action shifted toward implementing and strengthening the new UN Framework Convention on Climate Change (UNFCCC), which had been established there. In part, this focus was due to the proliferation of principles, rules, institutions, and processes that Rio created, allowing the G7 to delegate global climate leadership to the broader multilateral system about which hope was very strong. This delegation continued throughout 1993–1996, bringing less G7 summit action on climate change, if still at a level notably higher than before 1989. The G7's responsibilities on climate came more as reinforcing actions taken toward the UNFCCC process. Only at Halifax in 1995 did sustainable development make a prominent appearance on the G7 agenda, with some focus on climate change within.

Then in 1997, at the Denver Summit of the Eight, the G7 began to add initiatives independent of the UN system. G7 leaders noted in their final declaration their determination to "take the lead and show seriousness of purpose" in strengthening international efforts to combat climate change (G8 1997). With negotiations intensifying in the lead-up to the UNFCCC Conference of the Parties (COP) in December, the Denver Summit assumed strong leadership in pushing the UN toward concrete and workable agreements on climate mitigation strategies. This push culminated in the drafting of the UN's Kyoto Protocol in Japan later that year.

The number of climate commitments generated by the G7 during this period reflects this trend. Of the 433 commitments generated by G7 leaders from 1993 to 1997, only 27 or 6 percent of the total dealt with climate change. The highest number came at the 1997 Denver Summit, when the leaders made nine such climate commitments, or 6 percent of Denver's total 145. This compares with the four years of the preceding 1989–1992 period, when the leaders made 23 commitments or 10 percent of the total 233 commitments made.

Tokyo, 1993

At the 1993 Tokyo Summit, foremost on the leaders' agenda was a global strategy for improving macroeconomic policy, structural reforms to stimulate growth and job creation, preparing for the Uruguay Round of trade negotiations for the General Agreement on Tariffs and Trade (GATT), and producing aid packages to assist a democratically reforming Russia and other Eastern European economies in transition. The escalating conflict in Bosnia and Kosovo also weighed heavily on the leaders' minds, prompting them to issue a political declaration focused largely on the rapidly deteriorating situation in the former Yugoslavia. Thus the environment and climate change received very little attention, a neglect compounded by the decision not to host a G7 environment ministerial meeting in the lead-up (Kokotsis 1999).

Preparations

On the road to Tokyo, the Japanese, Americans, and Canadians pressed with some success to make the summit a more effective and productive forum for implementing global environmental commitments. Just prior to the summit, the Japanese hosts achieved a global consensus that sustainable development required environmental conservation. Japan reiterated its commitment, expressed by Prime Minister Kiichi Miyazawa at Rio the year before, to expand its environmental aid to developing countries by US$0.7 billion over a five-year period (Japan, Ministry of Foreign Affairs 1994). Tied to this financial commitment was the idea that increases in official development assistance would be provided in accordance with the principle that environmental conservation and development should be pursued together.

Under newly elected president Bill Clinton, the United States announced in April 1993 its intention to reduce its greenhouse gas emissions to 1990 levels by 2000. It further offered $25 million for country studies to provide the analytical groundwork for actions to address climate change in developing countries (Schneider 1992). Clinton also declared that the US would sign the UNFCCC (Stevens 1993). As the Denver Summit approached, on June 14, Clinton created the President's Council on Sustainable Development to "develop specific policy recommendations for a national strategy for sustainable development that can be implemented by the public and private sectors" (US Department of State 1993). The council was also charged with creating the national action plan to fulfil America's commitments under the Rio accord (Stevens 1993). Given Vice-President Al Gore's vision to equalize environmental and economic goals in every decision of the Clinton administration, the US now, more than ever, clearly needed to work closely with other governments in order to protect the global environment.

Although deference to the UN Rio regime on climate was supported by most summit leaders, Canada, now led by Progressive Conservative prime minister Kim Campbell, believed that the G7 should continue addressing the problem. Immediately prior to Tokyo, Canada noted that although the UN Commission on Sustainable Development (UNCSD) would play an important role, the G7 summit was the most immediate forum where economic stewardship would be expressed, frameworks would be selected, and instructions would be delivered to the multilateral organizations that implemented the relevant environmental conventions. Canada therefore prepared several environmental initiatives heading into the Tokyo Summit, including a commitment to deal with environmental matters at the GATT after the Uruguay Round, and the creation of a permanent, annualized G7 environment ministers' meeting.

The position that the G7 should continue leading on environmental issues was broadly accepted by nongovernmental organizations (NGOs). The Environmental Defense Fund in Washington DC noted in July 1993 that G7 leadership on issues of environmental degradation was urgently needed (Hajost et al. 1993). It also called upon G7 leaders to include environmental concerns in their economic decision making. This sentiment was supported by a group of 80 German environment and development NGOs in their letter to the G7 leaders prior to the Tokyo Summit (*Der Bund — die Tageszeitung* 1993).

Issues relating to the current economic recession took priority for other G7 leaders, particularly the United Kingdom, which seemed to have lost any interest in environmental issues in the lead-up to Tokyo (Maddox 1993). While the British government agreed to adopt a few environmental regulations set by the European Commission, including eco-labelling, it strongly opposed the commission's plans for a carbon tax on burning fossil fuels. The British government argued that individual countries should decide how to meet their targets for stabilizing carbon dioxide emissions as agreed to at the Earth Summit. This sentiment resulted in a sharp divide at the meeting of European Commission environment ministers in July, with some ministers concerned that the British attitude would prevent the commission from ratifying the UNFCCC.

At the Summit

As attention at Tokyo focused on job creation, growth, trade, and reform in Russia, less attention was placed on environmental concerns and climate change. Economic issues pervaded the agenda, along with the difficulties faced by the ex-communist economies in transition. However, there was some progress on integrating economic and environmental considerations and an acceptance of the goal of sustainable development (Kirton and Richardson 1995b). Moreover, there was a consensus on environmental issues for the first time in three years. As US under secretary of state Joan Spero noted at the end of the summit, "this is the first time in three years that the environment has not been a contentious issue in discussion at the summit." This was largely attributable to a change in approach by the new Clinton administration. According to Spero, "the President made a very lengthy intervention on the environment in the heads of state meeting, discussing both what his administration is doing within the United States and emphasizing an interest in cooperating with other G-7 on international environmental issues" (White House 1993).

Results

However, with only one of 16 sections devoted to environmental issues, the Tokyo communiqué delivered only four commitments related to climate change, the most robust of which was a promise to publish national action plans by the end of 1993. Although the leaders endorsed the Global Environment Facility (GEF), they failed to specify replenishment amounts. Focus had now shifted to the UNFCCC process, where the leaders welcomed its progress and renewed their "determination to secure environmentally sustainable development through an effective followup of the fruits of the UNCED [UN Conference on Environment and Development]" (G7 1993).

Dimensions of Performance

Across the dimensions of performance, the 1993 Tokyo Summit did very little for climate change (see Appendix A-1).

Deliberation
In the leaders' final declaration, climate change specifically took 3.1 percent of the declaration (see Appendix C-1).

Direction Setting
In the key paragraph the leaders confirmed their collective support for the UN process, noting their determination to secure environmentally sustainable development through an effective follow-up of the UN process.

Decision Making

Tokyo's leaders made only four climate commitments, down from seven the year before (see Appendix D-1). The most pronounced was the pledge to publish national action plans by the end of 1993, a commitment first articulated at Munich and reiterated at Tokyo. Second, leaders endorsed their support for the GEF. Third, leaders endorsed the work of the UNCSD, thereby renewing their determination to follow up the "fruits of UNCED." Fourth, leaders indicated their support for seeking internationally agreed arrangements on the management, conservation, and sustainable development of forests (G7 1993). Although these commitments lacked tangible targets and timetables on climate mitigation, they did indicate that global environmental issues remained relevant for the heads. Moreover, while limited in scope, climate change was the topic most addressed within the environment portion of the Tokyo declaration through commitments to publish national action plans, incorporate environmental appraisals into project plans of development banks, and establish international mechanisms for forest management. Tokyo failed to deliver hard targets, but at the core of these issues was the need to reduce emissions and combat climate change.

Delivery

Delivery of Tokyo's two assessed climate commitments averaged +0.57, the second highest level thus far after the +0.71 the year before (see Appendix E-1).

Development of Global Governance

Apart from the firm timeline on the delivery of national action plans, all other commitments adopted at the 1993 summit related to encouraging and preparing to improve the work of already established mechanisms and institutions. The UN system and its affiliated bodies created through the Rio process to tackle global environment and sustainable development lay at the heart of these institutions. This resulted, however, in decreased momentum after Rio with the G7 leaders losing their interest in assuming global environmental leadership themselves (Bayne 2000b).

Causes of Performance

Shock-Activated Vulnerabilities

The achievements on climate change at Tokyo were due in part to the presence of shock-activated vulnerabilities. The first is the 11 accidents related to oil tanks and rigs that happened that year (see Appendix G-4). That same year G7 members produced 15,133 megatonnes of carbon dioxide emissions (see Appendix G-7).

Multilateral Organizational Failure

Achievements also arose due to the multilateral organizational failure coming from the absence of the first COP, which took place only in 1995. Yet the immediate afterglow of the historic, successful UN Earth Summit and its divided regime was sufficient for G7 leaders largely to take a climate leadership sabbatical of their own, and merely support the UN in 1993.

Predominant Equalizing Capability

The internally equalizing capability of G7 members also led to achievements at the Tokyo Summit (see Appendix I). A now climate-committed United States, Japan, and Russia experienced a rise in their gross domestic product (GDP).

Common Principles

Tokyo's achievements were also made possible because of the shared common principles of democracy by G7 members (see Appendix J).

Political Cohesion

The one leader who sought G7 leadership, Canada's Kim Campbell, was new to the summit, came from a small country and had low political control. In contrast America's Bill Clinton, with high capability and control, preferred to support the UN. John Major's UK, with medium levels of both, was opposed to the G7 initiatives beyond the Rio regime (see Appendix K-1).

Constructed Club Participation

Achievements were also made at the Tokyo Summit because of its constricted participation. It was attended by only the nine members with no outside guests (see Appendix L-1).

Naples, 1994

With the fallout of the 1993 economic recession lingering, economic recovery and reducing high unemployment rates across G7 countries remained priorities for the 1994 Naples Summit. Russia's inclusion in the summit's political discussions for the first time made leaders focus on economies in transition, with aid to developing countries and issues of nuclear safety also figuring prominently. Although issues related to the environment and climate change were addressed, these represented a direct response to deliberations and recommendations by the G7 environment ministers, who had met again that year.

Preparations

Following a gap in 1993, G7 environment ministers gathered for a second stand-alone meeting in Florence on March 12–13, 1994, having met during the 1992 Earth Summit just a few months after the German-hosted meeting. Initially, the Japanese had not been clear on whether they would attend the ministerial. With Sheila Copps, Canada's environment minister also serving as deputy prime minister in the new Liberal government, there was some concern that her schedule would not enable her to attend. In the end, the meeting attracted all G7 environment ministers. It focused on climate change, biodiversity, freshwater resources, forestry, and desertification. The approach was mainly on environmental protection as a vehicle for job creation. Ministers also discussed the impact of trade on the environment, innovative mechanisms for financing sustainable development, and environmental risks posed by nuclear power stations in Eastern Europe (Agenzia ANSA 1994). Discussions covered fiscal changes in environmental policy that could shift the tax burden from labour and capital to environmentally harmful areas including toxic emissions and nonrenewable energy resources. The ministers fell short of agreeing on eco-taxes or carbon taxes, largely due to opposition from Canada and the US. They did, nonetheless, agree to end subsidies on environmentally harmful activities in the developing world.

The environment ministers continued to support the Rio regime, expressing growing confidence in the newly established international environmental instruments including the UNFCCC and GEF (G7 1994). In support of these instruments, every G7 environment minister, with the exception of the head of the US Environmental Protection Agency (EPA), agreed to set emissions reduction targets at 1990 levels by 2000 (Agenzia ANSA 1994). The US, responsible for 23 percent of these emissions globally, agreed in principle but argued that the deadline for application of the 1990 standard should not come so soon.

Despite push-back by the Americans, the ministers discussed the need to adopt further reductions that would extend beyond stabilizing emissions at 1990 levels by 2000 (G7 1994). Discussions also focused on technology transfer to developing countries to reduce their greenhouse gases as well as the need to restructure and replenish the GEF for more effective sustainable management of the global commons.

At the conclusion of the meeting, Italian environment minister Valdo Spini (1994) noted that "with respect to the Climate Convention, a converging view emerged on the potential of 'joint implementation' schemes for transferring energy-saving technologies" to countries that were not members of the Organisation for Economic Co-operation and Development (OECD). As a message to the Americans, however, he emphasized that, "joint implementation should be used as a mechanism for further reduction of greenhouse gas emissions, not as a resort to help stabilize emission by the year 2000 at their 1990 levels."

In outlining this solid set of global priorities, the G7 environment ministers were well poised to deliver their recommendations to their leaders meeting in Naples early that summer.

At the Summit

As the summit arrived, it was clear that those issues slated as priorities in the lead-up to Naples would dominate the leaders' agenda: jobs and growth, ratification of the Uruguay Round, the forthcoming launch of the World Trade Organization (WTO) in January 1995, enhanced development assistance, debt of the poorest, closure of high-risk nuclear reactors, reform in Russia and other economies in transition, and cooperation against transnational crime and money laundering. The leaders reflected on reform of the Bretton Woods institutions of the International Monetary Fund (IMF) and the World Bank, with some leaders, including Germany's Helmut Kohl and Canada's Jean Chrétien, taking a more narrow view on institutional reforms, choosing to focus specifically on monetary issues, currency exchange transactions, and stability in money markets. America's Bill Clinton, on the other hand, took a much broader approach to such reforms, demonstrating a keen interest in future leadership challenges including migration, transnational crime, money laundering, the environment, and over-population.

Results

In the end, the leaders noted that the environment remained a "top priority for international cooperation" but pledged to continue to work through the UN process for the success of the Rio conventions at forthcoming conferences. The G7 further agreed to "speed up the implementation of our national action plans called for under the Rio Climate Treaty," recognizing the need "to develop steps for the post-2000 period" (G7 1994). And although the leaders once again endorsed the replenishment of the GEF, they failed to stipulate specifically what the financial contribution should be.

The commitment on accelerating the implementation of those national action plans, called for under the UNFCCC, produced the most resolve among the G7 leaders. Each G7 member was expected to report on its individual accomplishments at the Halifax Summit the following year, in addition to developing strategies for the post-2000 period. With the convention formally entering into force on March 21, 1994, each developed country was required to submit a detailed initial communication within six months of the convention's entry into force for that signatory. Fifteen such initial communications were received by the convention's secretariat, including submissions by all G7 members. The secretariat noted some inconsistencies in reporting structures, but nonetheless declared that "these communications by and large [met] high standards for completeness and transparency" (Independent NGO Evaluations of National Plans for Climate Change Mitigation: G7 Countries: Summit Meeting, cited in Kokotsis 1999, 74).

Dimensions of Performance

Climate change achievements were again limited at the 1994 summit in Naples (see Appendix A-1).

Deliberation

Although the statement issued at the environment ministers' meeting was presented to the leaders at the Naples Summit, only five of the 34 paragraphs in the leaders' communiqué were dedicated to the global environment. Climate change took 2.6 percent of the communiqué, a drop from the year before (see Appendix C-1).

Decision Making

Four concrete climate commitments were made, the same as Tokyo in 1993 (see Appendix D-1). The first was to speed up implementation of each country's national action plans called required by the UNFCCC and to develop emissions reduction strategies for the post-2000 period. The second was to report on achievements by the Halifax Summit the following year. The third was to develop steps for the post-2000 period. And the fourth—similar to the London, Munich, and Tokyo Summits—endorsed the replenishment of the GEF.

None of the pledges contained targets that would bind the G7 leaders to their climate commitments. The promise to produce reports on the implementation of their national action plans by the 1995 Halifax Summit was the only commitment with a defined deadline. And although the leaders recognized the need to replenish the GEF, they failed to commit to any specific amounts.

Delivery

Over the year following the Naples Summit, compliance with the two assessed commitments was a high +0.71, among the highest in summit history to date (see Appendix E-1).

Development of Global Governance

The Rio regime again received robust support at Naples. As at Tokyo the year before, the G7 supported the UNCSD's work in reviewing the progress in its implementation.

Causes of Performance

In terms of performance, the modest trend from prior summits continued, as Naples offered few specific pledges on the environment or climate change. Naples's performance was driven by four forces: the positive shock of a democratizing post-communist world that diverted leaders' attention, the perceived success of the UN system after Rio, a desire to streamline the summit agenda with Russia's growing participation, and the arrival of an institutionalized G7 environment ministers' forum as part of the G7 club. Given concerns over high employment rates and economies in transition, there was little time or interest left for the environment. The G7 instead deferred to the UN bodies recently created to address climate concerns. With the climate convention now into its second year, the general sentiment among leaders was that the UN had created the appropriate bodies and processes to address climate mitigation.

The ongoing general decline and focus on environmental concerns at Naples reflected the leaders' desire to simplify the summit format and provide a final declaration that was driven more authentically by leaders themselves.

However, keeping the environment agenda alive was the G7 environment ministers' forum, begun in 1992 and now institutionalized as an annual event in 1994. As environment ministers addressed a range of climate issues such as joint implementation strategies for transferring energy technologies to non-OECD countries and initiatives aimed at further reductions rather than mere stabilization, the ministers themselves were left to coordinate targets and timetables for emissions reductions. The institutionalization of the environment ministers' meetings began a process of a major expansion of G7/8 ministerial-level institutions in the 1990s, to deal with subjects of even greater domestic character. This effectively resulted in an allocation of responsibility to the ministers charged with the environment portfolio, thereby clearing space on the leaders' agenda to address those issues requiring their immediate attention. The leaders would thus feel obliged to deal with the climate change recommendations prepared by their ministers.

Shock-Activated Vulnerabilities

The achievements made at the Naples Summit were due in part to the shock-activated vulnerabilities that year. The first was the nine accidents associated with oil tanker and rigs that occurred in 1994, compounded by the 21 in the previous two years (see Appendix G-4). Additionally, in 1994 G7 members produced 15,233 megatonnes of carbon dioxide emissions (see Appendix G-7).

Predominant Equalizing Capability

The internally equalizing capability of G7 members also led to achievements at Naples. All members rebounded from the year before with increases in their GDP (see Appendix I).

Common Principles

The fourth cause was the shared common principles of democracy by G7 members (see Appendix J). All members scored high on the Polity IV ranking of democracies. Summit guest Russia scored on the positive side, although down two from two years before.

Political Cohesion

The level of political control was fairly low for host leader Silvio Berlusconi, who was up for re-election in April 1996 (see Appendix K-1). However, four leaders returned to the Naples Summit table, which added to the political cohesion that year (see Appendix K-2).

Constricted Club Participation

At Naples, leaders continued the trend of expanding Russia's participation, reducing the level of constricted club participation (see Appendix L-1).

Halifax, 1995

The Canadian-hosted G7 summit in Halifax on June 15–17, 1995, focused on two questions critical to the rapidly evolving global landscape. First, how can one assure that the global economy of the 21st century will provide sustainable development with good jobs, economic growth, and expanded trade to enhance the prosperity and well-being of all people? Second, what institutional changes are needed to meet these challenges?

Leaders also agreed to review progress on employment and labour standards and to continue supporting trade liberalization and cooperation among international organizations. They promised to report on their progress in implementing their national action plans required by the UNFCCC.

Sustainable development, international institutional reform, and related issues of trade and environment were to be key subjects of a leaders-driven, business-like summit. The first crucial question for Halifax asked how to encourage environmentally sustainable development through investing in appropriate technologies, improving energy efficiency, cleaning up polluted areas, and creating job through enhanced environmental protection. The second critical question addressed how sustainable development could best be fostered through institutional reform of international organizations including the IMF, the World Bank, the WTO, the OECD, and various UN agencies (Kirton and Richardson 1995b).

Preparations

Preliminary consultations among G7 governments during the autumn of 1994 and early winter of 1995 had confirmed the leaders' desire to focus on international institutional reform, particularly as 1995 marked the 50th anniversary of the Bretton Woods institutions. The restoration and acceleration of economic and employment growth among virtually all G7 countries in 1994–1995, the absence of serious exogenous economic crises in post-communist societies, the conclusion of the Uruguay Round, and the Mexican peso shock of December 20, 1994, further suggested that Halifax would indeed concentrate on these priorities, as specified due to Bill Clinton's initiative at Naples the year before.

Within Canada, the issue of the environment held undertones of national unity—the dominant if hidden focus for the summit as host Jean Chrétien was facing a referendum in the province of Quebec seeking its separation from Canada. The federal government knew that environmental protection was a common issue tying all Canadians together. In both Quebec and the rest of Canada the environment resonated well. The Canadian public and the Canadian government largely felt that more could be done. Climate change, or any environmental issue, involved an extensive degree of national unity and national leadership, because resolving such issues required resources and legislation at both the federal and provincial levels. Climate change also had federal-provincial and regional impacts through the oil and gas sector, carbon tax policies, and the speed of financing for such policies. The issue of climate change would come up at the COP at Berlin in March 1995, and at the Hamilton environment ministers' meeting a month later. The Germans were in a good position on climate change, but Canada and the rest of the G7 were struggling to meet their commitments.

Halifax sought a clear commitment to the biodiversity convention. With federal and provincial protected lands combined, Canada's territories made up 8–9 percent of the 12 percent targeted for protected areas. The federal government was negotiating with the provinces about doing more. So federal-provincial cooperation was excellent with regard to biodiversity. Quebec was the first province to amend its government liabilities legislation so that it could be an independent signatory to the agreement on the environment that was part of the North American Free Trade Agreement (NAFTA), which had come into effect in 1994.

In the lead-up to Halifax, the Canadian government undertook two crucial initiatives to increase action on environmental protection. First, the government appointed an environmental auditor general, who would ensure that Canada would be responsible for cleaning up its own messes. Second, the *Canadian Environmental Assessment Act* was implemented, which angered Quebec (which felt the federal government was overstepping into provincial issues) but significantly improved the environmental assessment regime. At the summit, Chrétien wanted to avoid the issue of the international transportation of hazardous waste, where Quebec premier Jacques Parizeau claimed Ottawa had invaded his jurisdiction.

The hot topic in Canada of fisheries and oceans was not to be discussed at Halifax, in order to avoid what came to be known as the Christmas tree approach, with every participant adorning the agenda with its own pet "ornaments" at this once-a-year event. Canadians cared deeply about this issue, but Chrétien felt that it could be left to the UN to handle through its Conference on Straddling Fish Stocks and Highly Migratory Fish Stocks. Canada also wanted to avoid any disagreements between Canada and the Europeans, who preferred to focus on the environmental implications of fisheries and oceans. In negotiating the 1994 communiqué for Naples, pushed by Danielle Mitterrand, the wife of the French president, and several environmental NGOs, the French had wanted a reference to a whaling-free zone in Antarctica. Japan was embarrassed by it. Canada felt it was unnecessary. Canadian sherpa Reid Morden said that in response Canada would propose a cod-fishing–free zone on the nose and tail of the Grand Banks of Newfoundland. The French withdrew. The Canadian team anticipated that there would be local attention to the issue at the port city of Halifax, but Chrétien preferred to keep the summit agenda simple and focused.

Nonetheless, the discussion of the environment at Naples was the starting point for negotiations at Halifax. The Naples communiqué listed two items to be discussed at Halifax—national action plans and institutional reform. Halifax sought to reinforce the Rio environmental commitments as part of both an international and a national plan.

Two key events fed into the Halifax preparatory process. The first was the convening in Berlin of the UNFCCC's first COP from March 28 to April 7, 1995—the most important meeting on climate change since the Earth Summit. Signatories to the convention adopted the "Berlin Mandate," which acknowledged that current commitments in the convention were inadequate, and called for the adoption of an emissions reduction protocol by 1997. All the signatories, including the G7 countries, agreed in Berlin that in order to ensure the fulfilment of current stabilization commitments, they would have to arrive at the next COP in 1996 with a list of additional measures that would allow them to consider reductions beyond 2000 (Canada, Environment Canada 1995).

The Hamilton Environment Ministerial
The second event was the environment ministers' meeting held in Hamilton, Ontario, on April 30 and May 1, 1995, scheduled right after the UNCSD meeting in New York that was attended by many of the ministers. These G7 ministerial meetings were now institutionalized. Hamilton would set the stage for Halifax, with discussions on two objectives. The first was the integration of economic and environmental issues. The ministers made some progress on rendering government operations and government policy more environmentally friendly.

They recognized that governments, as the largest single "business" within their respective countries, played a critical role in committing to environmentally sound management and "buy[ing] green" (see Canada, Department of Foreign Affairs and International Trade [DFAIT] 1995a). Such policies would benefit governments financially, too, by generating more well-paying jobs through policies that promoted sustainable resource management and investment in environmentally friendly infrastructure.

The second objective was institutional responsiveness. The environment ministers sought a consensus "on how existing institutions can be used more effectively to deal with sustainable development, as well as to identify the gaps in the institutional architecture that have to be filled" (Cappe 1995). In this context, they gave careful consideration to the sustainable development issues that the G7 could best address, what distinctive contributions it could make, and how it could organize itself to this end. They were thus not prepared to cede leadership on all issues to the institutions of the UN.

The preparations for Hamilton were dominated by the question of whether Canadian environment minister and host Sheila Copps should invite only the G7 members or include Russia as well. The Russians were taking every opportunity since Naples to push to become a full member by the time of the Halifax Summit, although they did not ask for an invitation to Hamilton. The Russians had not been particularly engaged in Rio commitments, and were not included in the GEF's burden sharing, unlike the G7 members. There was a possibility that the Russians might try to turn Hamilton into a donors' meeting with the demander at the table—a situation best avoided. Copps carefully considered the implications, and decided not to invite the Russians.

The G7 environment ministers wanted to have more substantive discussions than they had at Florence. Above all, any outputs from Hamilton should feed into the sherpa process. The ministerial meeting was intended to be an integral part of the proceedings at Halifax, unlike the Florence ministerial, which ultimately had no impact on the Naples Summit.

Chrétien supported Copps's efforts to put sustainable development at the forefront of the summit's environmental agenda. She had developed a good rapport with her colleagues at Florence, and seized the opportunity to provide environmental leadership by hosting the meeting. This was an opportunity for Copps to increase awareness about the climate change issue, as well as to increase her own profile and credibility among her colleagues. Indeed, there was some apprehension that Hamilton would prove to be a distraction, detracting from any progress that might come in Halifax.

The meeting, held at McMaster University, discussed three central themes. The first was environment-economic integration, with a focus on greening the government through government spending, subsidies, trade barriers, and competitiveness. The second theme was the environmental priority issues from the Naples Summit—climate change, biodiversity, and toxics management, with an intervention by Elizabeth Dowdeswell, head of the UN Environment Programme (UNEP), who spoke about UNEP's emerging role in these areas. The third theme was institutions, including strengthening and reinforcing existing environmental institutions, and the sustainable development component of international financial institutions (IFIs). The ministers largely avoided the complex technical issues of indicators, instead opting for broader, thematic language. Essentially, their goal was to identify a consensus position on these themes.

The ministers prepared three draft discussion papers, reflecting these three themes, and asked 35 individuals outside government to comment. The papers would be circulated internationally, for use at all levels among G7 colleagues. The Hamilton results would also be filtered through the sherpa process.

In preparing the chair's summary, the environment ministers agreed on the importance of moving forward on climate change. They reaffirmed their collective goal of stabilizing greenhouse gas emissions by 2005. In the section on climate change, the ministers reaffirmed their "determination to fulfill our existing obligations under the Convention and our intent to meet the ambitious timetable to follow-up to the Berlin Conference of the Parties" (G7 Environment Ministers 1995). Moreover, they noted for the first time the importance of an accountability review, declaring that it was "essential to have in place mechanisms to allow measurement and reporting on progress." To keep these initiatives on track, there was an emphasis on ongoing support by national governments to environmental institutions including the UNCSD due to their key role as "high-level global forums for setting broad policy directions on sustainable development" (DFAIT 1995b, 31; see also Canada, Standing Committee on Foreign Affairs and International Trade 1995). On accountability, the G7 ministers thus deferred to the UN.

Copps subsequently took those ministers who were interested on a tour of the Arctic. The topic of the Arctic had fallen off the ministerial agenda because the Canadian foreign ministry wanted to avoid any issues that might lead to the Russians demanding an invitation.

The NRTEE Contribution
Such critical and potentially historic topics engaged the energies of Canada's National Round Table on the Environment and the Economy (NRTEE). An advisory body to the Canadian prime minister on sustainable development, it had a legislated mandate to act as a catalyst for change in Canadian society. On February 27, 1995, it convened a workshop in Montreal to examine how to integrate sustainable development considerations more fully into the Halifax agenda, in ways consistent with, and supportive of, the priorities established by the G7 at Naples the year before (Kirton and Richardson 1995a, 4).

The workshop was designed primarily to provide members of the NRTEE's Task Force on Foreign Policy and Sustainability with the analytical background and current information regarding the G7 and the international sustainable development agenda required to prepare advice to the prime minister for approval at the NRTEE plenary on March 9. It was also designed to exchange views with policy makers about the role of the G7 summit system regarding sustainable development, with a focus on the Halifax Summit and the Hamilton environment ministerial. Among the participants were senior Canadian government officials, academics, NRTEE members, and representatives from NGOs and business. Prominent individuals from Canada's G7 partners in the United States, Europe, and Japan added international perspectives.

The NRTEE's final report to the prime minister suggested that the G7—which accounted for 38 percent of industrial carbon dioxide emissions—had both the ability and responsibility to lead in emissions reductions. It concluded that the summit process served as "an important forum for generating action on critical issues and, in particular, climate change, where energy production, consumption, and [carbon dioxide] emissions are at the core, and where forests play an important role as carbon sinks" (Richardson and Kirton 1995)

With the preparatory process now well under way, Chrétien and his senior summit planners continued to emphasize that Halifax allow the leaders to engage in "free and open discussion" and that they "[did] not want their conclusions precast by officials meeting months in advance" (Smith 1995). The sherpas' job, therefore, was not to prepare in detail the substance of the work for the leaders in Halifax, but to "create an environment in which the leaders have maximum opportunity to discuss what is on their minds."

As the summit approached, it became more apparent that two issues were on the leaders' minds. The first was the level of Russian participation at the summit. With Russian involvement increasing at summits over the past few years, the Naples Summit had established a formula that was set to be repeated in Halifax. Russia would be at the table for the second half of the summit for political-security discussions. Despite Russian president Boris Yeltsin's wish to participate in the economic discussions, the prevailing view was that Russia's involvement should remain limited to the second day of the summit.

The second key issue was a review of the Bretton Woods institutions as their 50th anniversary approached. When Bill Clinton paid an official state visit to Ottawa in February 1995, he and Chrétien discussed at length the international monetary system and its capacity to deal with shocks, as reflected by the recent Mexican peso crisis. They discussed whether the IMF, the World Bank, and the WTO were best suited to accommodate the enormous geopolitical and economic changes of the past 50 years. If they were not, would the Halifax Summit provide the appropriate venue for evaluating their continued contribution? Adding environmental considerations to IMF governance was, however, outside their vision.

Related to the issue of international institutional reform was the question of whether the current structure of international organizations charged with managing global environmental issues—primarily UNEP—were best equipped to deal with the evolving threats posed by biodiversity loss and climate change. With economic growth, poverty, debt reduction, trade, energy, migration, and nuclear safety all related directly to sustainable development, the current capacity of the international system to address these interrelated global issues was increasingly coming into question, particularly for Canada. Canada was trying to change the conventional thinking about sustainable development challenges. Among these challenges was the issue of whether to "parcel out" sustainable development issues or integrate them (Cappe 1995). Canada maintained that the reason for the lack of focus on sustainable development came from breaking it into component elements of global security, economics, trade, and the environment, obviating any capacity for dealing with sustainable development in an integrated way. Indeed, with Canada pushing to link sustainable development and environmental degradation within the global security context, as Halifax approached the question of which other G7 members were on board was still unresolved.

The Sherpa Meetings

The first sherpa meeting of the Halifax preparatory schedule, held on January 23–24, 1995, in Ottawa, focused on the IFIs, the institutional review, and Russian participation. The sherpas met again on March 23–25, in Vancouver, British Columbia, and divided their time into one session focused on Russia and another on the G7. They discussed five papers on six sub-themes. There was a sub-theme on poverty and development and another on the environment, but the sherpas decided to roll them into a single paper on sustainable development. They spent much of their time in Vancouver discussing the Nuclear Safety Working Group.

At the Summit

The plan for the Halifax Summit was to open on Thursday night, June 15, with an informal working dinner of the G7. On Friday, the leaders would discuss the economic agenda "at seven." Then they would have a working dinner "at eight" with Yeltsin. On Saturday, they would spend the full day on the political agenda and other issues at eight. Transnational issues could be part of those discussions. Chrétien was open to having one chair's statement, or one for the seven and one for the eight.

Jacques Chirac, having been recently elected president of France, arrived in Halifax in the wake of a domestic political crisis. A French peacekeeper had just been killed in the Balkans, and Chirac was clearly under pressure, both as the new president and as the leader of a country with peacekeepers now under siege, to react appropriately. Chrétien, always putting France first, immediately shifted the agenda to deal with this item at the opening dinner on June 15 and spontaneously decided to issue an unplanned separate summit statement that very evening on Bosnian peacekeeping.

The next day, Friday, G7 members knew they could not address every issue of IFI reform without overloading the agenda. The discussions thus focused on international financial issues, particularly reforms to the IMF and World Bank. Yet leaders reinforced their commitment to activities undertaken by UNCED on global environmental cooperation.

On June 17 the eight leaders discussed the United Nations, emphasizing its humanitarian relief and development assistance initiatives. They determined that the occasion of the UN's 50th anniversary would be an ideal opportunity to launch reforms and discuss the institution's future. UN reform took up about one third of the declaration.

Results

The Halifax Summit delivered on its goal of discussing the need for institutional reform to meet the challenges of the 21st century, including strengthening the global economy, promoting sustainable development, reducing poverty, safeguarding the environment, preventing and responding to crises, reinforcing coherence, building institutional efficiency, and promoting nuclear safety.

On the environment, the leaders outlined their intent to "show leadership in improving the environment" through the "appropriate mix of economic instruments, innovative accountability mechanisms, environmental impact assessments and voluntary measures" (G7 1995). To this end, leaders noted that "efforts must focus on pollution prevention, the 'polluter pays' principle, internalization of environmental costs, and the integration of environmental considerations into policy and decision making in all sectors."

On climate change, the leaders once again endorsed the UN process, noting the importance of keeping their Rio commitments and committing to "work with others to fulfil our existing obligations under the Climate Change Convention" (G7 1995). They further endorsed the UN process by committing to "meet the agreed ambitious timetable and objectives" to follow up the first Berlin COP.

Touted by many as an immense political success on the environment, others criticized the summit leadership for ignoring environmental concerns during their deliberations and dedicating only three of 50 clauses of the final communiqué to global environmental issues.

Dimensions of Performance

Domestic Political Management
Halifax was touted by many as "an immense personal success for Prime Minister Jean Chrétien, and a considerable achievement for his senior Canadian government officials" (Nankivell 1995). Others argued that it did not go far enough in reaching substantive commitments on climate change, as leaders focused instead on reform of the Bretton Woods

institutions, the conflict in Bosnia, global unemployment rates, trade liberalization, and France's resumption of nuclear testing.

Deliberation

Climate change took only 0.7 percent of the Halifax communiqué, the lowest level since 1986 and a sharp drop from 2.6 percent the year before (see Appendix C-1).

Direction Setting

G7 leaders agreed that they had an obligation to "show leadership in improving the environment" through the "appropriate mix of economic instruments, innovative accountability mechanisms, environmental impact assessment and voluntary measures" (G7 1995). To this end, they noted that efforts must focus on preventing pollution, integrating environmental considerations into their policy and decision making, and applying the "polluter pays" principle.

Decision Making

Halifax produced a total of 78 commitments, 13 of which were on the global environment. These included seven on climate change, up sharply from the year before (see Appendix D-1).

Noting that climate change "remains of major global importance," the G7 leaders committed to fulfilling their obligations under the climate change convention and meeting the timetable and objectives set out at the Berlin COP (G7 1995).

The leaders took an important lead in reviewing and reorienting the systems for creating incentives for environmentally positive behaviour. They also reaffirmed their support for the UNFCCC and rapidly moved toward implementing its national action plans, including sharing with other countries ways and means to improve energy-efficient and non-polluting production, with an emphasis on renewable forms of energy. Moreover, by agreeing to meet Berlin's targets, the leaders effectively committed themselves to a stringent protocol that would legally bind the G7 governments to implement reductions in greenhouse gas emissions beyond 2000. The G7 (1995) underlined the importance of meeting the climate change commitments made at Rio, with the "need to review and strengthen them."

Delivery

A year later, the two assessed commitments saw members comply at a level of +0.29, a sharp drop from +0.71 the year before (see Appendix E-1). However, as a testament to its commitment to accountability, the Canadian government released the first ever follow-up to its summit initiatives, which showed a committed G7 host keen to take stock of its global environmental commitments during its tenure as summit chair. The *Halifax Summit Legacy*, with its strong focus on IFI reform and the implementation of the Intergovernmental Panel of Forests, represented "the most aggressive, systematic, and thus far productive, followup in G7 history" (Kirton 1996; see also Canada 1996). This early model for assessing and reporting on the outcomes of commitments represented an important milestone in the evolution of summitry, and a key precedent in assessing and determining overall summit accountability.

Development of Global Governance

At Halifax, G7 leaders made no reference to the outside international organizations that deal with climate change (see Appendix F-5).

Causes of Performance

Driving Halifax's substantial environmental and climate change performance was shock-activated vulnerability, the multilateral organizational failure emphasized by the leaders facing the coming 21st-century demands, the high political control of the host and key partners, and the institutionalization of the compact G7 club with an invited Russia.

Shock-Activated Vulnerability
An acknowledged common vulnerability shared by the G7 leaders advanced sustainable development and climate change initiatives at Halifax. The ongoing and steady, decade-by-decade increase in the frequency, scale, and impact of environmental disasters, produced a clear recognition that economic, social, and security crises were rooted, in part, in such disasters (MacNeill 1995). This was supported by language in the final declaration that noted that "disasters and other crises complicate the development challenge and have exposed gaps in our institutional machinery" (G7 1995). As these trends were continuing and worsening, recognition was evolving at the highest political level that famine, refugee crises, the growing loss of arable land, species, and forests, as well as the collapse of fisheries on Canada's east coast and elsewhere, were being linked directly to ecological breakdown (MacNeill 1995).

These shared vulnerabilities induced the leaders to consider how the international community could increase its institutional capacity to deal with the negative environmental, social, and security impacts of these trends. Tied to this was the question of whether the international community could make its development, trade, energy, and other sectoral agencies "directly responsible and accountable for formulating policies and budgets that encourage development that is sustainable in the first place" (MacNeill 1995). If so, how could this best be done? The Halifax Summit opened the doors for these discussions to take place in an open, candid, and business-like manner among those global leaders best suited to promote the institutional reforms needed to make these changes.

This recognition of vulnerabilities was substantially politically constructed, as no single material shock spurred summit action in 1995. The collapse of the Canadian fishery due to high seas overfishing was a shock but of a cumulative, chronic kind. And the three oil tanker accidents were a sharp drop from nine the previous year (see Appendix G-4).

Multilateral Organizational Failure
Multilateral organizational failure mattered much in driving the summit's main theme. Chosen as the centrepiece subjects of a leaders-driven summit, sustainable development, international institutional reform, and the related issues of trade and environment formed the basis of the business-like summit meeting in Halifax.

The summit's second critical question was how to advance these initiatives through reform of the leading international institutions tasked with promoting sustainable development, primarily the IMF, WTO, World Bank, OECD, and other UN agencies. In this respect, Halifax delivered a number of concrete results that would steer these initiatives in the right direction.

However, on climate change specifically, multilateral success prevailed. The UNFCCC's first COP in Berlin 1995 was marked by great uncertainty regarding the means by which individual countries could combat greenhouse gas emissions. This gave rise to the Berlin Mandate, which set a two-year evaluation and analysis period, whereby a catalogue of instruments would be developed to match countries' climate change initiatives with their individual needs.

The Berlin Mandate boldly concluded that current actions on climate mitigation were "not adequate" and that a process was to begin to "enable it to take appropriate action for the period beyond 2000 ... through the adoption of a protocol or another legal instrument" (UNFCCC 1995). According to the mandate, this process would be guided by key principles, primarily the protection of the climate system for the "benefit of present and future generations," the promotion of sustainable development, and coverage of all greenhouse gases and their emissions, with special consideration to the situations of the least developed countries. Participants at Berlin recognized that the global nature of climate change required the "widest possible cooperation by all countries," thereby necessitating "an effective and appropriate international response."

Predominant Equalizing Capability
Another factor contributing to the success at Halifax was the predominant equalizing capacity of G7 members. All members experienced an increase in their GDP from 1994 to 1995 (see Appendix I).

Political Cohesion
A fourth spur of success was high domestic political cohesion. Host Jean Chrétien had a secure parliamentary majority, a divided opposition, and a general election at least two years away (see Appendix K-1). To be sure, he faced a looming referendum on separation in Quebec, which the separatists looked like they might well win. This separatist scare affected Chrétien's summit planning and execution in several fundamental ways, including emphasizing the distinctive national value of ecology that all Canadians—including Quebecers contemplating separation—deeply shared (Kirton 2007a). Although Chrétien had held several economic portfolios but not the environment one since he had entered cabinet in 1968, he had a deep personal belief in Canada's national parks, mountains, natural beauty, and aboriginal people, from which his environmental convictions flowed. A Quebecer, he was well acquainted with the value of clean hydro electricity and nuclear energy, as well as the danger of high seas overfishing off the Gaspésie.

US president Bill Clinton, who set the overall theme of the summit, was still midway through his first term, even if his Democratic Party had lost control of the House of Representatives in November 1994. His vice-president, Al Gore, was a strong advocate of environmental protection and climate change control. Germany's Helmut Kohl, the G7's veteran leader by far, was also an environmental advocate, as was his supportive public at home.

Driven by the host's desire to focus on how to encourage environmentally sustainable development, political control at the leaders' level factored heavily into the agenda. Chrétien's own focus for Halifax centred on job creation through improved environmental protection, advances in energy efficiency, and investments in the right technologies.

Compact Club Participation
A final cause of Halifax's solid climate change performance was the further integration of the now institutionalized environment ministerial process into the Halifax agenda and Russia's participation in the summit (see Appendix F-1). Through their success at Hamilton in identifying a G7 consensus position on several key environmental themes, having ministers tasked with the environment portfolio meet separately unburdened the leaders

at Halifax from having to spend much of their time agreeing on how to drive their climate commitments forward. By feeding directly into the leaders' process, ministers, sherpas, and summit planners ensured the G7 were well informed about their climate agreements heading into their summit deliberations, thereby creating a sense of reassurance as these issues came up for discussion (Kokotsis 1999). Thus the compact participation at the leaders' level was institutionalized at the ministerial level as well.

Lyon, 1996

The main thrust of the 1995 Halifax Summit continued at Lyon the following year. The French presidency identified the central theme of its summit as "making a success of globalization for the benefit of all" (Hajnal 1997, 221). Within this theme was a continued focus on reform of the international institutions and regional development banks as well as enhanced effectiveness of the UN and its affiliated institutions. Most importantly, however, was a strong, sustained focus on the effectiveness of aid and development. At the 50th anniversary of the UN General Assembly (UNGA), French president Jacques Chirac announced his intention to have the Lyon Summit concentrate on long-term aid solutions for the least developed countries. This aid would flow from various sources, including public, private, bilateral, and multilateral ones.

Canada wished the Halifax agenda to continue at Lyon. Canada's key issues were promoting sustainable development and environmental sustainability and reducing poverty. Canada encouraged France in the lead-up to Lyon to define its aid effectiveness issue more broadly, moving away from focusing on reforming the institutions that deliver aid to understanding the overall "philosophy of aid" and assessing whether the programs devised 40 years ago were still functioning appropriately. Tied to this aid philosophy would be an assessment of the link between aid and the environment and how environmental protection and sustainability could best be integrated into Lyon's development agenda.

Preparations

Continuing the cadence initiated in 1992, G7 environment ministers met for the fourth time on May 9–10 in Cabourg, France. Discussions focused primarily on three key themes: the innovative principle of the relationship between human health and the environment, international institutional arrangements and their commitments to the environment, and trade and the environment, with a view to the forthcoming ministerial meeting of the WTO in Singapore later that year (DFAIT 1996). Driving these deliberations was the invitation, for the first time, to key NGOs from around the world to join the environment ministers in a working session to discuss the impact of global economic decisions on key environmental issues (G7 Environment Ministers 1996).

For the first time, the G7 acknowledged the protection of public health through environmental policies. At Cabourg the environment ministers (1996) noted the "increasing evidence and concern that pollution at levels or concentrations below existing 'alert thresholds' can cause or contribute to chronic public health problems." This discussion focused on the broader link between health and the environment, noting the importance of coordination on research and risk analysis, but failed to link the negative impacts of climate change directly to human health. Instead, the ministers focused on precautionary principles and sound science in creating adequate solutions for health issues related to environmental degradation. Deferring to the work of other international institutions tasked with managing global health issues, the G7 proposed to convene a joint meeting with leaders of the World Health Organization, UNEP, the UNCSD, and the GEF to deepen their understanding of the issues.

Similar to the Hamilton ministerial the year before, sustainable development continued to be a major theme. The ministers' discussed promoting sustainable development through existing international environmental institutions, notably the UNEP and the UNCSD. These institutions "should form the basis of the institutional framework for the environment and sustainable development" (G7 Environment Ministers 1996). The ministers strongly supported the UNCSD, noting that it should "continue to be the high level global forum at which broad policy directions for sustainable development are set, and long-term strategic goals for sustainable development are identified and agreed upon." They agreed that the UNCSD was the key mechanism responsible for ensuring the coordination of the various environmental conventions, including the UNFCCC. Of particular note in the section on sustainable development was a reference to accountability: the ministers noted the "importance of compliance with the terms of international environmental agreements" and welcomed the offer by the US to host an international conference on this issue the following year (G7 Environment Ministers 1996). However, there were only two references directly on climate change, both within the context of the ongoing work of the UN.

Discussions linking trade liberalization and environmental protection similarly excluded any meaningful references to climate change. Only during background discussions did the ministers agree that climate change posed a threat to the global environment, and that it was their common responsibility to address the issue at the forthcoming UNGA special session for the five-year review of the Earth Summit. With less than six weeks to go before the G7 summit, there was little expectation that much of consequence would be achieved on climate change policies or mitigation strategies at Lyon. These discussions and decisions would be relegated instead to the UN's environmental institutions as well as the ongoing UNFCCC process.

At the Summit

As the summit unfolded, the central themes emerging on the leaders' agenda included growth, trade, and investment, as well as unemployment, North-South economic relations, and the integration of Russia and the countries of Central and Eastern Europe into the world economy.

Results

Building on the Halifax initiatives, Lyon called for enhancing the effectiveness of various institutions including the UN, IFIs, the regional development banks, and the WTO. A major achievement was reached on development, with the leaders meeting for the first time with the heads of the UN, the IMF, the World Bank, and the WTO in a post-summit session to elaborate ways to implement the "new partnership for development," with a particular focus on grants, aid flows, and concessional financing to the poorest countries in the world, especially those in sub-Saharan Africa (G7 1996c). No multilateral environmental or health organization was invited.

The leaders referred to the environment in their Lyon communiqué only in the context of broader development initiatives. They noted that sustainable development and environmental protection remained key tenets of the G7's New Partnership for Development. As anticipated during the preparatory process, the leaders made no direct references or commitments on climate change. Rather, the G7 recognized that the UN and its associated systems had a "fundamental role to play in ... protection of the environment" (G7 1996b).

Dimensions of Performance

At Lyon, G7 leaders made improvements on attention to climate change, advanced on the deliberative, delivery, and development of global governance dimensions of climate change, but made less progress in their decision-making dimension (see Appendix A-1).

Deliberation
Climate change took only +0.8 percent of the Lyon communiqué, a marginal rise from the +0.7 percent the year before (see Appendix C-1).

Decision Making
The leaders' deliberations resulted in a total of 128 commitments, only three of which climate change, down from seven the year before (see Appendix D-1). In noting that "protecting the environment is crucial in promoting sustainable development," the leaders recognized the importance of placing "top priority" on integrating environmental protection more completely into their respective governments' policies (G7 1996a).

Delivery
Lyon's commitments were complied with at an average level of +0.57, a major advance from +0.29 the year before (see Appendix E-1). The leaders further agreed, for the first time, to "assess compliance with international environmental agreements and consider options for enhancing compliance" (G7 1996a). Yet detailed measures and concrete action plans in this regard fell visibly short.

Development of Global Governance
In its development of global governance, the G7 focused fully on reinforcing the work of the UN (see Appendix F-5). The leaders noted that 1997 would be a "pivotal year for the environment" (G7 1996a). They renewed their commitment "to all agreements reached at Rio," pledging to work "for a successful outcome of the 1977 special session of the United Nations General Assembly which would lead to their better implementation." Anticipating a successful outcome of the next COP, the leaders committed to strong actions to promote sustainable development of forests as well as the negotiation of a "global, legally binding instrument on particular persistent organic pollutants." But they fell short on detailing the mechanics of such an instrument or offering a time frame for its implementation.

Causes of Performance

As the French presidency had noted early in the Lyon preparatory process, reform of the Bretton Woods institutions and aid to the heavily indebted poor countries would top the agenda, it came as little surprise to find that environmental protection and climate change fared poorly in the leaders' deliberations and decision making. Despite some countries' intention in advancing the sustainable development agenda, the preferences of the host prevailed. Lyon's agenda focused primarily on issues related to debt and development, as well as other global issues including humanitarian emergencies, nonproliferation, nuclear safety, and regional security.

Shock-Activated Vulnerability
One cause of this performance was shock-activated vulnerability. It took the form of the 10th anniversary of the Chernobyl explosion and the resulting diversion of attention to nuclear safety. Oil-related accidents remained low (see Appendix G-4).

Multilateral Organizational Failure
Multilateral organizational failure was masked by the hope that the forthcoming 1997 UNGA special session and COP would make the divided Rio regime a success (see Appendix H-3).

Predominant Equalizing Capability
Another factor was the predominant equalizing capability of G7 members. Inequality increased, as all experienced increases in their GDP, except for Germany and Japan whose GDP decreased (see Appendix I).

Political Cohesion
Host Jacques Chirac had strong political control, having to face re-election a full six years hence (see Appendix K-1). With such political security at home, he noted his intent early in the planning to focus his summit on debt and development initiatives.

Compact Club Participation
Finally, constricted participation was unusually low. It was reduced by the special nuclear safety summit in Moscow on the road to Lyon and the post-summit session with the four multilateral organizations, but no environmental ones (see Appendix L-1). Although the G7 noted the importance of incorporating environmental protection into their national policies and the need for better environmental compliance, the substantive work on climate strategies, mitigation, and timetables would be relegated to the UN process and its related institutions, with few, if any, concrete initiatives achieved in this regard at Lyon.

After the Summit: COP-2 Geneva

In late 1996, COP-2 met in Geneva. It noted—but not adopted— a ministerial declaration. Rifts between the members and non-members of the Organization for Petroleum Exporting Countries (OPEC) led to a compromise agreement that was driven primarily by Timothy Wirth, the lead US negotiator and under secretary for global affairs. That agreement accepted the scientific findings on climate change from the second assessment report published by the Intergovernmental Panel on Climate Change (IPCC). It also rejected the notion of uniform harmonized policies in favour of flexibility and called for legally binding mid-term targets (Oberthür and Ott 1999). This notable shift in the position of the American government from the previous year was largely precipitated by the conclusions reached in the IPCC's (1995) second assessment report, which concluded that there was a discernible human influence on the global climate. In an effort to strengthen its environmental profile, the Clinton administration for the first time committed to a set of targets. The Geneva meetings concluded with a decision by the Japanese government to host COP-3 in Kyoto the following year.

Denver, 1997

When the G7 leaders and those of the European Union and Russia assembled in Denver from June 20 to 22, 1997, the challenge of preserving the globe's natural environment in general, and controlling climate change in particular, was a major concern. In response, they shaped the Rio regime, as the divisive principles of the 1992 UN convention would be converted into a more precise, legally binding protocol created at the Kyoto COP at the end of the year. The protocol would oblige a little group of industrialized countries (so-called Annex I countries) to do a little bit for a little while to control their carbon emissions even as the much larger group of developing countries (non-Annex I countries) were free to pollute as they wished. The divisions created by Rio's principles of "special but differentiated" sustainable development and its focus on sources and not sinks assumed stronger, sharper legal definition and force. Moreover, the US, the world's largest climate power, would refuse to ratify the Kyoto Protocol and thus remain ever farther out of the now doubly divided Rio-Kyoto regime.

For the leaders at Denver to make a real contribution to controlling climate change in their two days of discussions, it was necessary to overcome a main constraint. They were distracted by the priority of promoting the status of Russia's Boris Yeltsin, in part by being trapped in more unproductive pageantry than at summits in the recent past (Clinton 2004). They thus produced the "Denver Summit of the Eight," so termed because the Russians took part alongside the other G7 participants from the outset. The main communiqué was issued by the eight leaders and covered a broad range of global and foreign policy issues, including Africa. A separate G7 statement on the economy was released, which did not include input from the Russians.

Denver's major environmental issue was the "Special Session of the UN General Assembly on Environment and Development: The Earth Summit Plus Five," which would take place immediately after the summit in New York on June 23–27, 1997. Denver's first challenge was to provide the high-level political impetus to move forward on climate change. The ultimate target was the third COP in Kyoto in December. The G7's challenge was to lead the world in agreeing to conclude a binding international agreement—a protocol to the 1992 UNFCCC that had been signed by 166 countries at Rio—to meet the existing commitment to stabilize carbon dioxide emissions at 1990 levels by 2000. The protocol ambitiously sought to set targets and timetables for actual reductions in greenhouse gas emissions in the coming decade. Indeed, it was adopted at Kyoto but did not enter into force until February 16, 2005.

Denver's second environmental challenge was producing the missing global forestry convention, initially agreed to at Houston in 1990 as a deliverable for 1992. This regime was required to fully implement the Rio commitments and deal seriously with carbon sinks, through a global convention to protect the world's rapidly dwindling forests. Denver's leaders thus needed to agree to provide the critical missing element to complete the network begun with the UN conventions on climate change, biodiversity, and desertification and the agreement on overfishing the high seas. Building on the accomplishments of the Intergovernmental Panel on Forests formed in 1995, and its report of April 1997, the Denver Summit and the UNGA special session were critical to securing an agreement to launch negotiations for an international convention that would provide a legally binding instrument to formalize a comprehensive and integrated approach to preserving the world's forests. It would promote sustainable forest management, provide financial assistance and technology transfer, protect sustainable global trade in forest products, and, above all, create carbon sinks, thereby enhancing the health of the world's forests and the living things that inhabit them.

The G7 leaders also attempted to advance work on a broad range of climate-related environmental issues, such as linking climate and human health and striking a more balanced and integrated relationship between development and the environment and between trade and the environment. Injecting ecologically sustainable development into the US initiative on Africa as the very foundation of that program would give life to that initiative. It was narrowly conceived to deliver development assistance and trade liberalization, but lacked an equally strong recognition of the ecological requirements needed for sustaining growth, and the broader demands of peace building and state capacity building if genuine development in sub-Saharan Africa was to take hold. Developing a trade-environment regime in the period following the WTO's Singapore ministerial meeting in December 1996 required catalyzing movement in areas where the WTO had stalled. Included here would be a decision to give life to a new global environment organization proposed by Germany's Helmut Kohl to work on an equal basis with the WTO to define a balanced, mutually reinforcing trade-environment regime (Biermann and Bauer 2005).

Preparations

As summit host, Bill Clinton had signalled that one of his central objectives was to provide the high-level impetus needed to develop the global conventions on climate change and biodiversity, so painfully secured at the Earth Summit in Rio five years prior.

Among the G7 members, Canada ranked environmental issues most prominently in its priorities in the lead-up to Denver, particularly matters relating to forests. It wanted agreement on a global forestry convention with either a strong start or nothing at all from the G7. It encountered opposition from the US and the Japanese, with all other summit members supporting the Canadian position. Canada would not accept US language in the final declaration that would put off forestry for several years. It wanted to avoid a rhetorical, watered-down statement and to retain its flexibility to deal with the issue at the UNGA special session. The Americans argued that neither the environmental NGOs nor business supported such flexibility.

The first sherpa meeting in Washington in May presented the initial opportunity to work from a draft document that the Americans had circulated the week before. It noted that there would be an economic declaration at seven, excluding the Russians, and a political communiqué for the eight. It was difficult to discern the Americans' strategy, as there was no single priority issue on the agenda they circulated. They were preoccupied with engaging the Russians in the G7 process.

However, two subjects were highlighted. The first, on the economic side, was Africa, with the summit sending a strong signal that a holistic approach was needed for this lagging region of the world that would embrace economics, finance, and trade. The second issue was the environment, with a particular focus on climate change, looking ahead to the Kyoto COP, as well as on freshwater, which was also emerging as a key theme for Denver. By this point, all seven leaders were expected to pledge their attendance at the UNGA special session immediately following their summit in Denver. The US also signalled it wanted to move ahead on UN reform, building off a report that Secretary General Kofi Annan was scheduled to release in July. The G7 wanted to give him strong support, but not so strong that there might be suspicion that the G7 had indeed orchestrated it.

The annual meetings of G7 environment ministers continued in the lead-up to Denver, with EPA administrator Carol Browner hosting a meeting in Miami on May 5–6. Taking as their primary theme the need to protect children's health, the ministers agreed on several issues: environmental risk assessments and standard setting, children's exposure to lead, microbiologically safe drinking water, air quality, the negative effects of tobacco smoke, and endocrine-disrupting chemicals. On climate change, they stressed the overwhelming scientific evidence linking the build-up of atmospheric greenhouse gases with changes in the global climate system. They noted that climate change was likely to have wide-ranging, negative impacts on human health, including "more intense air pollution, the spread of infectious diseases" and "significant loss of life" (G7 Environment Ministers 1997). The ministers further lent their support to COP-3 in Kyoto, stressing "their commitment to achieving a strong agreement for controlling greenhouse gases." Now recognizing that actions by the developed countries were not enough to meet the objectives of the climate convention, the ministers agreed to work with developing countries in assisting them to address the problem. To this extent, they supported an agreement allowing flexibility in meeting those targets, while recognizing that global systems for monitoring climate change and other environmental trends were also needed.

At the Summit

As the summit participants moved into Denver, there were still square brackets in the draft communiqué, reflecting disagreements to be resolved—primarily on climate change, the environment, the forestry convention, UN reform, and Africa. As UN reform would soon be resolved by the sherpas, and Africa by their assistants (known as sous-sherpas), the greatest disagreement remaining for the leaders to deal with at Denver was on climate change and forestry.

On climate change, the Europeans, led by Germany's Kohl, wanted a deal to commit to a reduction of 15 percent in emissions by 2010. But with many summit members failing to meet their Rio commitments, there was resistance to establishing firm targets. Canada did not want to subscribe to any targets or timetables until it had defined its Kyoto position. The US and Japan agreed. Canada had fallen short by 8.7 percent on its Rio targets. Some European members had similarly not met their commitments, with France and Spain doing less well than Canada. The Japanese did not speak up as they were hosting Kyoto and had their own internal difficulties to resolve.

On forests, Canada's strategy was to reach an agreement on a convention. But it calculated that it had a better chance of doing so in the large multilateral special session of the UNGA. With six summit members on board, one against (the US) and one waffling (the Japanese), a compromise solution had to be devised. The US wanted NGOs to be part of the dialogue, but were caught by an unholy alliance of business and NGOs in opposing a UN forestry convention. Canada suggested involving the NGOs by engaging them to help develop the first steps and an action program that would be reviewed the following year. With the US agreeing on this language, a last-minute compromise was eventually found for the forestry text in the final communiqué.

Results

In the opening chapeau of Denver's communiqué, the leaders emphasized that "cooperation to integrate Russia's economy into the global economic system represents one of our most important priorities," clearly outlining a shared commitment for further Russian involvement in the summit process (G7 1994). But despite their attention on Russia's summit participation, a considerable amount of the communiqué was devoted to environmental issues, climate change, and energy policy. A record eight sections were dedicated to environmental considerations, including stand-alone sections on the environment, the UNGA special session, climate change, forests, freshwater, oceans, desertification, environmental standards for export credit agencies, children's environmental health, and institutions deemed "essential to coordinating global efforts to protect the environment."

For the first time in this period of reinforcement, summit leaders noted their determination to "take the lead" in strengthening international efforts to confront climate change (G7 1997). This would require them to "forge a strong agreement" at Kyoto that contains "quantified and legally-binding emission targets." Leaders further reinforced this pledge by promising to commit to "meaningful, realistic and equitable targets" that would result in emissions reductions by 2010. This agreement would need to ensure "transparency and accountability" by establishing appropriate mechanisms for "monitoring and ensuring compliance among the parties."

At the end of the summit, when presenting the final communiqué, Clinton emphasized the global environment by noting that the G7 was determined to do its part "to protect our environment for future generations" (White House 1997a). "Among other measures," he said, "we recommitted ourselves to the principles of the Rio summit. We intend to reach an agreement in Kyoto to reduce greenhouse gas emissions to respond to the problem of global warming." At his closing press conference, he emphasized that economic objectives "would be pursued while improving, not undermining, the environment" (White House 1997b). On the forthcoming Kyoto conference more specifically, Clinton stressed that all G7 members were "bound to adopt legally-binding targets to reduce greenhouse gas emissions" and in so doing, the G7's economies would have to grow "while improving the environment" (White House 1997b).

Dimensions of Performance

In its key dimensions of performance, the Denver Summit was a substantial success, above all in shaping a doubly divided Rio-Kyoto UN regime (see Appendix A-1).

Deliberation
Given that the summit coincided with the fifth anniversary of the Earth Summit in Rio de Janeiro, global environmental issues were granted significant attention. Climate change took 1.6 percent of the communiqué, double the level the year before (see Appendix C-1).

In all, 20 of the summit's 89 paragraphs were dedicated to the environment and energy issues, with five focused specifically on climate change. The leaders emphasized that "overwhelming scientific evidence links the build-up of greenhouse gasses in the atmosphere to changes in the global climate system" (G7 1997). They further noted that "if current trends continue into the next century, unacceptable impacts on human health and the global environment are likely." Reversing these trends would thus require "sustained global effort over several decades."

Direction Setting
Denver's affirmation of the facts and causation of climate change was more resolute than it had been at prior summits because the leaders recognized the connection between

greenhouse gases and the changing climate system. They also noted their determination "to take the lead and show seriousness of purpose in strengthening international efforts to confront climate change" (G7 1997). The importance of forging "a strong agreement" with "quantified and legally-binding emissions targets" at Kyoto later that year was key. Other important direction-setting steps were the links of climate change to trade, development, and human health.

Decision Making

In a striking shift from prior summits, nearly one quarter of Denver's commitments (19 of 75) involved the global environment. Nine—almost half of the 19—focused specifically on climate change and its related issue of forests (see Appendix D-1). This was the second highest number of commitments the summit was to make before 2005. It contained one of the first summit commitments linking climate change and health (Kirton, Kulik, Bracht, et al. 2014). Yet Denver was unable to bridge the stark differences between the Americans and the Europeans on either climate change or a forestry convention. The Europeans were bound by a collective decision by the European Commission to reduce emissions by 15 percent below 1990 levels by 2010, following the commitments made at Rio. While the Europeans urged their summit partners to do likewise, the others argued they could not extend beyond the principle of quantified targets and reductions by 2010 (Bayne 1997).

Although they did not agree on firm goals, the leaders declared their intent to "commit to meaningful, realistic and equitable targets that would result in reductions of greenhouse gas emissions by 2010" (G7 1997). This would ultimately require them to "work together to enhance international efforts to further develop global systems for monitoring climate change and other environmental trends."

The Canadians and Europeans wanted the summit to call for a global forestry convention. But given the resistance of the Americans and Japanese, Denver fell short. The leaders agreed only to support a "practical Action Program" that would involve the implementation of national programs and capacity building for sustainable forest management (G7 1997).

Delivery

The delivery of the climate decisions was solid. Denver's two assessed commitments averaged compliance of +0.29, down from +0.57 the year before but in line with the evolving norm (see Appendix E-1). To boost compliance, leaders added a remit mandate, in the form of requesting a report to assess progress in implementing the forest action program to be delivered by their next summit.

Development of Global Governance

In developing global governance, G7 leaders again focused overwhelmingly on the UN. The communiqué addressed the US need to include NGOs by stating that the G7 would work with environmental groups "to build consensus on an international [forestry] agreement with appropriately high international standards to achieve these goals" at the UNGA special session (G7 1997).

Causes of Performance

The causes of Denver's performance in shaping the doubly divided Rio-Kyoto UN regime were similar to those at work the year before.

Shock-Activated Vulnerability

Oil tanker accidents spiked to 10 in 1997, which spurred Denver's advance (see Appendix G-4). But there was no other severe shock to inspire the G7 to cut a different path than the one set by the UN.

Multilateral Organizational Failure

Prospective multilateral organizational failure was thus left to dominate, as the looming UNGA special session and Kyoto COP would lead the G7 directly to reinforce the divided Rio regime (see Appendix H-2).

Predominant Equalizing Capability

The overall predominant equalizing capability of G7 members was also a factor (see Appendix I). However, five of the members experienced a drop in their GDP, including France, Germany, Italy, Japan, and the EU.

Political Cohesion

In his political control, host Bill Clinton was fairly secure, not facing election until November 2000 (see Appendix K-1). Seven leaders returned to the summit table, adding to the group's political cohesion (see Appendix K-2).

Compact Club Participation

In 1997 club participation expanded with the official inclusion of Russia for the first time, which distracted the leaders' attention from climate change (see Appendix L-1).

Assessment

In the view of leading outside analysts, the 1997 Denver Summit was notable for what it initiated rather than for what it achieved (Bayne 1997). It had a heavy agenda, embracing new and traditional economic issues, political-security concerns, institutional reform, and Russian integration into the world economy, and produced progress reports on infectious diseases, terrorism, drugs, nuclear safety, money laundering, financial crime, and corruption. It even added new issues of human cloning and the international space station. The question thus turned to how much attention global environmental issues would receive.

Yet real progress on the environment came. At a minimum, Denver induced the leaders of the world's major powers to attend the UNGA special session immediately following the G7 summit—a meeting that gave high-level attention to the need to build upon the Rio regime. It further provided the critical catalyst to transform the promises of Rio into binding commitments in the Kyoto Protocol that would ensure countries took the hard decisions and concrete actions required to preserve the world's climate.

The summit advanced the process of creating the critical missing component of the world's constellation of environmental regimes—a global convention on forests. Once the UNGA special session convened, the prevailing consensus among participants was that the Denver communiqué language had been extremely helpful, particularly in moving the US forward. However, the Denver Summit gave an ambiguous response to the German call for a world environmental organization that would ensure compliance. Instead, it agreed on measures to strengthen existing environmental institutions, particularly UNEP.

After the Summit: COP-3 and the Kyoto Protocol

Following intense negotiations spanning 12 days, the Kyoto Protocol was adopted in December 1997. It introduced, for the first time, legally binding obligations and targets for greenhouse gas emissions (UNFCCC 1998). In all, 37 industrialized countries agreed to binding reductions of between 6 percent and 8 percent below 1990 levels by 2008–2012, a phase commonly recognized as the first commitment period. Kyoto placed a heavier burden on developed countries under the principle of common but differentiated responsibilities, while enabling developing countries to participate voluntarily through flexibility mechanisms designed to allow them to meet their emissions reduction goals with reduced impact on their economies.

The Kyoto Protocol was heralded as an important first step toward a truly global regime to reduce and stabilize greenhouse gas emissions and provide the necessary architecture for future international agreements on climate change.

However, following Kyoto's adoption it became increasingly apparent that countries remained deeply divided on whether a sufficient number of members would ratify the treaty and acknowledge its emissions reduction requirements. While the US proposed stabilizing emissions but not cutting them, the EU called for a 15 percent reduction. In the end, the US would be required to reduce its emissions by 7 percent below 1990 levels, pitting the Clinton administration against members of the US Congress who rejected these levels and vowed to veto Kyoto's ratification. Citing potential damage to the US economy, the Senate also rejected the agreement as it excluded key carbon emitters from the developing world, primarily India and China, from complying with the new standards (Oberthür and Ott 1999).

Divisions were further exacerbated once countries recognized that although the overall goal of Kyoto was to reduce emissions, in the end it enabled the creation of mechanisms for avoiding reductions. For example, generating credits from carbon sinks would eventually result in more atmospheric carbon emissions, all in the name of reducing emissions.

Conclusion

Over the period of G7 climate governance from 1993 to 1997 the G7 reinforced the divided UN control regime, flowing directly from the cadence created by the 1992 Earth Summit and resulting UNFCCC. This support was due largely to the prominence of the Earth Summit and its institutions and processes and the absence of any severe climate shocks to urge the G7 to strike out on its own different, more effective course. It could leave global climate issues to the larger multilateral system. A lower overall level of summit deliberation on climate came as G7 attention was drawn primarily to the integration of Russia, economic growth, international trade, and the escalating conflict in Bosnia and Kosovo. Foremost on the G7's agenda was its new preoccupation with assisting and integrating into the democratic west the once communist countries in Central and Eastern Europe and the former Soviet Union. This continued to the Denver Summit of the Eight, designed to present Russia and Yeltsin as full summit participants, even though climate change emerged as a major issue. To be sure, the G7's success in this regard, and the end of the Soviet Union, its economy, and its centralized system, brought major gains in the real world of climate change control. But this lesson was not integrated into G7 climate governance, which reinforced the divided UN regime.

Halifax in 1995 made the environment one of its centrepiece themes and went the distance in addressing unsustainable development practices. Denver in 1997 promised to "take the lead and show seriousness of purpose" in strengthening efforts to combat climate change. Yet it did so primarily by creating a doubly divided Rio-Kyoto regime. With the UNGA special summit taking place immediately after the Denver Summit and the negotiations for the Kyoto Protocol looming in December, the G7 took a leading role in pushing the UN to make concrete agreements on climate change mitigation on the Rio platform.

This G7 emphasis on reinforcing the divided regime was driven primarily by four key causes.

First, the absence of severe climate shocks let the G7 leave global leadership to the UN, rather than act on its own in a more urgent and effective way. To be sure, a collective sense of environmental vulnerability was beginning to take root during this period of summit reinforcement. By 1995, the G7 was increasingly aware of the impact, scale, and frequency of global environmental disasters, forcing the leaders to recognize that economic, social, and security crises were connected directly to such disasters. No longer could the leaders of the world's seven most industrialized countries ignore the fact that environmental collapse was linked to global food shortages, water and refugee crises, and the dramatic loss of arable land, forests, and biodiversity. Reconsidering how the international community could increase its institutional capacity to deal with these negative social, economic, security, and environmental trends became a prominent theme of both the Halifax and Denver summits. Yet these slow, chronic cumulative trends allowed enough time for the G7 to safely leave the problem to the UN.

Second, multilateral organizational failure was a major cause, in a particular form. It appeared to be low in this period the UN rushed to create the principles, conventions, protocols, and institutions of its Rio regime. These appeared to be sufficient to meet the coming cumulative chronic challenge, especially if the G7 leaders would reinforce a regime that promised to work.

Third, the level of political control and commitment in the summit hosts was high in 1995 and 1997. As summit hosts, both Canada (in 1995) and the US (in 1997) had the capacity to play a major role in setting the environment agenda insofar as Jean Chrétien and Bill Clinton both "[raised] issues to public attention, [defined] the terms of public debate, and [mobilized] public opinion and constituency support through speeches, press conferences, and other media events" (Vig 1994, 72). The result at both summits was a sustained and constructive dialogue by the leaders on sustainability and climate mitigation, and a commitment by both hosts to implement their international commitments on the environment. In the case of the 1995 Halifax Summit, this commitment came through Canada's *Halifax Summit Legacy*, whereby Canada's accountability to environmental commitments was documented, for the first time in the summit's 20-year history. This implementation trend continued at the US-hosted 1997 Denver Summit. Indeed, in 1996 the US had committed that "by the end of 1997, the State Department [would] host a conference on strategies to improve compliance with international environmental agreements—to ensure that those agreements yield lasting results, not just promises" (Christopher 1996).

Beyond Chrétien's personal attachment to environmental issues, global environmental protection was a top policy priority for the Canadian government in 1995. Canadian politicians knew they would not go wrong in associating Canada's name with certain environmental causes. By engaging in advocacy at the summits, Canada tried to develop an international consensus. Moreover, public opinion polls throughout the 1990s found that Canadians consistently ranked environmental protection high. By 1995, public opinion polls found that the environment continued to garner the strongest support among a list of 10 issues to be discussed at G7 summit, with 44 percent of Canadians in 1995 indicating their support for environmental considerations as a priority issue at the G7 (Goldfarb Consultants 1996). Chrétien thus had the support of his constituency to push for including environmental issues on the summit agenda and following through on the collective agreements reached.

Fourth, the compact participation of the G7 summit was lowered, with Russia's increasing involvement and the G7 leaders' preoccupation with making it a member of a new Group of Eight. Yet this task was never linked to climate change control, and thus divided the G7's agenda and divert attention from climate control. This division appeared at the ministerial level, both in the old G7 bodies for finance, foreign affairs, and trade and in the new environmental one.

The institutionalization of the G7 environment ministerial process, beginning in Germany before the 1992 Munich Summit, continuing in Florence in 1994, Hamilton in 1995, and Cabourg in 1996 reinforced the Rio regime. Indeed, those meetings had begun for and at the Earth Summit, in order to make Rio succeed. They hastened the G7 leaders' recognition of the importance of keeping the spirit of Rio alive, by continuing with the sustained implementation of the Rio commitments on climate change. Every G7 communiqué after Rio (and every G7 environment ministerial statement) endorsed the implementation of the Rio conventions, including the one on climate change.

References

Agenzia ANSA (1994). "From Tokyo to Naples, the Stages: The Florence Meeting on the Environment." ANSA Dossier (folder), Rome.

Bayne, Nicholas (1997). "Impressions of the Denver Summit." G8 Research Group, Denver. http://www.g8.utoronto.ca/evaluations/1997denver/impression/index.html (January 2015).

Bayne, Nicholas (2000). *Hanging In There: The G7 and G8 Summit in Maturity and Renewal.* Aldershot: Ashgate.

Biermann, Frank and Steffen Bauer, eds. (2005). *A World Environmental Organization: Solution or Threat for Effective International Environmental Governance.* Aldershot: Ashgate.

Canada (1996, February). 1995 Canada's Year as G7 Chair: The Halifax Summit Legacy. G7/G8 Research Collection in the John W. Graham Library, Trinity College, University of Toronto. Locator ID: 1.14.

Canada. Department of Foreign Affairs and International Trade (1995a). "Background Information: The Environment." Ottawa. http://www.g8.utoronto.ca/environment/1995hamilton/hamenvi7.html (January 2015).

Canada. Department of Foreign Affairs and International Trade (1995b). "The Halifax Summit, June 15–17, 1995: Background Information." Ottawa.

Canada. Department of Foreign Affairs and International Trade (1996). "The Lyon Summit, June 27–29, 1996: Background Information." Ottawa.

Canada. Environment Canada (1995). "Canada's National Action Plan on Climate Change." Ottawa. http://web.archive.org/web/20020420084029/http://www.ec.gc.ca/climate/resource/cnapcc/indexe.html (January 2015).

Canada. Standing Committee on Foreign Affairs and International Trade (1995). "From Bretton Woods to Halifax and Beyond: Towards a 21st Summit for the 21st Century." Report of the House of Commons Standing Committee on Foreign Affairs and International Trade on the Issues of International Financial Institutions for the Agenda of the June 1995 G7 Halifax Summit, House of Commons, Ottawa. http://www.g8.utoronto.ca/governmental/hc25/index.html (January 2015).

Cappe, Mel (1995). "From Hamilton to Halifax." In *The Halifax Summit, Sustainable Development, and International Institutional Reform*, John J. Kirton and Sarah Richardson, eds. Ottawa: National Round Table on the Environment and the Economy. http://www.g8.utoronto.ca/scholar/kirton199503/cappe/document.html (January 2015).

Christopher, Warren (1996). "American Diplomacy and the Global Environment Challenges of the 21st Century." Speech to alumni and faculty of Stanford University, April 9, Palo Alto CA. http://1997-2001.state.gov/wviww/global/oes/speech.html (January 2015).

Clinton, Bill (2004). *My Life*. New York: Knopf.

Der Bund — die Tageszeitung (1993). "Auf dem Gipfel der Tatenlosigkeit." July 7.

G7 (1993). "Tokyo Summit Political Declaration: Striving for a More Secure and Humane World." Tokyo, July 8. http://www.g8.utoronto.ca/summit/1993tokyo/political.html (January 2015).

G7 (1994). "G7 Communiqué." Naples, July 9. http://www.g9.utoronto.ca/summit/1994naples/communique (January 2015).

G7 (1995). "Halifax Summit Communiqué." Halifax, June 16. http://www.g8.utoronto.ca/summit/1995halifax/communique/index.html (January 2015).

G7 (1996a). "Chairman's Statement: Toward Greater Security and Stability in a More Cooperative World." June 29, Lyon. http://www.g8.utoronto.ca/summit/1996lyon/chair.html (January 2015).

G7 (1996b). "Economic Communiqué: Making a Success of Globalization for the Benefit of All." Lyon, June 29. http://www.g8.utoronto.ca/summit/1996lyon/chair.html (January 2015).

G7 (1996c). "A New Partnership for Development." Released after a Work Session with the Secretary-General of United Nations, the International Monetary Fund Managing Director, the President of the World Bank and the World Trade Organization Director-General, Lyon, June 29. http://www.g8.utoronto.ca/summit/1996lyon/partner.html (January 2015).

G7 (1997). "Communiqué." Denver, June 22. http://www.g8.utoronto.ca/summit/1997denver/g8final.htm (January 2015).

G7 Environment Ministers (1995). "Chairperson's Highlights." Hamilton, Canada, May 1.

G7 Environment Ministers (1996). "Chairman's Summary." Cabourg, France, May 10. http://www.g8.utoronto.ca/environment/1996cabourg/summary_index.html (January 2015).

G7 Environment Ministers (1997). "Chair's Summary." Miami, May 6. http://www.g8.utoronto.ca/environment/1997miami/summary.html (January 2015).

G8 (1997). "Communiqué." June 22, Denver. http://www.g8.utoronto.ca/summit/1997denver/g8final.htm (January 2015).

Goldfarb Consultants (1996). "The Goldfarb Report 1996." Toronto.

Hajnal, Peter I., ed. (1997). *International Information: Documents, Publications, and Electronic Information of International Governmental Organizations*. Englewood CO: Libraries Unlimited.

Hajost, Scott, Bruce Rich, and Todd Goldman (1993). "Struggles of Developing Nations a Job for G7: G7, World Bank Bear Burden of Environmental Reform." *Christian Science Monitor*, July 6, p. 19.

Intergovernmental Panel on Climate Change (1995). "IPCC Second Assessment: Climate Change 1995." Geneva. http://www.ipcc.ch/pdf/climate-changes-1995/ipcc-2nd-assessment/2nd-assessment-en.pdf (January 2015).

Japan. Ministry of Foreign Affairs (1994). "Japan's ODA Annual Report (Summary) 1994: The 40th Anniversary of Japan's ODA: Accomplishments and Challenges — 3. Perspectives of Japan's ODA Towards the 21st Century." Tokyo. http://www.mofa.go.jp/policy/oda/summary/1994/3.html (January 2015).

Kirton, John J. (1996). "The G7 Has Finally Reached Adulthood." *Financial Post*, June 22.

Kirton, John J. (2007). *Canadian Foreign Policy in a Changing World*. Toronto: Thomson Nelson.

Kirton, John J., Julia Kulik, Caroline Bracht, et al. (2014). "Connecting Climate Change and Health Through Global Summitry." *World Medical and Health Policy* 6(1): 73-100. doi: 10.1002/wmh3.83.

Kirton, John J. and Sarah Richardson (1995a). "The Halifax Summit, Sustainable Development and International Institutional Reform: Preliminary Discussion Paper and Background Material." February 27, National Round Table on the Environment and the Economy, Ottawa.

Kirton, John J. and Sarah Richardson (1995b). "Introduction: The Halifax Summit, Sustainable Development, and International Institutional Reform." In *The Halifax Summit, Sustainable Development, and International Institutional Reform*, John J. Kirton and Sarah Richardson, eds. Ottawa: National Round Table on the Environment and the Economy. http://www.g8.utoronto.ca/scholar/kirton199503/rt95ind.htm (January 2015).

Kokotsis, Eleanore (1999). *Keeping International Commitments: Compliance, Credibility, and the G7, 1988–1995*. New York: Garland.

MacNeill, Jim (1995). "UN Agencies and the OECD." In *The Halifax Summit, Sustainable Development, and International Institutional Reform*, John J. Kirton and Sarah Richardson, eds. Ottawa: National Round Table on the Environment and the Economy. http://www.g8.utoronto.ca/scholar/kirton199503/rt95mac.htm (January 2015).

Maddox, Bronwen (1993). "The World on His Shoulders." *Financial Times*, June 30.

Nankivell, Neville (1995). "Summit Produced Real Substance." *Financial Post*, June 30.

Oberthür, Sebastian and Hermann Ott (1999). *The Kyoto Protocol: International Climate Policy for the 21st Century*. Berlin: Springer.

Richardson, Sarah and John J. Kirton (1995). "Conclusion." In *The Halifax Summit, Sustainable Development, and International Institutional Reform*, John J. Kirton and Sarah Richardson, eds. Ottawa: National Round Table on the Environment and the Economy. http://www.g8.utoronto.ca/scholar/kirton199503/conclusion (January 2015).

Schneider, Keith (1992). "Gore Bringing Environmental Policy to Fore." December 15. http://www.nytimes.com/1992/12/15/us/the-transition-gore-bringing-environmental-policy-to-fore.html (January 2015).

Smith, Gordon (1995). "Canada and the Halifax Summit." In *The Halifax Summit, Sustainable Development, and International Institutional Reform*, John J. Kirton and Sarah Richardson, eds. Ottawa: National Round Table on the Environment and the Economy. http://www.g8.utoronto.ca/scholar/kirton199503/smith/document.html (January 2015).

Spini, Valdo (1994). "The Floreince Meeting on the Environment." G7 Napoli Summit '94 (dossier), Agenzia ANSA, Rome.

Stevens, William (1993). "Gore Promises U.S. Leadership on Sustainable Development Path." *New York Times*, June 15. http://www.nytimes.com/1993/06/15/news/gore-promises-us-leadership-on-sustainable-development-path.html (January 2015).

United Nations Framework Convention on Climate Change (1995). "Decisions Adopted by the Conference of the Parties." Report of the Conference of the Parties on Its First Session, Held at Berlin from 28 March to 7 April 1995. FCCC/CP/1995/7/Add.1, June 6, Berlin. http://unfccc.int/resource/docs/cop1/07a01.pdf (January 2015).

United Nations Framework Convention on Climate Change (1998). "Report of the Conference of the Parties on Its Third Session, Held at Kyoto from 1 to 11 December 1997." FCCC/CP/1997/7, March 24, Kyoto. http://unfccc.int/resource/docs/cop3/07.pdf (January 2015).

United States Department of State (1993). "Group of Seven (G7) 1993 Economic Summit. Fact Sheet: Global Environmental Issues." *US Department of State Dispatch Supplement* 4(3). http://dosfan.lib.uic.edu/ERC/briefing/dispatch/1993/html/Dispatchv4Sup3.html (January 2015).

Vig, Norman J. (1994). "Presidential Leadership and the Environment: From Reagan to Bush to Clinton." In *Environmental Policy in the 1990s: Toward a New Agenda*, Norman J. Vig and Michael E. Kraft, eds. Washington DC: CQ Press, pp. 71–95.

White House (1993). "Press Briefing by Counselor to the President David Gergen, Under-Secretary of State Joan Spero, Under-Secretary of Treasury Larry Summers, and Special Assistant to the President Bob Fauver." Tokyo, July 9. http://www.presidency.ucsb.edu/ws/?pid=60165 (January 2015).

White House (1997a). "Remarks by President Clinton in Presentation of the Final Communiqué of the Denver Summit of the Eight." Denver, June 22. http://www.g8.utoronto.ca/summit/1997denver/denclintp.htm (January 2015).

White House (1997b). "Transcript: President Clinton's Final Denver Summit Press Conference." Denver, June 22. http://www.g8.utoronto.ca/summit/1997denver/clint22.htm (January 2015).

Chapter 7
Retreat, 1998–2004

After the Group of Seven (G7) reinforced United Nations leadership on climate change from 1993 to 1997, it retreated from action until 2004, as it left global climate governance to a UN regime now equipped with the Kyoto Protocol agreed in December 1997. After the thrust of the Denver Summit of the Eight in June 1997 to shape the protocol, subsequent summits offered modest efforts and results on climate change. Now, with Russia at the table, the Group of Eight (G8) at the 1998 Birmingham and 1999 Cologne summits focused primarily on implementing the Kyoto Protocol and other commitments from the UN Framework Convention on Climate Change (UNFCCC) established at the Earth Summit in Rio de Janeiro in 1992. The 2000 Okinawa and 2001 Genoa summits did likewise, adding some discussion of energy efficiency and the role of developing countries in mitigating climate change. The 2002 Kananaskis Summit ignored climate change, apart from peripheral mention in its Africa Action Plan. After a brief return at Evian in 2003, climate change virtually disappeared at United States president George W. Bush's Sea Island Summit in 2004.

This period also marked a retreat in a second way—from an easy acceptance of the principles, norms, and rules of the Rio-Kyoto regime. Starting in 1998, G8 governors increasingly recognized that all major carbon-producing powers must be included in an effective climate control regime. The failure of the UN process to accept this argument led the G8 to retreat increasingly from global climate leadership, in support of the UN regime.

A systematic review of the G8 summit's public deliberation, decision making, and delivery on climate change confirms this notable retreat (see Appendix A-1). From 2002 to 2004, the issue's share of the communiqué dropped to the lowest level since 1986. The G8 produced only 30 climate decisions from 1998 to 2004, including just one in 2002, compared to 39 in the previous seven years from 1991 to 1997. While these were delivered at a strong annual average compliance score of +0.55, despite only −0.22 in 1999, this was largely because of the high compliance with the few and unambitious commitments from 2002 to 2004. It compared with the +0.50 compliance level from the previous seven years, where compliance every year had been at least +0.29.

The key cause of this G8 retreat was the UN's assumption of global climate change leadership after the 1997 Conference of the Parties (COP) at Kyoto, with a newly legalized doubly divided regime that had not yet failed. G8 retreat was further fuelled by the diversionary impact of the Asian-turned-global financial crisis from 1997 to 2002, and the al Qaeda terrorist shocks to the US on September 11, 2001 (Kirton 2013a). Both consumed the attention of G8 leaders at their summits, were not consciously linked to climate change in any way, and thus crowded out less compelling, demands—especially those that, in the view of most G8 members, the UN had under its control. The globally predominant overall and specialized climate power of the G8 members eroded, given the fast growth of the emerging powers led by China and India, whose leaders appeared at the G8 only once, in 2003. The arrival of Vladimir Putin as Russian president in 2000 and George Bush as US president in 2001 began an era of high domestic political cohesion, but one with leaders who had less personal commitment to the climate control cause and incrementally less national like-mindedness overall.

Finally, the G8's compact, club-like participation continued to be compromised by the need to elevate the status of Russia and socialize it into the norms of the G7 club, and by a 2003 effort at outreach that gave the rising climate powers no privileged, focused place. At the finance ministers' level, G7 governance was supplemented after 1999 by the broader, increasingly institutionalized Group of 20 (G20), where the established G8 powers and the emerging ones had an equal place. But the G20 focused on the 1997 financial crisis and the 2001 terrorist one, and did very little to supplement global climate governance, either by spurring the G8 or UN or by acting on its own.

Birmingham, 1998

The Birmingham Summit in 1998 was the first summit with the Russians as full members of an ongoing Group of Eight and the first where the leaders met alone, without their finance and foreign ministers (Hodges 1999). This resulted in a more limited, focused agenda, as many traditional issues were delegated to other G7/8 ministerials or working groups.

Preparations

The summit was thus preceded by more intense preparations within the sherpa network and other ministerial forums. In addition to the regular foreign and finance ministers' meetings, interior ministers met to discuss crime in Washington in December 1997, employment and finance ministers met in London in February 1998, energy ministers met in Moscow in March, and environment ministers met at Leeds Castle, England, in April (Bayne 1998).

At the first preparatory sherpa meeting, held in mid January 1998, environment issues were not discussed. There was a general feeling that Kyoto had bought the G8 members some time, and that further action could await a follow-on UN conference in Buenos Aires later in 1998. The real decisions would need to be made there.

As the G8 preparatory process proceeded, the US saw the global environment as an important component of the Birmingham Summit. Climate change should arise as part of the discussions on the global economy, environment, and sustainable development, taking into account how North-South cooperation encouraged development on a sustainable basis.

For Canada, environmental issues ranked sixth on Foreign Minister Lloyd Axworthy's list of foreign policy priorities set in the spring of 1998 (Smith 2008a). He felt, in a preliminary way, that Canada must be more active on global warming, with a focus on water, forests, and fish issues that had to be addressed internationally. There was also interest in the Arctic Council, which was to be the topic of a national foreign policy forum that year. Of potential relevance were issues of Pacific salmon (with a view to preventative diplomacy) and the environmental clean-up of abandoned US military bases in Canada.

Canada knew that the US wanted to discuss the environment at Birmingham. Although Congress was considering the Kyoto Protocol and there was speculation that the Kyoto Protocol would be signed, there was little likelihood that it would be ratified. The G8 would therefore consider components of congressional dissatisfaction, the key element being the absence of any participation by developing countries. Congress demanded that America's major trade competitors act on an equal footing to control their carbon emissions. Given the increasing prominence of major developing countries as major polluters, G8 members would discuss a strategy for securing their participation in the Kyoto commitments.

When G8 environment ministers convened at Leeds Castle on April 3–5, 1998, they discussed the next steps in addressing climate change, including the need for G8 members to take the lead in signing the Kyoto Protocol. Those next steps also included setting an agenda for the next COP in Buenos Aires (which included detailed plans for trading emissions credits), developing a strong and effective compliance regime for monitoring international efforts to reduce emissions, and pressing for greater consideration of climate change issues by international institutions including the World Bank and the International Monetary Fund (IMF).

Building on the outcome of Leeds Castle, as the Birmingham agenda evolved, the summit would have three broad climate-related themes. The first, "Sustainable Growth in a Global Economy," considered what the G8 should do for financial stability, trade development, the environment, and climate change in light of the financial crisis that had begun in 1997. Second, and related to the first, it would consider how the G8 could mitigate climate change in ways that minimized its effect on jobs and growth. Although the G8 leaders were expected to affirm their existing climate commitments, they would also look at flexibility mechanisms, including market-based emissions trading. Third, the G8 would look at how to engage the developing countries, notably China, Brazil, and Mexico, and secure voluntary commitments from them. The impact of the financial crisis, which had spread from Asia, and the inadequacies of the doubly divided Rio-Kyoto regime were evident here.

At the Summit

At Birmingham, the G8 leaders tried to focus on the three pre-set themes of money, jobs, and crime. Yet while the detailed preparatory process had minimized the volume of extra items on the summit's agenda, two new political crises emerged in the week prior to Birmingham: new turmoil in Indonesia provoked by the Asian-turned-global financial crisis and the Indian decision to resume nuclear testing, arousing the fear that Pakistan would soon reciprocate and spark an Asian nuclear arms race of a particularly dangerous kind. Both topics were added to the leaders' discussions. Moreover, as public angst mounted over the instability in financial markets, the loss of jobs, and the world's poor falling even further behind, Birmingham became the scene of massive demonstrations, with upwards of 50,000 protesters demanding debt forgiveness for the poorest countries by 2000. Little time was left for leaders to add their own imprint on climate change, beyond endorsing what the preparatory process and the momentum from Denver had produced.

To be sure, expecting a difficult exchange on climate change as at Denver the year before, the British government under Prime Minister Tony Blair allowed plenty of time for discussion on climate-related issues at Birmingham. The overall effect, however, was that none of the leaders was particularly disposed to such potentially lengthy and difficult deliberations. They opted instead to reach agreement for all to sign the Kyoto Protocol within a year, address the necessary domestic measures to meet their Kyoto commitments, and, further to US Congressional desires, engage the developing world in doing its part to reduce emissions. The prevailing consensus was that the tougher climate issues would best be left to the COP process in Buenos Aires later that year.

Results

Birmingham proved to be a successful summit on several crisis-driven subjects, and on debt relief for the poorest as well (Bayne 1998). On climate change it performed solidly, even if did little that was new. Few of the new linkages and directions affirmed at Leeds Castle were highlighted in the leaders' private conversation or in their communiqué. The G8's retreat from global climate governance had begun.

Dimensions of Performance

This retreat, however, is only partly visible in a systematic review of the dimensions of G8 climate performance (see Appendix A-1).

Deliberation

The communiqué released by the leaders at Birmingham on May 17 contained four paragraphs on climate change. The subject took up 5.3 percent of the communiqué—the highest total in eight years, since Houston in 1990 (see Appendix C-1).

Direction Setting

Yet these public deliberations produced few new principles and norms. Support for the UN's Rio-Kyoto regime was overwhelmingly approved. In paragraph 11 of the communiqué, the leaders endorsed, in general terms, the results of their environment ministers' meeting at Leeds Castle (G8 1998).

Decision Making

This reaffirmation of Rio was seen in the 10 climate commitments produced by the leaders, some with targets and timetables attached. This was more than the nine at Denver the year before. Leaders pledged to sign the Kyoto Protocol within the next year and urgently begin the ratification process, take the steps necessary at home to reduce greenhouse gas emissions significantly, continue working on flexibility mechanisms and sinks, develop rules for a trading and compliance regime, prepare for the Buenos Aires COP, and reach an agreement on Kyoto's Clean Development Mechanism (CDM), which supports emissions reduction projects that facilitate emission trading schemes. They also committed to seek ways to increase global participation, including from developing countries, to work with developing countries to achieve voluntary efforts and commitments, and to enhance efforts with them to promote technology development and diffusion. On forests, the leaders were more reserved but nonetheless pledged to assess the implementation of the G8 Action Plan (published the week prior) in 2000 and to support the UN's ongoing work on forests. Only in one commitment did they suggest the need for a new approach, in which all major carbon-producing powers must control their emissions.

Delivery

The delivery of these commitments was complete, as all members complied fully with all three commitments assessed over the following year (see Appendix E-1). This included their commitment to sign the Kyoto Protocol within the next year. The only two outstanding G8 members to the convention, the US and Russia, signed Kyoto in November 1998 and March 1999 respectively. The ratification needed to give the protocol legal force, however, was the next step—which the US never took.

Development of Global Governance

In its development of global governance, the G8's retreat to and from support for the Rio-Kyoto regime was strong. Inside the G8, members reinforced the development of global climate governance by hosting a meeting of environment ministers in April (see Appendix F-1). There were no references to outside institutions related to climate change (see Appendix F-5).

Causes of Performance

Shock-Activated Vulnerability

This retreat was caused by diversionary shocks from Asia, notably from Thailand, Indonesia, and Korea in finance and India with regard to nuclear weapons. With little time to consider new climate control directions, G8 leaders strongly supported the Rio-Kyoto regime in the immediate afterglow of what still seemed like the UN's great COP success in concluding the Kyoto Protocol six months before. With China and India's vibrant growth still relatively unaffected by the Asian financial crisis, and with their leaders absent from Birmingham, the G8 was less able to lead in a new, inclusive way. However, G8 leaders also made reference to climate vulnerability in their communiqué (see Appendix G-1). They declared that "the greatest environmental threat to our future prosperity remains climate change" (G8 1998).

Multilateral Organizational Failure

The G8 summit in Birmingham came at a time when the UN had given climate change substantial attention. The report published by the Intergovernmental Panel on Climate Change (IPCC) was two years old and there had not been a new or updated report since then (see Appendix H-1). The last COP was Kyoto, which most considered a success and hoped that the US would come on board (see Appendix H-2). The year before Birmingham, the UN also held what seemed like a successful fifth-year assessment of the Earth Summit (Rio+5) (see Appendix H-3). The UN had apparently succeeded on the climate change front and thus the G8 was not needed this year.

Predominant Equalizing Capability

Predominant equalizing capability changed, as Canada, Japan, and Russia experienced a decrease in their gross domestic product (GDP), with the other members experiencing a slight increase (see Appendix I).

Common Principles

There was also very little change in members' common democratic principles (see Appendix J). New member Russia, while democratizing, did not advance any further along the Policy IV regime authority index.

Political Cohesion

A politically secure Tony Bair as host, supported by US president Bill Clinton and Japanese prime minister Ryutaro Hashimoto, pushed the easier UN approach. They were supported by civil society, led by Jubilee 2000 at Birmingham, which pushed for and got G8 action on development as traditionally defined, in the form of debt relief for the poorest, rather than the newer approach of sustainable development or climate change control in its own right. Moreover, constricted participation was reduced at this first summit with Russia as a regular participant. The G8 had to respond to Russia's priorities, which did not include climate change.

After the Summit

Diversionary shocks, the Kyoto afterglow, and the civil society pressure on responsive leaders led Birmingham to support the Rio-Kyoto regime, but to retreat from setting out in new directions of its own. This approach was reflected in, and confirmed by, the UN's COP meeting later that year.

This fourth COP took place in Buenos Aires from November 2 to 13, 1998. Attended by 180 countries, including all the G8 ones, the parties agreed to a two-year plan for advancing the agenda outlined in the Kyoto Protocol. Under the Buenos Aires Action Plan, the parties agreed to reach decisions on several key issues by the end of 2000, including rules and guidelines for market-based mechanisms including international emissions trading, joint implementation, and the CDM; compliance rules and procedures, consequences for noncompliance; development and transfer of environmentally friendly technologies; and consideration of the negative impacts of climate change. The action plan would further work on transferring climate-friendly technologies to developing countries and address the special needs and concerns of countries affected both by global warming and by the economic implications of response measures (UN 1999).

The parties also agreed to begin a pilot phase for emissions reductions and carbon sinks, allowing the least developed countries additional opportunities to develop projects and gain experience that would enable them to participate in the CDM. They agreed to define, measure, and verify categories of carbon sinks. They instructed the IPCC to undertake a comprehensive study on land use and forestry activities to be completed by spring 2000.

Finally, the Buenos Aires Action Plan set a firm deadline for adopting rules on emission trading and other market-based tools agreed to at Kyoto. These would enable countries to meet environmental targets faster and at a lower cost (see Earth Negotiations Bulletin [ENB] 1998). With all G8 members adopting this plan of action, Buenos Aires provided a promising platform for advancing key climate-related issues, thereby delegating authority away from the G8 to the broader UN process. By setting the end of 2000 as the deadline for its key deliverables, it also encouraged the G8 to retreat from global climate governance of any kind when it would meet the next year, in Cologne, Germany, in the spring of 1999.

Cologne, 1999

The German-hosted G8 Cologne Summit in 1999 featured an unusually comprehensive and demanding agenda. It was the first time the preparations for the summit occurred while most of its members were engaged in a war in Europe (in Kosovo), and the first summit at which the G8 had to conclude the peace and plan a postwar reconstruction effort there. It also took place as a still fragile global economy was emerging from a two-year steady succession of financial crises that had taken the Asian financial crisis global and threatened at one point even to engulf the vibrant US economy (Kirton 2013a). Consequently, leaders had to make critical decisions on the shape of an international financial system appropriate to the needs of an increasingly globalized world and its anxious citizens. Russia participated in all the discussions except those on finance and economics (see Kirton 1999a).

Cologne met both these central challenges, and did so in ways that strengthened its mission of forwarding the principles of democratic governance, market-oriented economies, and now social protection in a rapidly changing world. However, on environmental issues and climate mitigation more specifically, little was accomplished, with the G8 again leaving these issues to the UN.

Preparations

The lead-up to the Cologne Summit featured meetings among G8 foreign, finance, trade, and labour ministers as well as G8 counter-terrorism experts. Only passing references to global environmental issues were made during the G8 foreign ministers' meeting on June 9–10, 1999. They noted their concern for people continuing to "live in poverty while serious environmental degradation persists" (G8 Foreign Ministers 1999). At their meeting in Schwerin, Germany, on March 26–28, the G8 environment ministers released a more detailed and comprehensive communiqué, linking globalization and environmental protection, trade and the environment, the mitigation of climate change, and the environment and security. They stressed the need for G8 members to show political will to take domestic measures to cope with climate change and share "best practices" in climate mitigation (G8 Environment Ministers 1999). Conclusions and recommendations reached at this meeting would form the basis of the leaders' climate discussions in Cologne.

At the Summit

With the aftermath of the financial crisis still strong, the leaders focused heavily on financial matters at Cologne—how to manage risk in a globalized financial system, the appropriate order of financial liberalization, significant increases in global unemployment levels, and the implementation of more prudent regulatory systems. A key concern was how to make Russia's Boris Yeltsin, who had reluctantly and retroactively approved the G7-led military intervention in Kosovo, feel like a full and valued member of the club.

Results

Cologne produced three key initiatives. The first was the Köln Debt Initiative, prepared by the G7 finance ministers (1999) to assist the poorest countries achieve a more enduring solution in managing their unsustainable debt burdens. The second initiative, the Köln Charter: Aims and Ambitions for Lifelong Learning, aimed at responding to the challenges of globalization, through "adaptability, employability and the management of change" (G8 1999). Third, the Stability Pact for South Eastern Europe was written and presented by the foreign ministers of the European Union, the European Commission, all the southeastern European states, various G8 countries, and international institutions and banks.

In the end, little more than lip service was paid to global environmental issues, including those related to climate change, despite pollution increasingly being touted as one of the few political problems that inevitably demanded international cooperation. Whether through the international ban on certain emissions or taxation on the international trade of emission quotas, the Cologne Summit missed an important opportunity to produce the clear and organized political response needed to address issues related to global climate mitigation. Its failure to set a strong foundation or emissions reductions meant that climate mitigation had to be addressed by other international forums, primarily the UN process.

Dimensions of Performance

For the first time a German-hosted summit produced little advance or innovation on climate change (see Appendix A-1).

Deliberation
In its deliberation, climate change took only 1.3 percent of the communiqué at Cologne, a sharp drop from 5.3 percent the year before (see Appendix C-1).

Direction Setting
The central objective of the Köln Debt Initiative was to focus on reducing poverty by releasing resources for investment in health, education, and social needs through major changes to the framework for heavily indebted poor countries (HIPCs). Yet the environment and climate change were left out of this new link.

Decision Making
In its decision making, Cologne made only four commitments on climate change, a sharp drop from the 10 the year before (see Appendix D-1).

Delivery
The delivery of these commitments was low. The one assessed commitment on climate change secured compliance of only −0.22. This was the lowest annual score ever for G7/8 delivery of its commitments on climate change to date (see Appendix E-1).

Development of Global Governance
What little attention and action on climate change arose focused on forwarding the UN regime. The leaders committed to support sustainable development through efforts to build a "coherent global and environmentally responsive framework of multilateral agreements and institutions" (G8 1999). Their agreement on environmental considerations being fully considered in multilateral agreements at the World Trade Organization (WTO), Organisation for Economic Co-operation and Development (OECD), and multilateral development banks offered no concrete commitments. The leaders did agree, however, that the potentially deadly Chernobyl nuclear reactors would be finally closed—a full 14 years after the 1986 explosion—by 2000 and that host Gerhard Schroeder would be dispatched to Ukraine to get the job done.

On climate change specifically, there was support for the early entry into force of the Kyoto Protocol later that year, as well as for a strong and effective Kyoto mechanism and compliance regime. G8 leaders (1999) underlined the importance of reducing "greenhouse gas emissions through rational and efficient use of energy and through other cost-effective means." But they failed to specify how or when they would implement these domestic measures, deferring instead to the UNFCCC process. Finally, although the issue of developing country participation in limiting greenhouse gas emissions was noted, with the G8 stressing the need to support their efforts through financial mechanisms, technology transfer, and capacity building, the leaders again fell short on making any firm financial commitments or identifying specific actions that would help mitigate the developing world's carbon emissions.

Causes of Performance

This low but UN-supportive performance on climate change was driven above all by the diversion of G8 leaders' attention to several shocks in unlinked fields. The first was the global phase of the Asian financial crisis with the collapse of G8 member Russia in August 1998, the collapse of the hedge fund Long Term Capital Management in the US in September, and the collapse of Brazil in October. The second was the war in Kosovo. And the third was the remembered shock of the Chernobyl nuclear explosion in nearby Ukraine.

In the face of such consuming crises, climate change could safely be left to a UN regime that appeared to work and whose critical decisions would come only at the end of 2000, after another G8 summit had been held. G8 leadership in support of the UN could thus be left to the Okinawa Summit next year.

Political cohesion was also waning. America's Clinton had entered his last two, lame-duck years. Russia's Yeltsin's personal performance suggested he too might soon have to leave. And the G8's compact, club-like participation continued to be compromised by the need to make Russia behave as a responsible member.

After the Summit

At COP-5, held in Bonn from October 25 to November 5, 1999, the UN worked to facilitate agreements on technology transfers, flexibility mechanisms, and compliance rules and procedures. However, it was unable to reach a final consensus on several issues, including how developing countries would best mitigate their greenhouse gas reductions through common but differentiated responsibilities. There arose a stalemate in the consultation process and a suspension of deliberations until the next COP (ENB 1999). With this came a new opportunity for G8 leadership on climate change.

Okinawa, 2000

The G8's Okinawa Summit, on July 21–23, 2000, faced several formidable challenges. The first was maintaining Japan's historic record, unique among G7 members, of hosting relatively successful summits (Bayne 2000b, 195; Putnam and Bayne 1987, 270). The second was living up to the standard set by the previous year's high-performing German-hosted summit at Cologne, at least on issues other than climate change (Kirton et al. 2001, 285–94). The third was ensuring a successful global economic recovery, beginning in Japan itself, after the 1997–99 global financial crisis (Kaiser et al. 2000). Other challenges included modernizing the international financial system to cope with the rapidly spreading digital "new economy," fostering a more humane, cohesive, and credible form of globalization, and designing a more coherent approach to global governance (Kirton 2001b, 143–67). Also arising was the need to broaden the process of G8 deliberation and decision making to involve the world's emerging market economies, its developing countries, and many civil society actors. And because the 2000 summit took place in Asia, where the Cold War had not yet ended, arms control, regional security, conflict prevention, and human security were important concerns.

Perhaps most importantly, the Okinawa Summit marked the start of the new millennium, the second 25 years of G7 operation, the 10th anniversary of the end of the European-centred Cold War, and the fifth anniversary of the G7's concern with globalization. The time was thus ripe for leaders to reflect on how G8-guided governance in the twenty-first century could avoid the manifold disasters brought by the older approaches to global governance that had dominated the twentieth century, end the continuing Cold War in Asia, and make globalization work for the poor as well as the prosperous everywhere in the world.

This comprehensive, compelling array of challenges left little space for Okinawa's leaders to govern climate change. Their earlier retreat continued.

Preparations

Preparations for Okinawa generally followed the same process as in 1998 for Birmingham and 1999 for Cologne, with a series of ministerial meetings held in the lead-up to the summit, including those by foreign and finance ministers (Bayne 2000a).

There were, however, three important modifications to the summit process and format. First, efforts were made to include other non-G7 countries both in preparations and with the G8 heads themselves, as leaders from three African countries and Thailand met a group of five summit leaders (excluding America's Bill Clinton, Germany's Gerhard Schroeder, and Russia's Vladimir Putin) en route to Okinawa. Japanese prime minister Keizo Obuchi's larger initiative to include China, however, did not succeed. Second, the summit was to offer, for the first time, official facilities for nongovernmental organizations (NGOs), marking the beginning of a long-term relationship between civil society representatives and the G8 at the annual summit gatherings. Third, there was an unprecedented set of other ministerial meetings, including the first thematically focused stand-alone meeting of G8 foreign ministers (held in Berlin in December 1999) and the first meeting of G7 education ministers (Kirton 2000; Hodges et al. 1999; Hajnal 1999).

The G8 environment ministers met in Otsu, Japan, from April 7 to 9, 2000. Most of the commitments they made were to ensure that the results of the next COP would help promote the early entry into force of the Kyoto Protocol. By noting their resolve to take the "political leadership necessary for the success of COP6," they signalled that little would be done at the summit itself (G8 Environment Ministers 2000).

G7 finance ministers met ten days prior to the summit, issuing reports on the international financial architecture, information and communications technologies (ICT), and money laundering. The week before the summit, G8 foreign ministers met again to produce their main declaration covering the environment, conflict prevention, small arms trade, terrorism, crime, the illicit trade in diamonds, and regional issues.

As the summit neared, tensions mounted in determining whether ICT or development would serve as the overriding theme. Competing as subjects for tangible results were a new generation of digital technology, debt relief for the poorest, multilateral trade liberalization after the disruption of the WTO's ministerial meeting in Seattle in 1999, and the Korean Peninsula.

Obuchi had issued very clear instructions to his summit team in the autumn of 1999 to put ICT forward as a leading topic. He felt very strongly that it was a future-oriented theme that ICT could bring happiness to ordinary people everywhere, and would be a driving force in Asia's economic recovery (Obuchi 2000). It was consistent with his initial idea to focus on human security, including peace of mind and caring for the poor, in an age of globalization. A possible double-edged sword, ICT could both foster economic growth while generating an unacceptable "digital divide."

On April 2, three months before the summit, Obuchi suffered a stroke that left him in a coma until his death on May 14. He was replaced by Yoshiri Mori (2000), secretary general of the Liberal Democratic Party, who repeatedly said he intended to follow Obuchi's lead. In a pre-summit tour of G8 countries, Mori advanced the ICT theme even more strongly, and received the approval of his colleagues.

On the eve of the summit, the sherpas faced protracted, deep-seated disagreements over food safety, trade liberalization, the rules governing electronic commerce, and the looming "Rio+10" review. On the subject of the 10th anniversary of the Earth Summit in Rio de Janeiro, as at Denver in 1997, the basic split centred on a European versus North American disagreement over the need to keep the Kyoto Protocol's targets for greenhouse gas emissions reductions. Canada and the US, as geographically large, hydrocarbon-rich countries with widely dispersed populations and many regions without year-round sunshine, preferred to relax the pace of implementation or opt out of their obligations. This was especially true as they knew their efforts could be overwhelmed by the unlimited emission increases of unbound countries in the now robustly growing developing world. By contrast, the Europeans, benefiting from their compact geography, the closure of dirty East German industry, and Britain's move to replace coal with natural gas, found it relatively easier to meet their targets, particularly given the all-EU regional bubble that the G8 members in the North American Free Trade Agreement lacked.

Some of the changes required by Kyoto required G8 governments to implement domestic legislation. But some G8 leaders claimed they were having trouble persuading their legislatures to pass the required laws. Those who had secured such ratification regarded this as a routine challenge. They remained unimpressed with the will, skill, and leadership of their colleagues in countries that had not. They further noted that similar challenges of overcoming legislative resistance arose on other G8 issues, such as securing funds to support reform in Russia.

On the climate-related issues of forests and oceans, there was some pressure within Japan for a strong statement on protecting forests, although this view was not widely shared. Japan wanted to take a direct approach by focusing on concrete, if limited, action. It argued that there were many treaties on the environment that had been signed but not ratified, suggesting that a similar push to secure the UN forest convention agreed to at the 1990 Houston Summit might be the wrong approach. Japan thus wanted to see more specific action, for example in curbing illegal international logging.

As the summit neared, there was not much discussion on the environment or climate change. For Canada, Okinawa presented an important opportunity to reach consensus on ratification of the Kyoto Protocol by 2002. But without a national strategy for climate change in the US, moving the ratification process forward would be extremely difficult. The most that could be expected from Okinawa were supportive words about Kyoto ratification, without tying leaders to any specific dates.

At the Summit

The Okinawa Summit itself unfolded according to plan, with few spontaneous initiatives from the leaders. The outstanding issues over ICT and global health were easily resolved. The leaders focused mostly on dealing with Russia's new president, Vladimir Putin, and celebrating Bill Clinton's contribution as this would be his last summit.

Results

The preamble of the leaders' final communiqué was replete with philosophical overtones, as the leaders marked the dawn of the new millennium and "foundations for a brighter world in the 21st century" (G8 2000). They reaffirmed the importance of building "on our basic principles and values" in all their endeavours, with the hope of achieving "greater prosperity, deeper peace of mind and greater stability" as they entered the new century.

With their attention so heavily focused on increasing prosperity, little room was left for concrete deliberations or decisions on climate change (G8 2000). In a seven-paragraph section on the environment, the leaders noted their intent on assisting "in mitigating the problems of climate change and air pollution." Here, leaders referred to the increased use of renewable energy sources as a way of improving the quality of life. They invited stakeholders to join a task force to prepare recommendations on the use of renewables in the developing world for consideration at their next summit.

Further action on climate change was deferred to the UN process, as the G8 committed to "close co-operation" in resolving "all major outstanding issues" on the Kyoto Protocol at the next COP-6 (G8 2000). Focus shifted to 2002, when summit leaders would endeavour "to prepare a future-oriented agenda for Rio+10" (G8 2000).

Okinawa thus did little on climate change. It offered a noncommittal acknowledgement that in order to achieve the Kyoto goals, strong domestic actions and supplemental flexibility mechanisms were needed (Bayne 2000a). No specifics on how this would be done were offered.

The G8 did, however, take some modest strides in creating a task force on the use of renewable energy in developing countries. Although the leaders recognized the need to "encourage and facilitate" investment in renewables as a means of "elevating the level of energy supply and distribution" in the developing world, the leaders made no financial pledges to support their efforts to invest in sustainable energy practices (G8 2000).

Okinawa also specifically mentioned forests and oceans. Japan was successful in achieving its goal of action on illegal logging, reinforced by Greenpeace, which issued daily communications during the summit.

Yet many of these commitments were "high on rhetoric and low on precise measures and identifiable resources" (Bayne 2000a). On climate change, Okinawa produced few ambitious results, deferring instead to the UNFCCC. Although noting in their communiqué the importance of resolving all major issues related to the Kyoto Protocol "as soon as possible," G8 leaders offered no solutions, handing the responsibility over to the COP-6 process scheduled for later that year at the Hague, Netherlands.

Dimensions of Performance

Domestic Political Management
With the Okinawa Summit costing a record $750 million, the world's media focused on the lavish expenditures, which seemed disproportionate to the results achieved. They thus offered sceptical or grudging judgements of the substantive accomplishments at summit's end.

Deliberation
In its public deliberation, Okinawa gave climate change only 1.6 percent of its final communiqué. This was only a slight rise from the low 1.3 percent the year before (see Appendix A-1).

Direction Setting
The summit resolved a tension in the preparatory process over whether the main message of Okinawa would be a celebration of globalization guided by the rich alone, with the new ICT revolution at its core, or of a broader set of values, including concern for those on the other side of the digital, knowledge, and health divides. In the end, Japan's Okinawa Summit offered the latter: a much broader, more inclusive vision, thereby forming the foundation for the focus on poverty reduction and development in Africa at subsequent summits. Yet neither here nor in its overall themes—increasing prosperity, peace of mind, and global stability "in a world of ever-intensifying globalisation"—was there any link to climate change (G8 2000).

Decision Making
Okinawa produced four commitments on climate change, the same as the year before (see Appendix D-1).

Delivery
In its delivery, the one assessed climate commitment had a compliance score of +0.44, a sharp rise from the −0.22 the year before (see Appendix E-1).

Development of Global Governance
In its development of global governance, the G8 leaders again focused heavily on supporting the work of the UN. However, their communiqué added: "We must engage in a new partnership with non-G8 countries, particularly developing countries, international organisations and civil society" (G8 2000). Inside the G8, the leaders launched a multistakeholder task force to prepare concrete recommendations for consideration at their next summit regarding "sound ways to better encourage the use of renewables in developing countries."

Causes of Performance

Shock-activated vulnerability was low, just as the Okinawa's climate performance was. The recent Asian-turned-global financial crisis, with its aftershocks in Turkey and Argentina, diverted attention to financial and economic concerns. The arrival of the new millennium aroused the anniversary memories of many wars from the past (Welch 2005). There were only four major oil-related accidents in 2000, down from six the year before (see Appendix G-4).

Multilateral organizational performance was still assumed to be substantial, amid hopes for a successful COP at the end of the year. The G8's global predominance and internal equality was eroding, driven by the vibrant growth in China and India and the continuing stagnation in Japan. Common democratic principles were cast in some doubt as Putin replaced Yeltsin in Russia. Political cohesion was lowered by the sudden replacement of Obuchi by Mori, the lame-duck status of Clinton and the change of leadership in Russia. Civil society pressure, newly reinforced by its institutionalized presence on the summit site, diverted attention from the environment to development, with the exception of the advance on illegal logging. And compact participation was lowered by the presence on the margins of leaders and heads of multilateral organizations concerned with issues other than environmental ones.

After the Summit

The COP-6 meeting in the Hague on November 13–25, 2000, was a resounding failure. Disagreements on carbon sinks, flexibility mechanisms, and a compliance regime pitted the EU, Japan, and Russia against the United States and Canada. Their inability to compromise resulted in the suspension of the negotiations without an agreement. The meeting produced a declaration, on behalf of Jan Pronk, the Dutch minister of the environment and chair, to continue the session in Bonn in July 2001. This failure was a disappointment to the 7,000 delegates from 182 countries whose ambition was to finalize any outstanding details by the time of the Rio+10 Summit in 2002, when the Kyoto Protocol was slated to enter into force (Davenport 2008; ENB 2000).

The lack of G8 climate leadership at Okinawa was an important cause of the failure of COP-6. However, the UN's failure at the Hague did little to inspire the G8 to reverse its retreat on climate change at Genoa in 2001.

Genoa, 2001

The Genoa Summit in 2001 took place against the backdrop of an antiglobalization movement rapidly gaining momentum, the anticipation of violent protest activity and terrorist attacks, a massive police and military presence, unsightly and daunting physical barriers around the main media and delegation centres, and deserted city streets and a boarded-up downtown core. Genoa resembled a "city under siege, even before the riots began" (Bayne 2001). The heavily anticipated violent demonstrations and riots did break out in the streets of Genoa, killing one protester, injuring more than 200 others, resulting in more than $50 million of damage, and prompting the G8 leaders to issue a statement in response (see G8 2001b). Nonetheless, the leaders launched significant initiatives on Africa, infectious diseases, education, ICT, and trade, making Genoa one of the more influential summits to date.

Genoa also featured the recently elected George W. Bush, representing a shift to the right in the United States. In an era of increasing globalization, the challenge of Genoa was to socialize the new, unilateralist, Republican president into the world beyond North America and into the plurilateral G8 fold. With a narrow electoral mandate, a divided Congress, and a fundamental philosophical shift from his predecessor Bill Clinton and political rival Al Gore, Bush came to Genoa strongly resisting his G8 counterparts on the issue that would remain the most divisive of the summit—climate change, including the ratification of the Kyoto Protocol. This division led the G8 to retreat further. But it also introduced, from Bush himself, the new principle that all consequential carbon-producing powers must control their carbon, rather than just the few OECD ones as under the Rio-Kyoto regime. This principle would form the basis of the new inclusive, unified regime that the G8 would pioneer when it returned to global climate leadership in 2005.

Preparations

The Italian hosts prepared their summit with lead-up ministerial meetings for finance, foreign, energy, education, environment, and justice. From the beginning they brought in many outsiders from civil society, trade union organizations, non-G8 countries, and international organizations. They embraced the United Nations, in order to counter any feeling that the G8 opposed its work. Indeed, they invited Secretary General Kofi Annan to Genoa and made a particular effort to ensure that he felt comfortable there (Kirton 2001a).

At an early stage the Italian hosts signalled their three key themes—poverty reduction, the environment, and conflict prevention. It appeared that climate change might return to the top tier. But they also sought to launch another round of multilateral trade negotiations after the failed WTO ministerial at Seattle in 1999 and to address the global economy, given the expected sharp slowdown in the US and its impact on Europe and emerging economies (*Reuters* 2001). Finally, on the eve of the summit, the Italians settled on four major issues: managing the economic situation; restarting trade negotiations; sharing growth among all countries within and beyond the G8, and responding to emergencies, particularly in the realm of poverty (Kirton 2001a).

On March 2–4, the G8 environment ministers met in Trieste to define an agenda for Genoa. They committed themselves to reaching an accord on cutting greenhouse gas emissions during the UN's "resumed COP 6" (G8 Environment Ministers 2001).

But by the end of March, the Bush administration firmly opposed the Kyoto Protocol. The president was "unequivocal" in rejecting Kyoto as it "exempts the developing nations around the world, and it is not in the United States' economic best interest" (White House 2001). Bush refused to impose mandatory carbon emissions reductions on US power plants and would not consider listing carbon dioxide as a pollutant under the clean air act (Kluger 2001; Coon 2001). On March 28, Christie Whitman, head of the US Environmental Protection Agency, said, "We have no interest in implementing that treaty" (*Associated Press* 2001). Environmental issues and climate change thus fell off the priority list.

During the two weeks leading up to the summit, the Bush administration released a series of multimillion dollar studies and initiatives, promising a $120 million, three-year investment in research on the natural carbon cycle, climate modelling, and the link between atmospheric chemistry and climate. Research funds were also allocated to study how to capture carbon dioxide from fossil fuel combustion plants as well as the effects of forestry practices in Belize and Brazil. These initiatives were widely regarded in the international community as inadequate and a way to stall action on Kyoto by pushing for more scientific research.

The response of G8 partners was mixed. The EU showed strong solidarity on ratifying Kyoto during its June 2001 summit in Göteborg, Sweden. Margot Wallström, the EU's environment commissioner, noted that, "EU ministers have ... confirmed that they stand firm behind the Kyoto Protocol and are ready to proceed with the ratification of it" (EU 2001). EU president Romano Prodi (2001) said, "There is no bigger common challenge for our planet than *climate change*. We remain deeply concerned about the US position on the Kyoto protocol. We simply *cannot* go back on the 10 years of international efforts."

As they arrived at Genoa, France's Jacques Chirac, Germany's Gerhard Schroeder, and the UK's Tony Blair stood strong in their support of ratifying Kyoto. With France boasting about its emissions at 4 percent below 1990 levels, the UK announced it was set to cut its emissions by 12.5 percent below 1990 levels by 2010 (an impressive 7.3 percent above the 5.2 percent target reached at Kyoto). Despite their strong positions, these leaders did not denounce Bush's unilateral repudiation of the Kyoto Protocol, seeking instead a compromise that would encourage American participation.

Yet despite the EU's defence, Italy—now led by Silvio Berlusconi, who had just won the May election—shifted its position in the immediate lead-up to Genoa, defending the US's support for more conclusive scientific research. Similarly, while initially showing strong support for ratifying Kyoto, Japan's Junichiro Koizumi now indicated that Japan might not ratify Kyoto should the United States not be on board. Koizumi did note, however, that he would discuss Kyoto during his bilateral meetings with Bush, Chirac, and Blair at the summit, offering to mediate by urging the US to reconsider Kyoto while seeking more flexibility from the EU.

Canada's Jean Chrétien as well as his ministers of the environment and natural resources initially affirmed their commitment to Kyoto and their determination to meet Canada's targets. Just weeks prior to Genoa, Canada (2000) allocated $1.1 billion over five years toward climate and clean air initiatives as part of the Government of Canada Action Plan 2000 on Climate Change. These initiatives included investments in effective transportation through fuel-cell technology development, the marketing of low-emissions vehicles, and emissions reductions in the government's own facilities by 31 percent below 1990 levels by 2010.

Canada then modified its position on Kyoto significantly. It stated that Canada would not ratify the final agreement should the terms be unacceptable. Environment Minister David Anderson noted, "Ratification is impossible to consider until we know what the rules are. We'll face that decision when it comes" (MacKinnon 2001). One of the key sticking points was carbon sinks. Canada said it would seek credit when Canadian forests and farmlands were removing atmospheric carbon dioxide.

At the Summit

Attending the G8 summit for their first time, America's Bush and Japan's Koizumi were lively participants. Britain's Blair and Canada's Chrétien had both been recently re-elected. Blair was particularly active on issues relating to Africa. Chrétien attracted interest as Canada's summit host the following year. France's Chirac was "outspoken and individualistic" (Bayne 2001). Germany's Schroeder made little obvious impact. The EU's Prodi—a former prime minister of Italy—seemed heavily preoccupied by the riots in Genoa.

Climate change was one of the most divisive issues discussed (Blair 2010, 550–51) . Belgium's Guy Verhofstadt, serving as head of the European Council, condemned Bush's climate policy to his face at great length. But in the face of this onslaught, Bush remained unmoved.

Results

Genoa failed to solve the deadlock on the Kyoto Protocol. The communiqué frankly acknowledged the current "disagreement on the Kyoto Protocol and its ratification" (G8 2001a). The leaders did commit to "working intensively together to meet [their] common objective." Moreover, all agreed on the need to stabilize atmospheric greenhouse gas concentrations through "effective and sustainable action" consistent with the "ultimate objective of the UN Framework Convention on Climate Change." Specific references to policy flexibility and "co-operation on climate-related science and research" reflected the

American position. The US indicated that it was working on alternative approaches to Kyoto, but that they would not be ready until the COP in Morocco the following year. The UN process was thus used to justify a US position and a G8 delay. The communiqué did point to the importance of including others in the Kyoto process, demonstrating support for the US position on the inclusion of developing countries in the climate debate, and adopting the core principle of the future unified regime.

On the innovative, multistakeholder Renewable Energy Task Force launched at Okinawa at the initiative of the Clinton-Gore administration, the communiqué did nothing more than "thank all those who participated in the work," without taking a position on it. Further action was assigned to G8 energy ministers. This reflected US fear that the task force's report—containing a wealth of conclusions and recommendations based on technical expertise from the International Energy Agency (IEA)—would inhibit its national energy strategy.

Genoa forged a common recognition by the leaders that the issue of climate change was bigger than the Kyoto Protocol, and that the most effective way forward was to build on the UNFCCC (G8 2001a). The resumption of the COP-6 negotiations took place in Bonn in tandem with the Genoa Summit. By deferring any unresolved divergences on emissions targets and timetables to the COP process, the G8 once again effectively conceded any further resolution on the climate debate to the UN.

Dimensions of Performance

A systematic review of the key dimensions of G8 performance confirms Genoa's further retreat on climate change, now reinforced by the arrival of George W. Bush (see Appendix A-1).

Domestic Political Management
In its domestic political management, Genoa was a clear failure. Negative media reports blamed the leaders for the escalation in violence and for achieving very little of substance (Jeffery 2001).

Deliberation
Genoa's public deliberation on climate change was solid, at 5.2 percent of the communiqué, up from only 1.6 percent at Okinawa the year before (see Appendix A-1). This returned the G8 to the level at Birmingham in 1998.

Direction Setting
In its direction setting, Genoa produced one striking advance: its affirmation of the importance of including others in the Kyoto process. Noting the disagreement over ratifying the Kyoto Protocol, the leaders affirmed their commitment to work intensively to meet their common objective. To that end, the G8 stated they would participate "constructively in the resumed Sixth Conference of the Parties in Bonn" through deepened discussions "among the G8 and with other countries" (G8 2001a).

Decision Making
In its decision making, Genoa produced four climate commitments, tied with Okinawa the year before (see Appendix D-1)

Delivery
The delivery of these commitments was low, at 0, a sharp drop from the +0.44 the year before (see Appendix E-1).

Development of Global Governance
The development of global governance yet again focused on support for the UN regime. The G8's retreat was confirmed by its termination of its own Renewable Energy Task Force after its brief one-year life.

Causes of Performance

Genoa's low performance was driven by the lack of any climate-related shocks, as there were only three major oil tanker accidents in 2001 (see Appendix G-4). The local shock of the violence and death at the summit had a diversionary effect. Multilateral organizational failure was now sufficiently and visible strong to have the G8 affirm the need to include others in the control regime, but not to induce the Europeans to abandon their support for the Rio-Kyoto regime, whose obligations they found relatively easy to meet.

Domestic political control was high, serving as a key catalyst in generating the division, retreat, and resulting compromise at Genoa. Given the leaders' exceptional domestic popularity and personal commitment to the G8, they were thus confident they could lead both domestically and abroad. Fresh electoral mandates, often with strong legislative majorities, were held by an anti-Kyoto Bush, whose popularity remained intact, by pro-Kyoto Blair, and by an adjusting and compromising Berlusconi and Chrétien. Koizumi enjoyed unprecedented popularity, even as his upper house elections on July 29 loomed. Putin was the first Russian leader to enter the presidency through the ballot box and maintained a firm grip at home that Yeltsin often seemed to lack.

Constricted participation was lowered by the presence at the dinner table of four African leaders—South Africa's Thabo Mbeki, Nigeria's Olusegun Obasanjo, Senegal's Abdoulaye Wade, and Algeria's Abdelaziz Bouteflika—who demanded development assistance rather than climate change control. UN secretary general Kofi Annan was also invited, and received support for the UN's Rio-Kyoto regime.

After the Summit

Genoa seemed, on the surface, to generate little success in resolving internal G8 divergences, thus falling short on achieving specific commitments that would significantly advance the Kyoto Protocol. However, through the ability, capacity, and political will of the G8 leaders to state their respective positions at the summit table, a compromise was found that appealed to all sides, ultimately securing the minimum number of votes for Kyoto to take effect. The consensus reached by the COP-6 participants at Bonn in July 2001 was therefore largely driven by the ability of the G8 to strike a satisfactory deal at Genoa. It thus served to prolong faith in the UN regime.

There was therefore some hope that the 2001 Bonn climate meeting could save Kyoto by inducing the Americans to accept the European plan proposed at the initial COP-6 at the Hague in November 2000. The G8 leaders had to lead the way. They did, as the impact of their high-level political discussions on climate change became apparent. The consensus of 178 countries at Bonn was driven by the ability of the world's top decision makers at Genoa to state their respective positions. In exchange for Canada and Japan receiving increased credit for carbon sinks, the EU was able to impose greater penalties for countries that did not meet their targets. In the end, the leaders of Japan, Canada, and Russia agreed to ratify the protocol, providing the minimum number of weighted votes required for it to take effect as international law (ENB 2001a).

For its part, the United States still resisted ratification but agreed in principle to further discussions through the UN framework, committing to present its "made-in-America" climate change action plan at COP-7 in Morocco at the end of 2001. The big push at Marrakech was to finalize the operational and technical details of the Bonn Agreement (ENB 2001b). The parties were intent on resolving all the outstanding issues, including eligibility requirements for financial mechanisms, a compliance regime, allowances for carbon sinks, and reporting and review. Agreeing on these issues was critical for the protocol to come into force. On the last day of COP-7, a global package deal was put together that could finally be accepted by all the parties so the protocol could be ratified by the time of the Rio+10 Summit in September 2002 in Johannesburg. Known as the Marrakech Accord, it laid out the decisions on rules and procedures for operationalizing the protocol (see Davenport 2008, 56). Russia's commitment to host a global conference on climate change in 2003 ensured that continued efforts would move the climate agenda forward, and, in so doing, seek convergence on the most divisive issues (G8 Research Group 2001).

Kananaskis, 2002

The 2002 G8 summit, held at Kananaskis in the secluded Canadian Rocky Mountains, allowed leaders to have their first meeting after the terrorist attacks of September 11, 2001, in an intimate and informal gathering, free of the violence that had afflicted Genoa the year before. With peaceful rallies in nearby Calgary attracting fewer than 2,000 people and only two demonstrators arrested, the leaders could focus on their three priorities of sustaining global growth, combating terrorism, and reducing poverty in Africa. To succeed on their centrepiece African agenda, they needed to create consensus not only among themselves, but also among the African leaders who, for the first time, joined them as full partners for the summit's second day. Kananaskis was one of the shortest summits in G7/G8 history, so the leaders had little time to reach the right consensus at the right time (Kirton 2002b).

Kananaskis also brought Russia more fully into the new G8 club, by deciding that Russia would host the summit in 2006. With its revised hosting sequence specified, the G8 thus signalled that it intended to remain in business as an international institution, with its current membership, for the next ten years (Kirton 2002c). In this context, there was little room to advance climate change control. Indeed, leaders did not even commit to attend the forthcoming World Summit on Sustainable Development (WSSD) in South Africa—Rio+10—where climate change would be front and centre on the policy agenda.

Preparations

As host, the Canadians introduced several distinctive features into the Kananaskis format. With the summit held in the remote Rockies, civil society and most of the media were kept about 100 kilometres away, in the Alberta capital of Calgary. This produced extremely high security costs for the leaders and their delegations—between $300 million and $500 million. The small size of the resort, however, meant limited accommodation, which served to encourage an "informal and spontaneous atmosphere" (Bayne 2002).

To prepare for Kananaskis, several G8 ministerial groups met, including labour and employment ministers, justice and interior ministers (with a special emphasis on terrorism), energy ministers, and environment ministers.

The energy ministers met on May 2–3, 2002 in Detroit. They concluded that energy security, economic growth, environmental protection, and sustainable development were all supported by improved energy efficiency and that diversification of energy sources, largely through renewables, could "help countries address climate change by reducing the greenhouse gas intensity of energy production and use" (G8 Energy Ministers 2002). Energy science and technology were also emphasized, particularly within the context of renewable energy and clean energy technologies. The ministers affirmed their collective support for the upcoming Rio+10 Summit, committing to "providing constructive and substantial input on energy to the World Summit on Sustainable Development and other similar fora."

The environment ministers met in Banff, not far from Kananaskis, on April 12–14, 2002, specifically to advance preparations for Rio+10 (Risbud 2006). They affirmed their collective support for the UN process by noting their determination to take "strong actions, in fulfillment of our commitments under the UN Framework Convention on Climate Change" (G8 Environment Ministers 2002). A section of their communiqué was dedicated to international environmental governance. They welcomed the leadership role taken by the United Nations Environment Programme (UNEP) in sustainable development governance. The statement on the WSSD showed the G8's ongoing and expanding support of the relevant UN bodies charged with managing global sustainable development.

At the Summit

At Kananaskis, the informal, isolated setting allowed for personal and spontaneous discussions among the leaders. This resulted in their agreement to allow Russia to host the G8 in 2006 and to the creation of the Global Partnership against the Spread of Weapons and Materials of Mass Destruction. But the second day was consumed by the more formal meeting on the G8 Africa Action Plan with South Africa's Mbeki, Nigeria's Obasanjo, Senegal's Wade, and Algeria's Bouteflika. There was no opportunity for the leaders to address climate change seriously, which was not a priority for any of the leaders or linked to any of the issues on the agenda.

As chair and summit host, Chrétien had the satisfaction of "seeing the summit work as he planned it, but he seemed ill at ease in his public appearances" (Bayne 2002). Blair and Chirac "were active and effective players," with Bush "an energetic participant." Putin seemed to hold his own, and the chemistry between him and Bush was quite amicable. Schroeder clearly played a key role in the decision for Russia to host in 2006, as Germany had to delay its turn.

With only a day and a half for deliberations (or about 30 hours from beginning to end), time for reaching consensus was tight. On day one, the leaders discussed sustaining global growth, advancing the newly launched Doha round of WTO negotiations, and ensuring universal primary education by 2015. In the wake of September 11, discussions that evening promptly turned to terrorism, with the leaders charting a path for keeping weapons of mass destruction (WMD) out of terrorists' hands and protecting the global transportation system.

The big push came on day two, with the G8 leaders outlining a new vision for African development. It was one that reinforced Africa's own priorities—good governance, peace and security, knowledge and health, trade and investment, agriculture, and water as a fundamental public good.

Results

Given ongoing antagonism over the Kyoto Protocol, the leaders' limited discussions on the environment focused almost exclusively on how to deliver sustainable development. Preparations for Rio+10 were in trouble, largely because of diverging opinions among the G8 leaders, and even the presence of Kofi Annan failed to produce a firm commitment from the leaders to participate in Johannesburg. Chirac spoke at great length about sustainable development and said publicly that he had urged all his G8 colleagues to attend. But, in the end, only he and Blair indicated they would.

On climate change specifically, nothing of consequence was done. There was only one fleeting reference to the issue in the chair's summary, recognizing simply that "climate change is a pressing issue that requires a global solution" (G8 2002). Details of what that global solution might be were not offered, making Kananaskis devoid of any important commitments on the climate debate. And despite a number of other, more substantive issue-specific G8 statements on Africa, WMD, transport security, and Russia's role in the G8, no stand-alone document was issued on climate change or, more broadly, the environment or sustainable development. This signalled once again, the deferral of this highly contentious issue from the G8 to the UN and the COP process.

Dimensions of Performance

The climate performance of the Kananaskis Summit was very low across all dimensions of governance (see Appendix A-1).

Domestic Political Management
In its domestic political management, host Jean Chrétien received favourable media treatment, but not at all due to summit action on climate change.

Deliberation
In its public deliberation, the G8 gave climate change only 0.2 percent of the communiqué, a sharp drop from the year before and the lowest since the 1986 Tokyo Summit (see Appendix A-1).

Direction Setting
In its direction setting, no new principles on climate change were affirmed by the G8 leadership.

Decision Making
In its decision making, the G8 made only one commitment on climate change, the lowest level by far since the absence of any at the Toronto Summit in 1988 (see Appendix D-1).

Delivery
In its delivery of this one decision, however, summit members did comply at a very high level of +0.89, up significantly from only 0 the year before (see Appendix E-1).

Development of Global Governance
The overall lack in the G8's development of global governance on climate change was evidenced by its implicit support, once again, for the UN's Rio-Kyoto regime.

Causes of Performance

The key cause of this very low climate performance was the ultimate diversionary shock of the direct, deadly terrorist attacks on the US on September 11, 2001. This made George W. Bush somewhat reluctant even to attend the summit or to deal with issues other than those directly related to terrorism. Climate change was not on his mind.

Bush's resistance to climate change commitments, seen at the Genoa Summit, was now emboldened by the domestic popularity and G8-shared preoccupation brought by September 11. Chrétien was thus able to get Bush to come to a G8 summit held next door and make it work for Africa as well as terrorism but climate change was out of reach.

The UN's multilateral process still seemed to offer promise, while the G8 devoted its energies elsewhere.

Political cohesion was high, but leaders were personally and professionally committed to issues other than climate change. Chrétien was a summit veteran whose involvement began back in 1978 in Bonn, where he served as finance minister under Pierre Trudeau. Chrétien designed Kananaskis to bring out his "determined, no-nonsense, business-like best" (Kirton 2002b). Yet despite his high level of domestic political capital, his continuity in hosting again after the 1995 Halifax Summit, his commitment to the environment there, and the first-place rank of the environment in Canadian public opinion, as it had been for over a decade, the environment and climate change virtually disappeared.

Bush, whose domestic popularity had been buoyed by his response to September 11, showed no sign of flexibility. He remained willing to take unilateral action on behalf of the US without much regard for his G8 counterparts (Afionis 2009). Without the support or consent of the US, the G8 was forced to continue to delegate climate change discussions and negotiations to the UNFCCC process.

The reduced compact participation also took its toll. The presence of the African and UN leaders on the second day of the summit elevated development on the agenda, with minimal environmental sustainability built in.

Kananaskis thus "did what summits are intended to do: the heads acted to resolve issues that had not been settled at lower levels" (Bayne 2002). The overall results of Kananaskis therefore were the G8 Africa Plan, a $1 billion commitment to replenish the HIPC trust fund, a negotiated program to clean up nuclear and chemical materials in the former Soviet Union, and the decision for Russia to host its first G8 summit. But summit leaders made no serious attempt to address climate change.

After the Summit

When COP-8 convened in New Delhi, India, from October 23 to November 1, 2002, little of substance was accomplished. Parties often stuck to their entrenched positions and deferred many substantive issues to future COP meetings. The US veto on progress prevailed, both at the G8 and the UN.

Evian, 2003

The 29th annual G8 summit took place in Evian-les-Bains, France, on June 1–3, 2003. It was an unusually suspense-filled event, due to tensions between France and the US over Iraq and doubts about whether George Bush would even attend, how long he would stay, whether he and the French hosts would reinforce or repair their recent UN-bred divisions, whether two lead-up meetings in St. Petersburg with the "East" and Evian with the "South" would energize or exhaust the G8, and whether continental European protesters would again consume the summit by violence, destruction, and death. Yet Evian also launched a new summit cycle with Russia now as a full member, had unprecedented outreach to countries around the world, sought to build a democratic Iraq and to contain North Korea and India–Pakistan, addressed the epidemic of Chinese-created severe acute respiratory syndrome (SARS), and brought better corporate governance, economic growth, and clean water to the world.

Evian was the longest and largest summit to date, extending over five days across two countries and involving more than one third of the world's leaders. It produced an unprecedented number of documents and action plans, including one on marine environment and tanker safety and another on science and technology for sustainable development. Although these action plans revived G8 action on sustainable development and the global environment, little time was left for the leaders themselves to mobilize any funds or establish any concrete follow-up mechanisms that would ensure that these climate-related commitments were met.

Preparations

In the lead-up to the summit, debates over the US invasion of Iraq in March 2003 pitted the US, and a supportive UK, Japan, and Italy, against an opposed France, Germany, Russia, and Canada, "raising doubts about the future of the transatlantic relationship" itself (Bayne 2003). The military conflict subsided before the summit, allowing attention to shift to Iraq's rebuilding. Weaker economic growth prospects and a tumbling US dollar placed new demands on the G8. Host Jacques Chirac thus declared that Evian should focus on restoring confidence.

The G8's Africa group continued its work, with chair and former IMF managing director Michel Camdessus insisting "on keeping it distinct from the main sherpa process" (Bayne 2003). There were meetings of G8 ministers for foreign affairs, finance, development, justice and the interior (chiefly on terrorism-related issues), and the environment.

From April 25 to 27, 2003, the environment ministers met in Paris and discussed environmental issues, primarily as they related to Africa, sustainable production and consumption, and oceans and maritime safety. References to climate change addressed facilitating research and development in clean, efficient energy use, although no firm commitments were made. Support was again given to the UNFCCC process. Ministers agreed that reductions in greenhouse gas emissions would "play an important role in achieving the ultimate objective of the United Nations Framework Convention on Climate Change" (G8 Environment Ministers 2003).

The four-day combined sherpa and sous-sherpa meeting at Evian in mid May was very tense. The Anglo-American coalition offered many proposals, but the French insisted on having their way. It was agreed that the G8 leaders would start in St. Petersburg with an entirely ceremonial meeting with leaders of the EU and the Commonwealth of Independent States (CIS). They would depart on the morning of June 1 for a meeting of 27 world leaders in Evian in the early afternoon. In the late afternoon, the G8 leaders would huddle alone to discuss Africa and the New Partnership for Africa's Development (NEPAD), and then meet their African guests. The next morning, the G8 summit proper would start and end one-and-a-half days later with a chair's summary and a briefing by the chair and several other delegations. All knew the world would be waiting to see if the Americans stayed until the end (Kirton and Panova 2003). In fact, Bush left early on the last day of the summit to go to the Middle East.

Substantively, it was agreed that Evian would produce documents on seven subjects: the Global Partnership; famine; freshwater, marine/oceans issues, and tanker safety; HIV/AIDS and SARS; counter-terrorism capacity building; and a responsible market economy. Also possible were agreements on corruption and transparency (related to the responsible market economy, perhaps as an appendix or addendum); trade and development (a Doha-related text); health; science and technology for sustainable development; and security of transport. There was also a possible agreement on nonproliferation and WMD. Two weeks before the summit started, only half of its potential deliverables had been agreed to.

Environmental issues would fall under corporate social responsibility and specific items on science and technology and on water and maritime safety. The centrepiece was expected to be science and technology for sustainable development, in the form of a paper on breakthrough developments in science and technology and in research and development. The British, supported by the Japanese, drove this process. On water, Evian would leverage the EU's pledge of €1 billion, by inspiring donations from other countries. The summit would also call for strong measures on maritime safety, and perhaps even a pledge, resisted by Japan, to phase out or otherwise deal with tankers with single hulls.

At the Summit

The summit took place in Evian's Hotel Royal, so the G8 leaders and their delegations could be housed in one venue. The main media centre was at Publier, five kilometres from Evian, with the press accommodated in various ski resorts, some as far as 50 kilometres away. Non-G8 leaders stayed across Lake Geneva, in Lausanne, Switzerland. As many as 15,000 police and troops successfully protected the Evian region, resulting, however, in significant protest activity across the border in Switzerland, with "destructive rioting over several days in Geneva and Lausanne, which the Swiss police found hard to control" (Bayne 2003). At the summit itself, G8 leaders met with 11 leaders from the developing world—China, India, Brazil, Mexico, Nigeria, South Africa, Algeria, Senegal, Egypt, Malaysia, and Saudi Arabia—as well as Switzerland, in appreciation for its close cooperation during summit preparations (G8 2003a). The heads of the UN, IMF, World Bank, and WTO also attended.

With the exception of Greek prime minister Konstantinos Simitis, representing the Greek EU presidency, the same group of G8 leaders at Evian had met at the two previous summits. Following rifts over Iraq, the leaders seemed eager to put the past behind them, at least for the summit's duration. Bush and Putin had begun the process of reconciliation, by declaring at a bilateral in St. Petersburg on June 1 that their relationship was strengthened by recent troubles (US Department of State 2003). Chirac took the opportunity to stress his common ideas with Blair, primarily on Africa. Schroeder, Koizumi, Berlusconi, and Chrétien "were less prominent, but all seemed to mingle and no one appeared discontented or left out" (Bayne 2003). The climax came just before noon on the second day, at the bilateral between Bush and Chirac who showed they could work together, through smiles, embraces, and the use of first names.

Results

With the divisive legacy of the Iraq war overcome and personal bonds restored, the G8 summit released a torrent of 16 documents, containing a record 207 concrete commitments—the highest in summit history. Of these, 63 commitments fell in the political-security sphere (primarily relating to terrorism), with 38 commitments on economic issues, 21 on development, and only four on climate change. Commitments on climate change came in the sustainable development section, linking science and technology as a means of providing "cleaner, more efficient energy for the fight against air pollution and climate change" (G8 2003a).

Despite the high number of commitments on an unusually wide array of issues, few bore the leaders' personal stamp, or were backed by the financial resources, mandates, and follow-up mechanisms that G8 leaders alone can provide. Reconciliation had come too late to do more. Even the $3 billion that Evian tried to lock in for the Global Fund to Fight AIDS, Tuberculosis, and Malaria launched at Genoa had to await for a subsequent European summit and donations from Japan, Russia, and others, before the pledged money would become real (see Kirton and Kokotsis 2003). The communiqué noted the need for major new funding several times, but failed to identify how, when, or from where it would be obtained.

Evian also offered almost no requests or remits that bound the G8 to return to any issues at its US-hosted summit the following year. As the summit ended without any hint of where or when that summit would be held, the Americans were left with a completely free hand for designing and delivering it (Kirton and Kokotsis 2003). It would thus be up to George Bush, as host during in a presidential election year, to determine if and how the G8 could be used to forge effective climate and environmental governance in the years ahead. A comprehensive poll of American and European attitudes found that a majority of Americans actually favoured ratifying the Kyoto Protocol (see Ikenberry 2003).

Dimensions of Performance

On climate change, performance at Evian was solid across most dimensions, primarily in its direction setting, decision making, delivery, and, to a lesser extent, in its development of global governance and deliberation (see Appendix A-1).

Deliberation
In its public deliberation, the G8 allocated only 0.3 percent of the communiqué to climate change. This was a marginal advance from the 0.2 percent the year before at Kananaskis, but still down significantly from 5.2 percent two years prior at Genoa (see Appendix A-1).

Direction Setting
In its direction setting, the leaders noted the importance of enhancing sustainable development in their opening chapeau, but offered no direct references to climate change.

Decision Making
The Evian Summit produced four commitments on climate change, up from one the year before (see Appendix D-1). They made 69 commitments on the environment as a whole, largely in the action plans on water, marine environment and tanker safety, and science and technology for sustainable development. There were commitments to assisting "as a

priority" countries that paid attention to clean water policy and to "give high priority in Official Development Aid allocation to sound water and sanitation projects" (G8 2003d). The action plan on the marine environment and tanker safety promoted measures to manage fisheries and protect the oceans. The action plan on science and technology for sustainable development promised cooperation in "global observation, cleaner energy, agriculture and biodiversity" (G8 2003c). References to finding "appropriate methodologies" for rapid innovation and clean technologies—necessary in mitigating climate change—were relegated to other international forums, including the UNFCCC, the IEA, the UN Economic Commission for Europe, and the Expert Group on Technology Transfer. That document also noted that the leaders would "discuss various aspects of the global climate change problem at the World Conference on Climate Change" in Moscow in September 2003 (G8 2003c). Most of the activities mentioned relied on existing programs and institutions and sought to involve developing countries.

These action plans represented the first advance on many environmental issues for several years, but had been achieved only by avoiding controversial issues such as ratifying the Kyoto Protocol. The chair's summary did mention, however, that those G8 members that had ratified Kyoto "reaffirm[ed] their determination to see it enter into force" (G8 2003a). That was a decision that ultimately rested on the Russians.

Delivery
Delivery of the G8's climate decisions was high, as leaders complied with the two core assessed commitments at a level of +0.88 during the year following the conclusion of the Evian Summit (see Appendix E-1).

Development of Global Governance
The G8's development of global governance continued to support the UN and its Rio-Kyoto regime. However, its endorsement of the World Climate Conference in Moscow offered some hope for a new approach.

Causes of Performance

To a considerable extent, Evian fell victim to the diversionary effect of the old-style security shock of the U.S.-led attack on Iraq in March 2003. However, a climate-specific shock arose from the massive, visible spill from the oil tanker *Prestige* that ran aground off the coast of Spain—one of the four major tanker accidents that year (see Appendix G-4). A direct result was a section in the action plan on marine environment and tanker safety that said the G8 would take the lead in "accelerating the phasing out of single hulled tankers," addressing "through appropriate measures the special risks posed by the carriage of the heaviest grades of oil in single hulled tankers," and speeding up "the adoption of guidelines on places of refuge for vessels in distress"(G8 2003b).

Multilateral organizational failure at the UN Security Council (UNSC) exacerbated the divisions among G8 members over Iraq. However, these divisions were set aside late in the Evian preparatory process, when a UNSC resolution passed with the support of all G8 members the week before the summit began.

Evian's small advances were propelled by the highly experienced and politically secure G8 leaders there (Kirton and Kokotsis 2003). Its constricted participation was low, given its unprecedented thrust toward outreach. This left little time for the G8 leaders to meet alone to discuss and decide their many serious issues such as climate change, especially as George Bush left Evian a day early for the Middle East.

After the Summit

In the meantime, climate change control continued to rely on the work of the UN process. COP-9, held in Milan on December 1–12, 2003, was intended to deal with the few unfinished issues that would assist in cleaning up the text of the Kyoto Protocol and end the deadlock between the developed and developing countries (ENB 2003). It failed, as the obligations of the developing countries continued to be troublesome. Without either the Australians or Russians ratifying the protocol, the UN was unable to bring it into legal force (Dessai 2001). This UN failure was left for the G8 to address at Sea Island in 2004, should it so choose.

Sea Island, 2004

George W. Bush hosted the 30th annual summit at Sea Island, Georgia, on June 8–10, 2004. As he prepared to do so, few observers thought he was likely to produce any significant successes (Kirton 2005a).

Yet in the preparatory process, the US increasingly listened to, learned from, and adjusted to its G8 partners. This was especially so on Africa, but not at all on climate change. Intensifying energy and terrorist shocks reminded America and its allies of their common vulnerability. The failure of the UN or US alone in response, the still predominant and equalizing capability of G8 members, and the fidelity of Sea Island's central agenda to the core G8 principles of promoting open democracy, individual liberty, and social advance also propelled Sea Island toward success. Yet the poor domestic political capital of Bush and most of his potential strong summit supporters, and the large number of visitors and issues to be dealt with in a very short time, were major constraints (Kirton 2005a).

On the environment, Italy pushed its initiative on a global earth observation system. Japan forwarded its "three Rs" proposal on reducing, reusing, and recycling. Bush resisted both, essentially eradicating environmental issues from the G8's discussions and results.

Preparations

Disappointed with his first encounter in Genoa in 2001, Bush was sceptical about the value of the G8, and even the need to hold the summit every year, including in 2004. Nonetheless, in the summer of 2003, the Americans confirmed that they would host a summit, in an informal setting on the Georgia coast as a very short meeting (Kirton 2005a). At the last sherpa meeting under the French presidency in November 2003, the Americans made it clear that they wanted no lead-up ministerial meetings (apart from those for finance and home affairs), no new money pledged, and very few—if any—guests. As themes for their summit, the Americans offered security, prosperity, and freedom. Africa and ecologically sustainable development were absent from the American plan.

However, once they assumed the G8 chair on January 1, 2004, the Americans mounted a very dense set of preparatory meetings of sherpas, foreign affairs sous-sherpas, political directors, finance deputies, and African personal representatives. G8 ministers of justice and home affairs met on May 10–11 to discuss terrorism issues; foreign ministers met in

Washington on May 14 and debated Iraq and Israel–Palestine, but produced no statement; and finance ministers met in New York on May 23 and issued an optimistic forecast for the world economy, despite concerns over rising oil prices. Bush deployed his ministers in G8 forums on his four summit priorities: transport security and terrorism (including terrorist finance), the Greater Middle East Initiative and its delicate Iraqi link, the nonproliferation of WMD, and private sector-led development. In an unusual move, he even met with G8 foreign ministers at their Washington gathering. Yet G8 environment ministers did not meet, for the first time since 1993.

As the spring unfolded, the enthusiasm of America's G8 partners as well as actors in the US administration propelled ever more items onto the agenda. In the end, the agenda ambitiously covered the economic and political-security domains and reflected the distinctive priorities of all major G8 partners. But there were limits. There were pressures from France to return to Evian's science and technology for sustainable development, to upgrade the Global Environment Observation System of Systems, and to respond to Tony Blair's desire to address climate change directly. Japan, which had supported its US ally on the political/security agenda, was pushing its one key issue of the three Rs. It believed that the environment and growth may have seemed contradictory 30 years earlier, but were now fully compatible: growth required the environment, and science and technology bound the two together. But the decision made in the autumn of 2003 to leave sustainable development and Africa almost entirely to Britain's 2005 G8 presidency endured.

The leaders met in the exclusive resort area of Sea Island, while the media assembled in Savannah, 130 kilometres away. The few protesters who turned up were vastly outnumbered by police and security forces. The US made no attempt to engage NGOs, provided no facilities for them, and prohibited the circulation of their material within the media centre (Bayne 2004).

At the Summit

As the summit approached, the US sought involvement from regional leaders in its Middle East initiative. In the end, the G8 met with a mixed group from the Middle East and North Africa on the first day—Afghanistan, Algeria, Bahrain, Jordan, Turkey, Yemen, plus the new Iraqi president—and a strong African contingent on the second—Algeria, Ghana, Nigeria, Senegal, South Africa, and Uganda. These quests did not include the non-G8 countries needed to control climate change.

When the summit started, its leaders (with the exception of Japan's Junichiro Koizumi) had already met on June 6 at the 60th anniversary of the D-Day landings in Normandy. The official opening social dinner took place on the evening of June 8, after bilateral meetings earlier that day. Sessions began the following morning with attention focused on the world economy, trade, and entrepreneurship for development. Following a working lunch with regional leaders, the G8 leaders released their Middle East reform plan. They dedicated the afternoon to transport security and nonproliferation. Their working dinner that evening covered regional issues. This session, among the leaders alone, with neither a fixed agenda nor note takers, was a top-level political dialogue on the Middle East. The G8 then shifted to development on June 10, leading up to the working lunch with the Africans. While Canada's Paul Martin and Italy's Romano Prodi left before summit's end, some of the leaders met again at the state funeral for Ronald Reagan in Washington on June 11 (Bayne 2004). Most of the G8 leaders were thus effectively together from June 6 to 11, an unprecedented amount of time for the world's political elite to spend together.

As host, Bush came across as "energetic and determined, with a visceral commitment to see things through in Iraq," making his points "by the force, not the subtlety, of his arguments" (Bayne 2004). Chirac was typically individualistic in his demeanour, "going out of his way to distance himself from the U.S." By contrast, Schroeder was much more conciliatory toward Bush and appeared engaged on issues relating to the Middle East and debt relief. Blair was less visible than at prior summits, "perhaps holding his fire for Gleneagles in 2005." Koizumi pressed points of direct concern to Japan, particularly on North Korea and the environment, but supported Bush on Iraq. Martin drew on his lengthy experience as the only former finance minister within the group. Berlusconi, Putin, and Prodi "made no clear mark."

Results

The result of Sea Island was a summit of substantial success, above all on Africa and the Middle East. But it was a great failure on climate change, where very little was done. To no one's surprise, Bush's centrepiece issue, the Greater Middle East Initiative—promising sovereignty, security, development, and democratization—emerged as the clear summit victor. The UNSC's unanimous resolution to transfer sovereignty to Iraq went far in reuniting the G8 after bitter divisions during the past 18 months. The G8 followed with a bold declaration promising support for the principles of democracy, the rule of law, and human rights in the region, backed by a financial pledge (Kirton 2005a).

On Africa, the G8 secured $1 billion for peace support initiatives, an additional $1 billion for the HIPC trust fund, up to $200 million to eradicate polio by 2005, and $375 million to develop vaccines for HIV/AIDS. African leaders attended the G8 for the fourth year in a row, with the veterans Mbeki, Wade, Bouteflika, and Obasanjo joined by newcomers from Yoweri Museveni of Uganda and John Kufuor of Ghana for a lunchtime session with G8 leaders on the summit's third day.

Another area of accomplishment was WMD, where the G8 imposed a one-year moratorium on the export of materials that recipient states could use to acquire nuclear weapons. On regional security, the G8 produced a strong statement on "Gaza Withdrawal and the Road Ahead to Middle East Peace."

In other areas of traditional G8 summit action, including the environment and climate change, however, there were few results. In discussing the energy component of the world economy, the G8 began to explore solutions on energy efficiency, conservation, and alternatives to oil (G8 2004a). Moreover, the leaders expressed their concern with how the threat of terrorism could hurt or end the strong economic recovery by having an impact on energy prices. The action plan on science and technology largely approved existing work and set no new directions, apart from endorsing Japan's three Rs initiative.

On climate change, the British had sought plenty of time for the leaders to discuss this topic, expecting a difficult exchange as at Denver in 1997, but no one was in the mood (Kirton 2004b). Little mention was made of the rising problem of global warming apart from Chirac's disappointment with Bush's unwillingness to speak on the matter in detail. Thus, with little attention devoted to environmental issues, Blair reiterated his position that as host in 2005, his G8 summit would feature Africa and climate change as its centrepiece themes. On the latter issue, Blair and his summit planners would have to start from scratch (Kirton 2005a).

Dimensions of Performance

On many of the standard dimensions of summit performance, Sea Island set new highs overall (see Appendix A-1). Indeed, in all, with the exception of money mobilized, Sea Island was a major advance from the performance of almost all past summits (Kirton 2005a). But issues of the environment and climate change were almost entirely left out.

Domestic Political Management

In overall terms, Bush's summit won public approval and praise from both the G8 leaders and their guests. Only 500 civil society activists arrived, for activities that produced only 15 arrests on minor charges, with no bodily injury and no physical damage at all. With summit costs at only one third of what the French had spent at Evian, only 1,492 media members arrived to scrutinize the G8 and to report on the event, largely using information dispensed by the US administration alone.

Bush's popularity had been plummeting in the polls for the previous two months—with presidential rival John Kerry taking a six- to seven-point lead. Yet a Fox News poll released on June 10, the last day of the summit, showed that Bush had cut Kerry's lead to a statistically insignificant 2 percent (Fox News 2004). Climate change stood near the bottom of America's political concerns.

Few of Bush's G8 partners enjoyed the similar domestic political benefit of a bounce in the polls from spending time with him in the United States. Canada's Paul Martin, facing a general election on June 28, used his summit performance to reverse his slide in the polls. In Japan, Koizumi's approval dropped to where it had been before the strong surge brought by his pre-summit trip to North Korea. Blair returned to the UK to a devastating defeat in local and European elections. His Labour Party's 22.3 percent of the vote in the European parliament election was the lowest since before World War I. France's ruling Union pour un mouvement populaire received only 16 percent (compared to 30 percent for the opposition Socialists), Germany's ruling Social Democrats 21 percent, and Italy's ruling Forza Italia about the same (Kirton 2005a).

Deliberation

As a deliberative institution, the G8 issued a record 16 documents, often very lengthy and detailed ones covering 10 separate issue areas. However, climate change took only 0.3 percent of the final communiqué, the same very low level as the year before (see Appendix A-1).

Direction Setting

In setting normative directions and principles, Sea Island highlighted the themes of freedom and democracy, both in the opening passage of its chair's summary and throughout the other documents. Although no direct references were made to climate change in the summary, the stand-alone document on science and technology for sustainable development devoted a section on the "Reduce, Reuse, and Recycle ('3R') Initiative" to encourage more efficient use of resources and materials, as well as a section on "cleaner, more efficient energy," aimed at facilitating wider use of renewable energy and energy efficiency technologies (G8 2004b).

Decision Making

In its decision making, Sea Island generated a record 253 commitments in its individual documents, with some reiterated in the chair's summary. But only three dealt with climate change directly, down from four the year before (see Appendix D-1).

Delivery

Delivery of these decisions was high. The two assessed core climate commitments generated an overall compliance level of +0.89, almost tied with Evian's score (see Appendix E-1).

Development of Global Governance

Sea Island did very little to develop global governance. On other issues, however, it created or directed, 19 G8 or G8-centred institutions at the ministerial, official, and, importantly, civil society levels. Moreover, it issued more than 500 instructions, of both guidance and support, to a vast array of other international institutions.

Causes of Performance

Some observers suggest that America's best effort came when Ronald Reagan hosted at Williamsburg in 1983 (Nau 2004; see also Nau and Shambaugh 2004). Others saw Sea Island as performing well, in keeping with an upward trend starting with the 1990 Houston Summit and then the 1997 Denver Summit (Fauver 2003; Brainard 2004). That performance—strong overall, but weak on climate change—was driven by several forces from both inside and outside the G8 (Kirton 2005a).

Shock-Activated Vulnerability

The first force was America's rising vulnerability to the twin shocks of energy and terrorism. Just one week prior to the summit, world oil prices reached historic highs—daunting news as American oil imports continued to rise and gas prices at the pumps soared (see Appendix G-3). A rising number of troops and civilians were killed by terrorists in Iraq (Kirton, Roudev, et al. 2005). The discovery and use of sarin nerve gas against US forces in Iraq in May proved that the nightmare of terrorists deploying WMD remained a reality from which no G8 country was exempt. These diversionary shocks from terrorism and WMD, as at Kananaskis in 2002, overwhelmed the energy shock and its incentive for climate change control.

Multilateral Organizational Failure

The second cause was the overall failure of the UNSC to reach any real political-security arrangements on Iraq. After remaining deadlocked for so long, movement at the UNSC finally came the day prior to the summit, when a resolution unanimously gave "solid international backing for the transfer of power and the continued pressure of coalition troops" (Bayne 2004). This provided "a good foundation for G8 agreement on the broader Middle East initiative … and for other political decisions on nonproliferation, transport security and peace support in Africa." But it relegated the economic and development agenda to second place and erased any possibility of discussion on the environment or climate change. This was despite Britain's repeated attempts to keep environmental issues alive at the working level and thus available for an upgrade to the leaders' level should the occasion arise.

Political Cohesion

Political cohesion was mixed. Save for newcomer Paul Martin of Canada, the same leaders assembled for the fourth year in a row. But with elections looming for most of them, "domestic political factors placed greater constraints on the heads ... than they had faced in earlier years" (Bayne 2004). Martin would go to the polls on June 28; Prodi's term as European Commission president would expire later in the year; Berlusconi and Blair were expected to hold elections in 2005. Chirac and Schroeder, although electorally secure, were experiencing diminished political support at home. Bush was seeking re-election in November and was behind in the polls. He had little incentive to adjust on climate change, which ranked very low on his voters' list of concerns.

Constricted Participation

The fifth force was the low constricted participation due to the many guests from the broader Middle East and Africa, none of whom cared about climate change. Their presence and their priorities left little opportunity for Blair, Chirac, and Koizumi to raise the climate change issue they cared about.

After the Summit

COP-10, the last one before Kyoto's entry into force, was held in Buenos Aires on December 6–18, 2004. The signatories had agreed that Kyoto would enter into force on February 15, 2005, with binding targets for 37 industrialized countries to reduce their greenhouse gas emissions by 5 percent below 1990 levels between 2008 and 2012 (ENB 2004). Key issues at Buenos Aires remained mitigation and adaptation policy responses to climate change, with developing countries continuing to seek accommodation on targets and financial support from the developed world. Frustrated with the deadlock, the UK—which would hold the presidencies of both the G8 and the EU in 2005—began announcing its intention to move climate issues through those forums instead. It wanted to shift the climate debate away from the failing UN process back to the control of the G8, where like-minded, developed democracies could work collectively on a new regime for effective climate governance.

Conclusion

From 1998 to 2004, the G8 retreated from leadership on climate change, refraining from shaping a new regime or even supporting the increasingly ineffective UN one. The 1998 Birmingham and 1999 Cologne summits focused on the Kyoto Protocol and other commitments under the UNFCCC. Okinawa in 2000 did much the same, adding limited discussion on energy efficiency and the role of developing countries in climate mitigation. Genoa in 2001 forged a common recognition by the G8 that the most effective way forward on climate change was to build on the UNFCCC process. The 2002 Kananaskis Summit made only marginal mention of climate in its G8 Africa Action Plan. The limited restoration at Evian in 2003 was promptly followed by the disappearance at the otherwise successful Sea Island Summit in 2004.

This shift from G8 to UN leadership in global climate governance was primarily the result of six forces.

The first was the arrival of severe, nonstate shocks on a vulnerable America, with the September 11 terrorist assault and the old security shock of the American-led invasion of Iraq in 2003. Although these generated a new oil price shock, they diverted G8 attention from environment issues to security ones. Behind lay the G8's preoccupation with a wide range of new and emerging policy issues brought on by the complex challenges of globalization, accentuated above all by the Asian-turned-global financial crisis that began in 1997.

The second force was apparent multilateral organizational success due to the flourishing and the institutions and processes created by the 1992 Earth Summit in Rio de Janeiro. By systematically and continuously endorsing the Rio conventions in their final declarations, the G7 effectively supported the notion that the Earth Summit represented a landmark in meeting global environmental challenges through the creation of a global partnership on the environment and development (US Department of State 1992).

The third force was the absence of a committed G8 leader who was able and willing to push the climate agenda in the face of formidable American opposition after 2000 and the competing post-September 11 demands. There was no Helmut Schmidt to make the link from an oil shock to the need for climate change control. And the support of the lame-duck Clinton-Gore administration turned into adamant opposition in 2001 when the Bush administration arrived.

Fourth, the continued presence of the now firmly institutionalized G8 environment ministerial process, beginning in 1992 and continuing throughout this period, generated discussion, consensus, and commitment on climate mitigation strategies. Yet it increasingly reduced the need for the leaders to focus their attention on these issues at the summit and it disappeared altogether in Bush's year as host.

The fifth force was the increasing erosion of the compact participation of the leaders' club, as the summit included ever more guests who were concerned with issues other than climate change.

Sixth, the G8's inclusion of the most relevant powers on both the sources side and sinks side declined quite dramatically from 1998 to 2004. In the earlier phase of global climate governance, the G7 included all the consequential climate actors in its deliberations, including the United States. This was due to the small, flexible nature of the forum, and its status as a collection of the world's top emitters of carbon sources and containers of carbon sinks. During this phase of retreat, however, the G8 was ineffective in reaching out with sufficient speed to rising emitters and absorbers. It deferred to the hard law of the UN, with a divided, development-first regime that the US was unwilling to follow, and failed to bring the South's largest emitters into a plurilateral climate change dialogue centred within the G8. This ultimately created a seven-year lag in effective G8 climate governance, while UN leadership induced emerging economies to reach G8-like carbon emissions levels by 2005.

References

Afionis, Stavros (2009). "The Role of the G8/G20 in International Climate Change Negotiations." *In-Spire Journal of Law, Politics, and Societies* 4(2): 1–12. http://www.academia.edu/721847/The_Role_of_the_G-8_G-20_in_International_Climate_Change_Negotiations (January 2015).

Associated Press (2001). "US Won't Follow Climate Treaty Provisions, Whitman Says." *New York Times*, March 28. http://www.nytimes.com/2001/03/28/politics/28WHIT. html (January 2015).

Bayne, Nicholas (1998). "Impressions of the Birmingham Summit." May 19, G8 Research Group, Birmingham. http://www.g8.utoronto.ca/evaluations/1998birmingham/ impression/index.html (January 2015).

Bayne, Nicholas (2000a). "First Thoughts on the Okinawa Summit, 21–23 July 2000." July 27. http://www.g8.utoronto.ca/evaluations/2000okinawa/bayne.html (January 2015).

Bayne, Nicholas (2000b). *Hanging In There: The G7 and G8 Summit in Maturity and Renewal*. Aldershot: Ashgate.

Bayne, Nicholas (2001). "Impressions of the Genoa Summit, 20–22 July 2001." July 28, G8 Research Group, Genoa. http://www.g8.utoronto.ca/evaluations/2001genoa/assess_ summit_bayne.html (January 2015).

Bayne, Nicholas (2002). "Impressions of the Kananaskis Summit, 26–27 June 2002." Kananaskis, July 23. http://www.g8.utoronto.ca/evaluations/2002kananaskis/assess_ baynea.html (January 2015).

Bayne, Nicholas (2003). "Impressions of the Evian Summit, 1–3 June 2003." June 3. www. g8.utoronto.ca/evaluations/2003evian/assess_bayne030603.html (January 2015).

Bayne, Nicholas (2004). "Impressions of the 2004 Sea Island Summit." June 29. http:// www.g8.utoronto.ca/evaluations/2004seaisland/bayne2004.html (January 2015).

Blair, Tony (2010). *A Journey*. London: Random House.

Brainard, Lael (2004). "G7/8 Oral History: Interview." G8 Research Group, February 6. http://www.g8.utoronto.ca/oralhistory/ (January 2015).

Canada (2000). "Government of Canada Action Plan 2000 on Climate Change." Ottawa. http://env.chass.utoronto.ca/env200y/ESSAY2001/gofcdaplan_eng2.pdf (January 2015).

Coon, Charli E. (2001). "Why President Bush Is Right to Abandon the Kyoto Protocol." May 11, Heritage Foundation, Washington DC. http://www.heritage.org/research/ reports/2001/05/president-bush-right-to-abandon-kyoto-protocol (January 2015).

Davenport, Deborah (2008). "The International Dimension of Climate Policy." In *Turning Down the Heat: The Politics of Climate Policy in Affluent Democracies*, Hugh Compston and Ian Bailey, eds. Basingstoke: Palgrave Macmillan.

Dessai, Suraje (2001). "The Climate Regime from The Hague to Marrakech: Saving or Sinking the Kyoto Protocol?" Working Paper 12, December, Tyndall Centre for Climate Change Research. http://www.tyndall.ac.uk/content/climate-regime-hague-marrakech-saving-or-sinking-kyoto-protocol (January 2015).

Earth Negotiations Bulletin (1998, 16 November). "Report of the Fourth Conference of the Parties to the UN Framework Convention on Climate Change: 2–13 November 1998." 12(97). http://www.iisd.ca/download/pdf/enb1297e. pdf (January 2015).

Earth Negotiations Bulletin (1999, 8 November). "Report of the Fifth Conference of the Parties to the UN Framework Convention on Climate Change: 25 October–5 November 1999." 12(123). http://www.iisd.ca/download/pdf/enb12123e.pdf (January 2015).

Earth Negotiations Bulletin (2000, 27 November). "Report of the Sixth Conference of the Parties to the Framework Convention on Climate Change: 13–25 November 2000." 12(163). http://www.iisd.ca/download/pdf/enb12163e.pdf (January 2015).

Earth Negotiations Bulletin (2001a, 30 July). "Summary of the Resumed Sixth Session of the Conference of the Parties to the UN Framework Convention on Climate Change: 16–27 July 2001." 12(176). http://www.iisd.ca/download/pdf/enb12176e.pdf (January 2015).

Earth Negotiations Bulletin (2001b, 12 November). "Summary of the Seventh Session of the Conference of the Parties to the UN Framework Convention on Climate Change: 30 October–10 November 2001." 12(189). http://www.iisd.ca/download/pdf/enb12189e. pdf (January 2015).

Earth Negotiations Bulletin (2003, 15 December). "Summary of the Ninth Conference of the Parties to the UN Framework Convention on Climate Change: 1–12 December 2003." 12(231). http://www.iisd.ca/download/pdf/enb12231e. pdf (January 2015).

Earth Negotiations Bulletin (2004, 20 December). "Summary of the Tenth Conference of the Parties to the UN Framework Convention on Climate Change: 6–18 December." 12(260). http://www.iisd.ca/download/pdf/enb12260e.pdf (January 2015).

European Union (2001). "EU Reaction to the Speech by US President Bush on Climate Change." IP/01/821, Brussels, June 12. http://europa.eu/rapid/press-release_IP-01-821_en.htm (January 2015).

Fauver, Robert (2003). "G7/8 Oral History: Interview." G8 Research Group, March 13. http://www.g8.utoronto.ca/oralhistory/ (January 2015).

Fox News (2004). "Bush, Kerry Still Closely Matched." June 10. http://www.foxnews.com/story/2004/06/10/06102004-bush-kerry-still-closely-matched/ (January 2015).

G7 Finance Ministers (1999). "Report of G7 Finance Ministers on the Köln Debt Initiative to the Köln Economic Summit." Cologne, June 18. http://www.g8.utoronto.ca/finance/fm061899.htm (January 2015).

G8 (1998). "Communiqué." Birmingham, May 15. http://www.g8.utoronto.ca/summit/1998birmingham/finalcom.htm (January 2015).

G8 (1999). "Communiqué." Cologne, June 20. http://www.g8.utoronto.ca/summit/1999koln/finalcom.htm (January 2015).

G8 (2000). "G8 Communiqué Okinawa 2000." Okinawa, July 23. http://www.g8.utoronto.ca/summit/2000okinawa/finalcom.htm (January 2015).

G8 (2001a). "Communiqué." Genoa, July 22. http://www.g8.utoronto.ca/summit/2001genoa/finalcommunique.html (January 2015).

G8 (2001b). "Statement by the G8 Leaders (Death in Genoa)." Genoa, July 21. http://www.g8.utoronto.ca/summit/2001genoa/g8statement1.html (January 2015).

G8 (2002). "The Kananaskis Summit Chair's Summary." Kananaskis, June 27. http://www.g8.utoronto.ca/summit/2002kananaskis/summary.html (January 2015).

G8 (2003a). "Chair's Summary." Evian, June 3. http://www.g8.utoronto.ca/summit/2003evian/communique_en.html (January 2015).

G8 (2003b). "Marine Environment and Tanker Safety: A G8 Action Plan." Evian, June 3. http://www.g8.utoronto.ca/summit/2003evian/marine_en.html (January 2015).

G8 (2003c). "Science and Technology for Sustainable Development: A G8 Action Plan." Evian, June 2. http://www.g8.utoronto.ca/summit/2003evian/sustainable_development_en.html (January 2015).

G8 (2003d). "Water: A G8 Action Plan." Evian, June 2. http://www.g8.utoronto.ca/summit/2003evian/water_en.html (January 2015).

G8 (2004a). "Chair's Summary." Sea Island, June 10. http://www.g8.utoronto.ca/summit/2004seaisland/summary.html (January 2015).

G8 (2004b). "Science and Technology for Sustainable Development: "3R" Action Plan and Progress on Implementation." Sea Island. http://www.g8.utoronto.ca/summit/2004seaisland/sd.html (January 2015).

G8 Energy Ministers (2002). "G8 Energy Ministers Meeting: Co-chairs' Statement." Detroit, May 3. http://www.g8.utoronto.ca/energy/energy0702.html (January 2015).

G8 Environment Ministers (1999). "G8 Environment Ministers Communiqué." Schwerin, Germany, March 28. http://www.g8.utoronto.ca/environment/1999schwerin/communique.html (January 2015).

G8 Environment Ministers (2000). "G8 Environment Ministers Communiqué." Otsu, Japan, April 9. http://www.g8.utoronto.ca/environment/2000otsu/communique.html (January 2015).

G8 Environment Ministers (2001). "G8 Environment Ministers Communiqué." Trieste, Italy, March 4. http://www.g8.utoronto.ca/environment/2001trieste/communique.html (January 2015).

G8 Environment Ministers (2002). "Banff Ministerial Statement on the World Summit on Sustainable Development." Banff, April 14. http://www.g8.utoronto.ca/environment/020415.html (January 2015).

G8 Environment Ministers (2003). "G8 Environment Ministers Communiqué." Paris, April 27. http://www.g8.utoronto.ca/environment/2003paris/env_communique_april_2003_eng.html (January 2015).

G8 Foreign Ministers (1999). "Conclusions of the Meeting of the G8 Foreign Ministers." Cologne, June 10. http://www.g8.utoronto.ca/foreign/fm9906010.htm (January 2015).

G8 Research Group (2001). "From Okinawa 2000 to Genoa 2001: Issue Performance Assessment — Environment." http://www.g8.utoronto.ca/evaluations/2001genoa/assessment_environment.html (January 2015).

Hajnal, Peter I. (1999). *The G7/G8 System: Evolution, Role, and Documentation.* Aldershot: Ashgate.

Hodges, Michael R. (1999). "The G8 and the New Political Economy." In *The G8's Role in the New Millennium*, Michael R. Hodges, John J. Kirton, and Joseph P. Daniels, eds. Vol. 69–74. Aldershot: Ashgate, pp. 69–74.

Hodges, Michael R., John J. Kirton, and Joseph P. Daniels, eds. (1999). *The G8's Role in the New Millennium.* Aldershot: Ashgate.

Ikenberry, G. John (2003). *Strategic Reactions to American Preeminence: Great Power Politics in the Age of Unipolarity.* Washington DC: National Intelligence Council.

Jeffery, Simon (2001). "Protester Shot Dead in Genoa Riot." *Guardian*, July 20. http://www.theguardian.com/world/2001/jul/20/globalisation.usa (January 2015).

Kaiser, Karl, John J. Kirton, and Joseph P. Daniels, eds. (2000). *Shaping a New International Financial System: Challenges of Governance in a Globalizing World.* Aldershot: Ashgate.

Kirton, John J. (1999). "An Assessment of the 1999 Cologne G7/G8 Summit by Issue Area." June 20, G8 Research Group, Cologne. http://www.g8.utoronto.ca/evaluations/1999koln/issues/kolnperf.htm (January 2015).

Kirton, John J. (2000). "Creating Peace and Human Security: The G8 and Okinawa Summit Contribution." May 26, G8 Research Group. http://www.g8.utoronto.ca/scholar/kirton200002/ (January 2015).

Kirton, John J. (2001a). "Generating Genuine Global Governance: Prospects for the Genoa G8 Summit." July 15, G8 Research Group, Genoa. http://www.g8.utoronto.ca/evaluations/2001genoa/prospects_kirton.html (January 2015).

Kirton, John J. (2001b). "Guiding Global Economic Governance: The G20, the G7, and the International Monetary Fund at Century's Dawn." In *New Directions in Global Economic Governance: Managing Globalisation in the Twenty-First Century*, John J. Kirton and George M. von Furstenberg, eds. Aldershot: Ashgate, pp. 143–67.

Kirton, John J. (2002a). "The Promise of the Kananaskis Summit." *Calgary Herald*, June 26. http://www.g8.utoronto.ca/evaluations/2002kananaskis/assess_promise.html (January 2015).

Kirton, John J. (2002b). "A Summit of Historic Significance: A Gold Medal for the Kananaskis G8." *Calgary Herald*, June 27. http://www.g8.utoronto.ca/evaluations/2002kananaskis/assess_goldmedal.html (January 2015).

Kirton, John J. (2004). "Prospects for the G8 Sea Island Summit Seven Weeks Hence." Paper prepared for a seminar at Armstrong Atlantic State University, April 22, Savannah. http://www.g8.utoronto.ca/scholar/kirton2004/kirton_seaisland_040426.html (January 2015).

Kirton, John J. (2005). "America at the G8: From Vulnerability to Victory at the Sea Island Summit." In *New Perspectives on Global Governance: Why America Needs the G8*, Michele Fratianni, John J. Kirton, Alan M. Rugman, et al., eds. Aldershot: Ashgate, pp. 31–50.

Kirton, John J. (2013). *G20 Governance for a Globalized World*. Farnham: Ashgate.

Kirton, John J., Joseph P. Daniels, and Andreas Freytag, eds. (2001). *Guiding Global Order: G8 Governance in the Twenty-First Century*. Aldershot: Ashgate.

Kirton, John J. and Ella Kokotsis (2003). "Impressions of the G8 Evian Summit." Evian, June 3. http://www.g8.utoronto.ca/evaluations/2003evian/assess_kirton_kokotsis.html (January 2015).

Kirton, John J. and Victoria Panova (2003). "Coming Together: Prospects for the G8 Evian Summit." Conference on "Governing Globalization: Corporate, Public, and G8 Governance", May 27, Fontainebleau. http://www.g8.utoronto.ca/scholar/kirton2003/kirton_prospects_030520.html (January 2015).

Kirton, John J., Nikolai Roudev, Michael Lehan, et al. (2005). "Shocks from Terrorism to G8 Countries and Citizens: Major Incidents." G8 Research Group, July 20. http://www.g8.utoronto.ca/evaluations/factsheet/factsheet_terrorism.html (January 2015).

Kluger, Jeffrey (2001). "A Climate of Despair." *Time*, April 1. http://content.time.com/time/magazine/article/0,9171,104596,00.html (January 2015).

MacKinnon, Mark (2001). "Canada Threatens to Reject Kyoto Pact." *Globe and Mail*, July 19. http://www.theglobeandmail.com/news/national/canada-threatens-to-reject-kyoto-pact/article4150728/ (January 2015).

Mori, Yoshiro (2000). "Address by Prime Minister Yoshiro Mori at the Discussion Group on the Kyushu-Okinawa Summit, Okinawa Summit." June 5. http://www.ioc.u-tokyo.ac.jp/~worldjpn/documents/texts/summit/20000605.S1E.html (January 2015).

Nau, Henry (2004). "G7/8 Oral History: Interview." G8 Research Group, May 7. http://www.g8.utoronto.ca/oralhistory/nau040507.html (January 2015).

Nau, Henry and David Shambaugh, eds. (2004). *Divided Diplomacy and the Next Administration: Conservative and Liberal Alternatives*. Washington DC: Elliott School of International Affairs, George Washington University.

Obuchi, Keizo (2000). "Statement by Prime Minister Keizo Obuchi Discussion Group on the Kyushu-Okinawa Summit." February 28. http://www.mofa.go.jp/announce/announce/2000/2/228-2.html (January 2015).

Prodi, Romano (2001). "EU-US Summit, Göteborg 14 June 2001 Statement of Romano Prodi President of the European Commission." IP/01/849, Göteborg, June 15. http://europa.eu/rapid/press-release_IP-01-849_en.htm (January 2015).

Putnam, Robert and Nicholas Bayne (1987). *Hanging Together: Co-operation and Conflict in the Seven-Power Summit.* 2nd ed. London: Sage Publications.

Reuters (2001). "G8 Leaders to Tackle African Crisis at Summit." March 7.

Risbud, Sheila (2006). "Civil Society Engagement: A Case Study of the 2002 G8 Environment Ministers Meeting." In *Sustainability, Civil Society, and International Governance: Local, North American, and Global Contributions*, John J. Kirton and Peter I. Hajnal, eds. Aldershot: Ashgate, pp. 337–42.

Smith, Gordon (2008). "Canada's Foreign Policy Priorities." *Globe and Mail*, April 23.

United Nations (1999). "Report of the Conference of the Parties on Its Fourth Session, Held at Buenos Aires from 2 to 14 November 1998." Buenos Aires. http://unfccc.int/resource/docs/cop4/16.pdf (January 2015).

United States Department of State (1992). "US Environment Initiatives and the UN Conference on Environment and Development." *US Department of State Dispatch Supplement* 3(4). http://dosfan.lib.uic.edu/ERC/briefing/dispatch/1992/html/Dispatchv3Sup4.html (January 2015).

United States Department of State (2003). "Press Availability in St. Petersburg." Konstantin Palace, St. Petersburg, June 1. http://2001-2009.state.gov/p/eur/rls/rm/2003/21113.htm (January 2015).

Welch, David (2005). *Painful Choices: A Theory of Foreign Policy Change.* Princeton: Princeton University Press.

White House (2001). "Press Briefing by Ari Fleischer." Washington DC, March 28. http://www.presidency.ucsb.edu/ws/?pid=47500 (January 2015).

PART IV
Pioneering the Inclusive Global Regime

Chapter 8
Restoration, 2005–2007

The year 2005 marked a key turning point in global climate governance. It ended several years of the Group of Eight (G8) relying on an initially promising but increasingly failing United Nations system. It did so by restoring climate change to the forefront of the G8's own agenda and action. Under the strategic leadership of British prime minister Tony Blair, and over the strong initial resistance of United States president George W. Bush, Gleneagles was the first G8 summit to issue a separate document devoted entirely to climate change. It was the first to declare that climate change was a human-induced condition. It launched a new dialogue on climate change with the key non-G8, rapidly rising, emerging powers and included them in national efforts to control caution. It inspired more action from both the industrialized and emerging economies and focused G8 attention on a wide range of mitigation strategies including clean fuels and renewable energy innovation, awareness and adaptation, climate monitoring, and illegal logging (Bayne 2005a). It thus started the restoration of G8 global leadership on climate change and the process that a decade later had produced the basis of a new regime.

The period of restoration from 2005 to 2007 saw a rise in the centrality of climate issues at G8 summits. Communiqués now labelled climate change more important and urgent. In 1985, climate change had been considered one among other environmental "concerns" (Group of Seven [G7] 1985) A decade later in 1995, it had become a problem of "major global importance" (G7 1995). Ten years later, at Gleneagles, it was deemed a "serious and long-term challenge" (G8 2005b). And in 2007, it became "one of the major challenges for mankind" and an "urgent challenge" to be addressed (G8 2007a). As recognition of the climate challenge rose in prominence, so too did the G8's increasingly inclusive response from 2005 to 2007. This restoration of G8 leadership, now in an expanded, more inclusive form, arose at a time of growing UN failure to deliver an effective post-Kyoto regime that constrained all the consequential carbon-emitting and -absorbing powers in the world and thus the UN's failure to control the increasingly costly challenge faced by all.

Gleneagles, 2005

On July 6–8, 2005, G8 leaders assembled in Gleneagles, Scotland, for their 31st annual summit. They sought to make history on two highly ambitious global goals: financing democratic development in Africa and controlling climate change. Overall, they succeeded to an exceptional degree, producing new summit highs in domestic political management, deliberation, direction setting, and money mobilized, while performing strongly in decision making and the delivery of their commitments.

On climate change, some saw the Gleneagles Summit's performance as disappointing business as usual (Oosthoek 2005). In contrast, veteran summit analyst Nicholas Bayne (2010, 206) judged that "in climate change, Bush accepted for the first time that global warming was caused by human activity. The summit agreed to develop new technology to reduce greenhouse gas emissions and launched a dialogue embracing all the big emitters, including the United States and emerging powers like China. This enabled progress to restart in UN contexts, even before Bush left office." Overall, Gleneagles, with its score of A−, due in part to its performance on climate change, is one of only three A-level grades ever awarded, and the first since the energy-fuelled success of 1978. Robert Putman and Nicholas Bayne (2005b) awarded other British-hosted summits as follows: London in 1977 a grade of B−, London in 1984 a grade of C−, London in 1991 a grade of B−, and Tony Blair's Birmingham in 1998 a grade of B+.

This analysis shows that Gleneagles was a very strong success on climate change. Even as Gleneagles concluded the G8's half-decade campaign on propelling African development, it launched a new one on climate change control. Gleneagles produced new principles and a new process to build an inclusive beyond-Kyoto regime in which all consequential carbon powers and polluters would be constrained. It induced a hitherto resistant United States and the major developing countries to take carbon-constraining action, with the development, transfer, and use of clean technologies as the lead instruments. At Gleneagles, a Kyoto-unconstrained and climate-sceptic George Bush and his Kyoto-committed G8 partners joined the invited leaders of the rising, now top-tier climate change and energy powers of China, India, Brazil, South Africa, and Mexico to launch a new regime, as a foundation to build on in future years. Propelling these historic achievements were the forces highlighted by the concert equality model of G8 governance, with the addition of a more inclusive G8 that was becoming a global climate governance network hub. From the outside, G8 leaders were reminded of past shocks and present vulnerabilities by rising and volatile energy prices, driven by increasing demand from the economically growing, systemically significant powers and by political unrest in a terrorist-infected Middle East and an unstable Africa. These vulnerabilities were brought home by the galvanizing deadly terrorist attacks in London on July 7, the summit's first full day. In 2005, the greenhouse gas emissions of China became the first-ranked carbon polluter in the world. The UN-based multilateral organizations were struggling to deliver the Millennium Development Goals (MDGs), with their weak environmental content, or to meet the Kyoto Protocol's targets and induce the US to join.

The G8's global predominance and internal equality in capability were increased by the growing power of Russia and Canada, both Kyoto-committed, full-strength energy superpowers, with fiscal surpluses available to finance sustainable development abroad. It was diminished by the rapidly growing overall and ecological power of China, India, Brazil, South Africa, and Mexico outside the group. But as a result, those countries attended the G8 summit as a Group of Five (G5) for the first time, thus enhancing the predominance and equality of this new, expanded G8 plus G5 club. The G8's seminal mission and shared principles of open democracy, individual liberty, and social advance drove an African strategy now centred on democratic development and a climate change process that included the leading developing country democracies of India, Mexico, Brazil, and South Africa. Substantial domestic political capital and control came from host Tony Blair, just re-elected with a historic third majority government, from a secure Vladimir Putin from Russia and Junichiro Koizumi from Japan, and from a recently re-elected George Bush whose party controlled both chambers of Congress in the United States (Kirton 2005d, 2006c).

Yet success in restoring the G8's climate leadership, now in an expanded, more inclusive form, was ultimately driven and delivered by the exceptional vision, skill and experience of Blair, hosting the G8 for a second time. He brought a long-term strategy, an ambitious agenda locked in at a very early stage, and a dogged determination to stick to his major objectives to the end. To back his overall "British bulldog" approach, he constructed an innovative preparatory process featuring many regular and creatively blended G8 ministerials, the multistakeholder Commission for Africa, an energetic burst of bilateral summitry, and a creative constellation of business leaders. Its climax came with the unprecedented mobilization of hundreds of thousands of voters and hundreds of millions of viewers through the "Make Poverty History" campaign and the Live Eight concert (Hajnal 2008). Also important was the momentum offered by the unprecedented continuity and good compliance with the commitments these same leaders had made at the summit hosted by Bush the previous year. This summit club significantly included Russia, which Blair had welcomed as a full member of the newly designed, leaders-only G8 he had created and hosted in 1998. Blair could thus look to Russia, as the G8's incoming host in 2006, to implement and build on the innovative strategy that Gleneagles launched.

Preparations

Blair began with a long-term vision of how his summit could launch the principles and the process to create a new, expanded, ultimately effective regime for climate change control. One of his two central objectives was to "at least establish the principle of a new global deal on climate change—to include the US and China—to follow the expiration of the Kyoto Protocol in 2012" (Blair 2010, 548). This involved creating the new "G8+5" format to enable "the largest global players—or most of them—to gather, albeit informally, at the only non-regional global political meeting outside a formal UN or WTO [World Trade Organization] structure". In Blair's view, "the G8+5 was a crucial forum in which debate and discussion between the main emitters could happen reasonably informally", and their agreement was necessary to produce an effective response (Blair 2010, 552). His first challenge was to change the attitude and approach of Bush, who distrusted summits, was sceptical of climate change, was prepared to stand alone, and was disliked by many in the G8.

For his summit, Blair chose a date just after Britain would assume the presidency of the Council of the European Union. Blair could thus keep his G8 summit participants small and speak with the full weight of a now expanded EU-25 behind him. He selected a secure location where he could recreate the informality of Sea Island, Kananaskis, and Commonwealth Heads of Government meetings, with the leaders far enough away from Genoa-like European radicals but in a spot large enough to accommodate his G5 guests, responsible civil society representatives and the world's media. By choosing to hold the first G8 summit in Scotland, he could also claim to be supporting Gordon Brown, his chancellor of the exchequer, political partner, and now rival, who came from there.

Blair declared his priority agenda at an unusually early stage. At the final sherpa meeting of the French presidency in the autumn of 2003, the British announced that Gleneagles, almost two years later, would focus on Africa, climate change, and a third theme to be defined in the spring of 2005 by international events at that time. Blair was convinced of the importance of his generation of politicians making positive advances on climate change, and strongly supported the scientific findings linking global warming to carbon dioxide emissions. The British sought to start a process, by securing agreement on the science, on the need for action by the G8 and by Kyoto-unbound systemically significant climate powers from the developing world, and on the key principles on which a twenty-first century control regime beyond Kyoto would be built. Blair's focus on climate mitigation was timely, given the coming into force of the Kyoto Protocol on February 16, 2005, following its ratification by the Russian Federation in November 2004. Blair was careful, however, to note that his government supported economic growth as a means of providing prosperity, but that this growth would have to be sustainable. He was vigilant in wanting G8 leaders to produce credible outcomes within their means to keep. Other G8 countries shared Blair's approach, notably Paul Martin's Canada, which wanted some acknowledgement at Gleneagles that human activity contributed to climate change. The Canadians hoped that Gleneagles would represent a summit of anticipation and coordination, particularly given the forthcoming climate-related sessions at the UN's Millennium Review Summit in New York in September, and at the WTO Doha ministerial in Hong Kong and the Montreal Conference of the Parties (COP) at the end of the year.

Blair broke new ground by inviting carefully chosen outsiders to join G8 leaders for the discussion on climate change, seeing no point in having climate discussions at Gleneagles without the presence of the world's top emitters. Manmohan Singh of India and Luiz Inácio Lula da Silva of Brazil accepted their invitations very early. China's Hu Jintao agreed to come only during the third week in June. Blair also invited Vicente Fox of Mexico, thereby adding George Bush's fellow rancher and Rio Grande neighbour, and a systemically significant emerging power that, along with Canada, was a net exporter of oil and a major supplier to an ever thirstier United States. Together with Britain's Commonwealth colleague Thabo Mbeki from South Africa, these "plus five" powers would meet with their G8 partners to discuss climate change alone in a one-and-a-half hour session on the third day of the summit. These five outreach partners would thus engage in a meaningful dialogue with their G8 colleagues and together arrive at common conclusions whose principles and processes would provide the foundation of a new regime.

An early draft of the relevant passage of the communiqué, leaked in March 2005, gave hope that Bush's US had accepted Blair's core objectives, as it noted that "the world is warming" and that humankind was partly responsible. Passages indicating that G8 members would pledge money for specific projects showed a sense of urgency and a willingness to undertake immediate action as a result. Yet a subsequently leaked draft suggested that a now engaged US president had forced the summit to relent on all of these points (Brown et al. 2005). Not being a signatory of the Kyoto Protocol, the US did not want any mention of it in the final communiqué. It was willing only to engage in additional scientific research on the issue.

Immediately after the sherpa meeting in mid June, phone calls flew back and forth, as officials rushed to save themselves from having to take to their leaders a heavily bracketed text that still needed to be negotiated. As June ended, the positions moved closer. But Bush's team was still reluctant to accept the basic scientific fact of human-induced global warming, or even acknowledge that "our world is warming." Optimism and pessimism about whether they would adjust arose in an uneasy balance. Those involved in climate change negotiations for many years judged that Gleneagles, with its limited time for leaders to deal with this subject, would accomplish little.

The British sought to make the connection between African development and climate change control. One link flowed from the fact that rising energy prices harmed not only an oil-thirsty and import-dependent America and its G8 partners, but also the poorer countries to a much greater extent. Indeed, the rise of energy prices to close to $60 a barrel cost Africa an additional $10 billion a year. Global warming was drying African and Chinese soils and thus bred desertification, crop failures, famine, and other effects that harmed the poor more than the rich who had the resources to adapt. G8 measures, such as a dedicated tax for development on passenger air travel, would help achieve both goals.

At the Summit

The summit opened with a dinner hosted by Queen Elizabeth. Just before the dinner, Tony and Cherie Blair had a drink alone with George and Laura Bush. After dinner, Blair worked to overcome Bush's still stiff resistance on climate change. The next morning, his first meeting was a bilateral with Bush, to see if the US president would be part of a process with the "express objective of reaching a new post-Kyoto deal" (Blair 2010, 557). Blair then met with Hu Jintao, who was very reluctant to adjust due to China's still low level of development.

This meeting was interrupted by news of the terrorist attacks in London. Despite the immediate fear that they might disrupt the summit, they had a strong galvanizing effect. In the chapeau of the final communiqué, the leaders affirmed that they had come to "Gleneagles to work to combat poverty and save and improve lives" (G8 2005a). Terrorists would not disrupt their work. In addition to intensifying their efforts to counter terrorism, they pledged to continue to do the same on "poverty and climate change."

On the summit's first full day, the G8 leaders, now with British sherpa Michael Jay substituting for Blair as chair, remained resolute in proceeding with their agenda. On day two, Blair returned from London, and the leaders placed energy issues at the top of their priority list. They first concentrated on the clear and present danger of rising energy prices, and underlying supply and demand. But they quickly realized that the solution lay in the direct connection to clean energy, conservation, climate change control, and intra-G8 cooperation with hydrocarbon-laden Russia and Canada. Russia's indication of energy security as the priority theme for its G8 summit the following year added even more credibility to the debate. The resulting Action Plan on Climate Change and Clean Energy affirmed several forward-reaching actions, including improving energy efficiency, diversifying the energy supply mix, promoting the use of cleaner fossil fuels, increasing commitments to research in energy technology, promoting the development of renewable energy, and managing the impact of climate change through better monitoring and data interpretation (G8 2005c).

On climate change itself, the British had pushed hard for a follow-up mechanism, which the Americans had resisted with equal force. But a new "G8 Plus Five Dialogue" was put in place. It had a three-part mandate: to address the strategic challenge of transforming the G8's energy systems to create a more secure and sustainable future, to monitor implementation of the commitments made in the Gleneagles Plan of Action, and to share best practices among participating governments. The leaders tasked their respective governments to engage in this dialogue and welcomed Japan's offer to receive a report at the summit it would host in 2008 (G8 2005b). This move sent a strong signal that the G8 was resuming its global leadership in climate change, now in an inclusive partnership with the world's emerging carbon powers and leading polluters, and starting with the classic component of energy at the core.

Results

At the end of their three-day summit, G8 leaders issued their final communiqué, making climate change the top priority, placing it immediately after their opening paragraph on the London terrorist attacks. All leaders agreed that "climate change is happening now, that human activity is contributing to it, and that it could affect every part of the globe." As a result, they "resolved to take urgent action to meet these challenges" (G8 2005a).

They referred to the Gleneagles Action Plan on Climate Change and Clean Energy as a key element in demonstrating their collective commitment to taking concrete measures to develop clean energy technologies and help vulnerable communities adapt to the impact of a changing climate. They welcomed the leaders of Brazil, China, India, Mexico, and South Africa, noting the importance of continued international cooperation on climate-related issues between the developed and developing world.

The G8's concentration on climate and energy issues marked the beginning of a new dialogue between the G8 and other countries with "significant energy needs" (G8 2005a). This new dialogue, reinforcing the G8's collective commitment in exploring how best to exchange technologies while reducing emissions, became a key element in bridging the North-South climate divide.

And on the issue of Kyoto, the G8 affirmed their commitment to continue to tackle climate change at the COP meeting in Montreal later that year, with those having ratified the protocol, continuing "to work to make it a success" (G8 2005a).

Dimensions of Performance

Domestic Political Management
In its domestic political management, Gleneagles issued no communiqué compliments on climate change (see Appendix B-1). However, in the domestic media, its performance was exceptionally strong in the approval it created for its British host (Kirton 2005b).

Deliberation
Gleneagles was a strong success in terms of deliberation. While Tony Blair was determined to concentrate his colleagues' attention on basic principles, rather than the fine print of detailed communiqués, in the end, Gleneagles produced 14 documents, among the highest to date. These included documents on Africa and climate change. The evolving British desire for comprehensiveness and detail had competed with an American preference, supported by the Russians, for a short general statement followed, if necessary, by a more detailed action plan.

Direction Setting
In its direction setting, Gleneagles was a strong success (see Appendix A-1). Its priority placement and democratic affirmations of climate change soared to new highs. In its substance, all G8 leaders, including Bush, affirmed Blair's desired principles that climate change was a scientific fact, caused by human activity and requiring urgent action from the G8 members as well as major developing countries. On causation, leaders acknowledged "that increased need and use of energy from fossil fuels, and other human activities, contribute in large part to increases in greenhouse gases associated with the warming of our Earth's surface" (G8 2005b).

Decision Making
Gleneagles was a strong success in terms of decision making. Driven by an American desire for measurable results, and G8 members' wish for clear, politically locked-in promises, Gleneagles produced a robust 29 specific, future-oriented, decisional commitments on climate change among its 212 overall, thus accounting for more than 10 per cent of the commitments reached in the final communiqué (see Appendix A-1).

Gleneagles also set a new summit record for money mobilized, as the estimated $205 billion attributable to the summit far surpassed the previous high of $50 billion raised at Kananaskis in 2002 (Kirton, Brady, et al. 2005). Although the most of that money came in the form of debt relief for the poorest countries and commitments to doubling official development assistance (ODA) by 2010, $3.5 million was earmarked for climate change initiatives in Africa, forest biodiversity, and legal timber procurement.

Delivery

In delivering these commitments, the performance at Gleneagles was very strong (G8 Research Group 2006a). Compliance with the core commitments assessed over the year following the summit averaged +0.80 (see Appendix E-1). Complete compliance came from the UK, Germany, France, and the EU, which the UK chaired at the time of the summit. A once reluctant US complied at a rate of +0.80.

The six core climate issues, including renewable energy, received complete compliance scores. Russia moved from negative overall compliance at the end of 2005 to compliance in the positive range, after assuming the responsibility of host in January 2006.

There had been a clear and credible desire to comply with these commitments from the start. They contained a large number and variety of compliance catalysts—setting specific targets and timetables, specifying agents responsible for implementing the commitments, generating remit mandates for reports to subsequent summits, and directing other international institutions to act.

Development of Global Governance

Gleneagles also performed very strongly in developing global governance. Inside the G8, it created five new G8-centred institutions to help put its new directions and decisions into effect. These were the Dialogue on Climate Change, Clean Energy, and Sustainable Development, the Global Bioenergy Partnership, the Working Group on Innovative Financing Mechanisms, the Experts Group on IPR [Intellectual Property Rights] Piracy and Counterfeiting, and the African Dialogue Follow-up Mechanism (Kirton 2005c).

Further to Blair's commitment to move his Gleneagles climate initiatives forward, the G8 energy and environment ministers, plus ministers from 12 other countries, met in London on November 1, 2005. They focused on short-term priorities in transitioning to a low carbon economy, the importance in developing new approaches to technology cooperation, and scaling up investments in clean energy technologies. Blair spoke at the conclusion of the conference, noting that the "evidence of climate change was getting stronger and even those who doubted it accepted there were concerns over energy security and supply" (Downing Street 2005). He acknowledged the importance of the Kyoto Protocol and added that the world needed to combine it with the need for growth with "a proper and responsible attitude" toward the environment. This follow-up a month before the December COP in Montreal further showed that the G8 was resuming global leadership, after having long ceded that role to the UN.

Causes of Performance

Fuelling these historic achievements were the forces highlighted by the concert equality model of G8 governance with the addition of a more inclusive G8 becoming a global climate governance network hub.

Shock-Activated Vulnerability

The first cause of the G8's restoration of global leadership on climate change, now in a more inclusive form, was the set of small shocks that reminded G8 leaders of their past and present compounding, interconnected vulnerabilities. From the outside, the leaders were reminded of past shocks and present vulnerabilities by rising and volatile energy prices, driven by increasing demand from the fast growing systemically significant powers, and political unrest in a terrorist-infected Middle East and unstable Africa. These vulnerabilities were brought home by the deadly terrorist blasts in London on the summit's first full day. The steady flow of coffins of American military service people killed by increasingly imported terrorist insurgents in Iraq and Afghanistan reached politically problematic levels for the US public and Congress, still sensitized by Vietnam. Similarly afflicted was the host UK.

A classic energy mini shock forced leaders to recall the dangers unleashed in 1973 and 1979 (see Appendix G-3). In the lead-up to Gleneagles, from January 1 to June 24, oil prices on the New York Mercantile Exchange rose 60 percent, compared to only 21 percent for the comparable period in the lead-up to Sea Island the year before. In the week from June 17 to June 24, just before the final sherpa meeting on July 1–2, oil prices reached new highs on four of the six trading days, ending at a new record of $68.09. Futures markets revealed that prices were expected to rise further in the coming months, and to stay at elevated levels in the years ahead. These increases took an immediate toll on growth prospects for the G8 and the global economy. Stock markets in the US, Europe, and Japan started to drop, as the steady rise of oil prices damaged prospects for corporate profits, growth in gross domestic product, and stable prices. Suddenly, memories were triggered of the stock market crash of October 1987.

On the supply side, the oil price surge stemmed from several factors. Unusually extreme weather, plausibly associated with climate change, closed offshore rigs in the Gulf of Mexico. The recent Asian tsunami that hit on December 26, 2004, still reminded many of the deadly costs of unpredictable extreme weather and natural forces and was a dramatic referent for the suffering in Africa. It was a spur to a Gleneagles-generated earth observation system that could measure, warn, and thus help defend against both naturally caused tsunamis and earthquakes and human-fuelled climate change.

Social unrest erupting in Norway, Nigeria, and elsewhere, resulting partly from insecure oil tanker routes and liquefied natural gas terminals to bring supplies in from abroad, added to the sense of threat. The nuclear component added to this tightly connected complex of reawakened energy-financial-terrorist-social shocks, as the US had not constructed a single new nuclear power plant since its Three Mile Island explosion in 1979. A new element arose on the demand side, as this first embryonic energy crisis of the twenty-first century was driven by the take-off in growth in the often newly democratizing, dynamic powers of the Gleneagles guests—Mexico, Brazil, India, South Africa, and China.

A succession of terrorist shocks reinforced the sense of urgency. On June 12, a bomb on a train in Russia, presumably set off by Chechen terrorists, injured several and reminded Russia of the ongoing threat from terrorists at home with Islamic affiliates in Afghanistan, Iraq, and the broader Middle East. A Chechen blast killed 10 Russian servicemen at the end of June, with a similar effect. New terrorist mini shocks arose in Nigeria, where al Qaeda-affiliated Islamic fundamentalists threatened G8 diplomatic posts.

And then, during the Gleneagles Summit itself, came the deadly al Qaeda-affiliated terrorist attack on the tube and buses in London, the capital city of the summit host, causing extensive death and instantly creating a strong sense of solidarity among all G8 and G5 leaders assembled at the summit. What the September 11 assaults had done for the

G20 finance ministers at their October 2001 meeting in Ottawa, the 7/7 attacks did more immediately for the new "G13" leaders at Gleneagles (see Kirton 2013a). It gave George Bush a particular incentive to cooperate with his G8 partners in order to find an adequate way out of what was quickly becoming his broader Middle East quagmire.

Moreover, beyond the sudden shocks, a critical threshold had been passed for the core, chronic, compounding climate vulnerability of greenhouse gas emissions. In 2005, for the first time, China surpassed the US as the leading emitter of greenhouse gas in the world (see Appendices G-7 and G-8). Brazil had surpassed Russia the year before. India had surpassed Japan in 1995. It was clear that climate change could not be controlled unless the Rio-Kyoto regime was replaced with one where all major carbon polluters were obliged to control their carbon.

Multilateral Organizational Failure
The 1944–1945 generation of multilateral organizations centred on the UN were clearly failing to generate an inclusive and hence effective regime. They were also failing to deliver a Kyoto Protocol that would meet its targets, and to induce the US to join. Nor did it look likely that they would deliver the MDGs, which were weak on environment anyway.

Predominant Equalizing Capability
The G8's global predominance and internal equality in terms of capability were increased by the growing power of Russia and Canada, both committed to Kyoto and full-strength energy superpowers, with fiscal surpluses available to finance development abroad (see Appendix I). But that capability was diminished by the rapidly growing overall and ecological power of China, India, Brazil, and Mexico on the outside.

Common Principles
The G8's seminal mission and shared principles of open democracy, individual liberty, and social advance drove a G8 African strategy now directed toward democratic development. They also propelled a climate change strategy that included the leading developing country democracies of India, Mexico, Brazil, and South Africa.

Political Cohesion
Political cohesion at home also helped. High domestic political capital and control came from host Tony Blair, just re-elected with a historic third majority government, from a secure Vladimir Putin and Junichiro Koizumi, and from a recently electorally refreshed George Bush, whose party controlled both chambers of Congress (Kirton 2005d). There was also high continuity, with Blair hosting his second G8 summit and having attended all since his first at Birmingham in 1998.

Also driving success was the confidence that came from the UK's historic achievement in hosting G7/8 summits, and in using the many other international institutions in which it had a leading place. According to Robert Putnam and Nicholas Bayne (1984, 1987), it tends to host above-average summits. Gleneagles was no exception, with a score of A–. The UK hosted more successful summits than the US, although not ones with the assured high success of Japan.

Where the UK most stood apart from its peers was in its consistently high compliance with the commitments made at the G7/8 summits over the earlier 30 years. It almost always ranked first as a faithful complier. This was due to its strong sense of good international citizenship, its skill in shaping commitments that conformed to British desires, and the ease of compliance in a parliamentary system that rarely experienced minority or coalition governments or deeply entrenched governing party factions. Britain thus went to Gleneagles with clean hands and much moral authority among its colleagues, as the most genuinely committed partner, in practice as well as on paper, of the club at the hub. Where it stood out most as host was in its pioneering steps in summit outreach, leading to permanent membership in an expanded G7 club. In 1977, it first welcomed the EU to the G7. In 1991, it held the first G7 dialogue with the reforming Soviet leader Mikhail Gorbachev. And in 1998, Tony Blair first welcomed Russia's Boris Yeltsin as a full, permanent member of a new G8 club.

Yet success on the chronic problems of African development and climate change was ultimately driven by an exceptional performance from inside the G8 itself. A highly skilled and experienced Blair brought an ambitious agenda locked in at a very early stage and a dogged determination to stick to his major objectives to the end. To back his "British bulldog" approach, he constructed an innovative preparatory process featuring many regular and creatively blended G8 ministerial meetings, the Commission for Africa, an energetic burst of bilateral summitry, and a creative constellation of business leaders. The climax came with the unprecedented mobilization of hundreds of thousands of voters and millions of viewers through the Make Poverty History campaign and the Live Eight concert. Also important was the momentum offered by the unprecedented continuity and good compliance with the commitments these same leaders made at Bush's Sea Island Summit the previous year. This "summer-camp club" significantly included Russia in the newly designed, leaders-only G8 Blair had created and hosted in 1998. He could thus look to Russia, as the incoming host, to implement and build on the bold new directions that Gleneagles launched.

Constricted Club Participation
This club-like cohesion allowed Blair to expand the participation of leaders at Gleneagles in a controlled way, to produce a still-constricted club. Each of the G5 leaders had attended a G8 summit before and most were members of the Commonwealth or the Asia Pacific Economic Cooperation (APEC) or Asia–Europe Meeting (ASEM) clubs.

After the Summit

The 11th COP and the first ever Meeting of the Parties (MOP) to the Kyoto Protocol was held in Montreal from November 28 to December 10, 2005. Chaired by Canadian environment minister Stéphane Dion, they pushed for the adoption of the protocol's operational and technical provisions, including the use of flexible mechanisms and options for future cooperation. Kyoto had entered into force in February 2005, making it a legal instrument, but the Marrakesh Accord stipulating the technical and operational details had not been adopted. There was thus much concern at Montreal that the protocol and its mechanisms would unravel and fail to curb emissions. Dion was successful in having the accord adopted without any major amendments—a move that would "breathe life" into the protocol and provide the basis for its implementation (Earth Negotiations Bulletin [ENB] 2005, 10).

Ongoing disputes occurred, however, on adaptation methods for developing countries. While the developing world called for immediate action to counter climate-related problems, the developed countries proceeded with caution, wary of the funding costs associated with moving too quickly. There was some dialogue between both sides but, in the end, there was no agreement on funding for such initiatives, stalling implementation further

(ENB 2005). Nor was there any agreement on improving the quality and quantity of Clean Development Mechanism (CDM) projects, or on pushing toward future commitments on innovation and technology transfer. Coupled with Canadian prime minister Paul Martin and former US president Bill Clinton's attack on the Bush administration for its lack of "a global conscience," the Montreal COP/MOP left several global climate mitigation challenges hanging in the balance (Smith 2008b; Bueckert 2005). The divided UN regime remained deadlocked.

St. Petersburg, 2006

The opulence and grandeur of the Constantine Palace in historic St. Petersburg was an appropriate venue for Vladimir Putin's first ever Russian-hosted G8 summit, taking place on July 15–17, 2006. The new G8 leadership on climate change launched the year before at Gleneagles provided an appropriate foundation for Russia's focus on global energy security. Although not surprising for a country interested in developing its energy sector, its decision to focus its summit on energy security tied in well with the G8's broader mandate of addressing environmental challenges, diversifying energy sources through renewables, encouraging investment in the energy sector, and expanding scientific and technological cooperation. Indeed, at the end of the Gleneagles Summit in 2005, Putin publicly signalled his choice of energy security as the priority theme for St. Petersburg, thus taking advantage of the concentration on, and success in, climate change control at Gleneagles and the components of energy efficiency, renewable energy and low-carbon, climate-friendly nuclear power (Kirton 2006c).

Preparations

Russia's skilled summit officials, many of whom had lived and worked in the West, had an intuitive and rich understanding of open markets lacking among many of their colleagues in the domestic departments of the Russian government. The unusually vibrant process of G8 expert and ministerial meetings to prepare for the summit—the latter featuring gatherings for energy as well as health and education—brought knowledge of established democracies' operations into the domestic Russian departments.

Russia's initial G8 proposals on energy security contained little recognition of the contribution that free and thus more efficient market mechanisms could make. However, by March 16, the statement of the G8 energy ministers (2006), drafted by the Russian host, proclaimed that "meeting energy security challenges will require reliance on market-oriented approaches." It also noted that to attract investment "it is essential for countries to have open and favourable investment regimes including stable and predictable regulations, clear tax laws, and efficient administrative procedures." Moreover, it recognized that the development of low carbon technologies was "crucial" and that the development of innovative energy technology solutions "will have longer term environmental, economic and energy security benefits."

Putin addressed the G8 energy ministers at the end of their meeting. He noted the socioeconomic and political risks associated with energy security, but emphasized that "insufficient energy resources are a serious obstacle to economic growth and their non-rational use can lead to environmental disasters" (Putin 2006). He referred to the important role the G8 could play in energy conservation, by searching for "new breakthrough technology," supported by "environmentally friendly energy resources." Putin said he would propose concrete initiatives on these issues at St. Petersburg and was prepared to "take full part in their practical implementation."

As such comments indicated, Russia's approach to energy security included a role for environmental protection and conservation. As early as February 2006, the newly institutionalized Civil 8 process, the Russian presidency's innovative initiative to engage global and Russian civil society and nongovernmental organizations (NGOs) in the summit process, had injected a strong ecological emphasis into the energy agenda. At the culminating Civil 8 plenary gathering in Moscow on July 3–4, Putin met—for the first time in G8 history—with more than 650 NGO leaders from around the world. In a fully public session, transmitted to the global community through the international media covering the event, he heard and responded frankly to the consensus advice of civil society on issues all across the summit's agenda, and beyond it as well, including nuclear energy. He made 17 public pledges to civil society that he would take new, specific actions at and outside the St. Petersburg Summit, in response to their advice.

Reinforced by such pressure, and the G8-generated global attention to Russia's approach to energy, Putin soon made a historic, if expensive, decision to re-route a new pipeline to protect the ecological integrity of Lake Baikal, the largest body of freshwater in the world. On a visit to the Russian embassy in Ottawa in April 2006, Russian sherpa Igor Shuvalov was quoted as saying that the G8 leaders "must examine the issue of energy security in the context of global climate change" (Campbell 2006).

By April, the G8's energy security action plan was coming together. It incorporated seven essential elements: increased transparency, stability, and predictability in energy markets; increased investment all along the energy chain; ways to enhance energy efficiency; issues related to energy poverty; diversification of energy sources; measures to mitigate climate change; and the physical safety of key elements of the world's energy system. The Russians noted the importance of the G8's role in linking climate change with emissions through new technologies (including renewables), as well as safe and reliable nuclear power.

Blair supported Russia's position. At a meeting with business leaders in June, a month before the St. Petersburg Summit, he said he would maintain the momentum from Gleneagles by raising climate change with his counterparts at the summit (BBC News 2006). Two days later he appointed a "special representative on the Gleneagles Dialogue" to lead international negotiations (Black 2006).

As St. Petersburg drew near, Russia signalled its broader support for a market-oriented approach in two important ways. On July 1, it made the ruble convertible, allowing money to be transferred in and out of Russia. On July 15, it allowed some shares of Rosneft, its state-owned oil firm, to be sold to Russian citizens and foreigners through trading freely on the London Stock Exchange and thus be subject to international regulatory, accounting, and corporate governance rules. It similarly planned to privatize, with majority foreign ownership, some of its electricity-generating firms.

At the Summit

At the St. Petersburg Summit itself, the negotiations among the sherpas and leaders on climate change saw little advance. US resistance to setting a shared, obligatory goal for increasing energy efficiency prevailed. So did the fall-back push to set country-specific targets. The result was an agreement to "consider" setting targets and reporting by year end

(*Asahi* 2006). Japan's Koizumi indicated he would take a strong lead on energy efficiency at the summit that Japan would host in 2008.

Results

As anticipated, G8 leaders delivered their ambitious and innovative priority agenda on energy. The summit's document on energy security emphasized transparency, efficiency and conservation, renewables, and climate change control. Leaders agreed that the need to protect the environment and tackle climate change was a serious challenge linked with several others, including high and volatile oil prices. They affirmed the importance of "ensuring sufficient, reliable and environmentally responsible supplies of energy at prices reflecting market fundamentals" (G8 2006).

The St. Petersburg Plan of Action on Global Energy Security reinforced the G8's commitment to implement and build on previous summit promises related to energy security (G8 2006). These promises included improving investment in the energy sector, enhancing energy efficiency technologies, diversifying the energy mix, protecting critical energy infrastructure, reducing energy poverty, and increasing attention to climate change and sustainable development.

On climate change, the leaders noted the importance of meeting their shared objective of reducing greenhouse gas emissions by delivering on their climate change commitments from Gleneagles. They confirmed their commitment to making the Kyoto Protocol a success through several approaches to deal with the "interrelated challenges of energy security, air pollution control, and reducing greenhouse gas associated with long-term global climate change" (G8 2006). Kyoto's binding emission reduction targets, however, and the exemption for developing economies (including China and India) remained the main points of contention for the Bush administration. Indeed, the action plan failed fully to bridge the divide between the US and its G8 partners. The text read "those of us committed to making the Kyoto Protocol a success underline the importance we attach to it." Still, leaders pledged to produce a report detailing their respective successes in implementing climate mitigation strategies by the time of Japan's G8 summit in 2008.

As the summit drew to a close, several G8 leaders, including Putin, Blair, and France's Jacques Chirac, mentioned climate change in their final press briefings. Putin drew on the link between energy security and environmental protection by noting that "environmental protection and reducing the effects of climate change are also important aspects of global energy security. In this respect I note the G8's readiness to fulfil all previous environmental commitments and to work out new, effective steps to protect the environment" (Russia's G8 Presidency 2006).

Chirac spoke extensively about climate mitigation, both privately with other G8 leaders and publicly with the press. He drew particular attention to the need to secure emissions reductions commitments for the post-2012 period. He said that "we cannot talk about energy security while there is no progress on climate change. Mankind is dancing on the edge of a volcano" (Louet 2006). In a clear rebuke to his colleagues, he also said that "it is our duty to act, before our citizens, before mankind, before future generations who risk paying a grave price due to the passivity of certain people." Blair distributed several documents to the press highlighting his government's work to implement its Gleneagles climate change commitments. He suggested to his G8 counterparts that the "Plus Five" members of China, India, Brazil, Mexico, and South Africa be made full members of the G8, in the hopes of forging a new round of emissions reduction targets beyond 2012 that would include developing countries and the United States (Elliott and Wintour 2006).

Dimensions of Performance

The St. Petersburg Summit made substantial advances on energy with climate change as a key component. Energy took centre stage as the first of the three priority subjects, and as the area where the G7/8 had the longest experience and greatest success (Kirton 2006b). Indeed, the most successful summits had come as a result of global energy governance, first as part of the large package deal for macroeconomic management forged at Bonn in 1978, then in response to the Iranian-generated second oil shock at Tokyo in 1979, and most recently in the agreement on climate change at Gleneagles in 2005.

Overall, St. Petersburg set new highs on some key dimensions of performance, with climate change and related issues playing a prominent place. It received grudging recognition from the leading media in the G8 members for its good work. Its documented deliberations approached the historic high set at Sea Island in 2004. Energy pervaded much of the discussions and statements. The summit also set several new directions, affirming a large number and broad array of democratic values as the foundation for its work in energy security, as well as in education, health, Africa, counterfeiting, corruption, counter-terrorism, nonproliferation, and the Middle East. It generated 114 commitments on energy security and its related links—representing more than one third of record high of 317 commitments reached. While the $4.4 billion in new money mobilized was far less than the $205 billion raised at Gleneagles, it was still above the sum raised at Evian in 2003 (G8 Research Group 2006b). St. Petersburg also created three new G8 bodies to help carry out its work, including those on energy, piracy and counterfeiting, and stabilization and reconstruction (for development, security, and diplomacy).

Domestic Political Management
In domestic political management, St. Petersburg's performance on climate change was poor. No communiqué compliments were produced (see Appendix B-1). But Putin's responsiveness to environmental issues in a high-profile G8 context at home represented a substantial advance.

Deliberation
St. Petersburg's performance on deliberation was strong. Its 1,533 words on climate change, taking 3.1 percent of the lengthy communiqués, were the second highest in G8 history thus far in their volume, if a sharp drop from Gleneagles the year before (see Appendix A-1).

Direction Setting
St. Petersburg also performed strongly on direction setting on the subject of climate change specifically. The priority placement of climate change was strong, with robust affirmations in the two references in the chair's summary. However, there were no references to democracy or human rights in the climate passages of the communiqué (see Appendix A-1).

Decision Making
In its decision making, St. Petersburg's performance was strong. Leaders made 20 commitments on climate change—the second highest number to date, but a drop from the 29 reached at Gleneagles the year before (see Appendix D-1).

Delivery
In the delivery of these decisions, performance at St. Petersburg was solid. Average compliance for the nine commitments assessed was +0.35 or 67 percent (see Appendix A-1).

Development of Global Governance
St. Petersburg's performance on the development of global governance was substantial. Leaders created the new G8 expert group on securing energy infrastructure (see Appendix F-2). The leaders guided outside international organizations 10 times on climate change, led by six references to the UN Framework Convention on Climate Change (UNFCCC) (see Appendix F-5).

Causes of Performance

The St. Petersburg leaders were pulled into a productive performance by profound, proliferating pressures from a precarious world. They faced familiar shocks that revealed their vulnerability in the fields of energy, terrorism, and health—largely the themes that Russia focused its summit on. By the time of St. Petersburg, all G8 leaders knew that they could not rely on the multilateral organizations of the UN system for solutions. Despite some progress on security and development at the World Summit in New York in September 2005, there was still no UN-based world energy organization and the Kyoto Protocol was still not ratified by the United States. Given shared vulnerability to the many clear and present global dangers, and without effective multilateral organizations, G8 leaders recognized they could count only on themselves. In response, the G8 brought equalizing capability among its members, with the least capable overall—host Russia and its Arctic neighbour Canada—standing as the only full-strength energy superpowers in the G8 club and world at large. The leaders also knew they had to mobilize their common commitment to open democracy, individual liberty, and social advance to produce a credible collective response to core global challenges (Kirton 2006c).

Shock-Activated Vulnerability
The first shock-activated vulnerability came again from the classic cause of high performance—energy. It arose from unprecedented high, sustained oil price spike that had its largest and most pervasive impact on oil-importing America, Japan, and Germany. On January 1, 2006, the St. Petersburg Summit year started at $74.88 per barrel, up substantially from the start of earlier years (see Appendix G-3). It soon soared to new heights in April, remained high until the end of June and then jumped. On July 6, a week before the G8 leaders would gather, it spiked to a new historic high of $82.88 a barrel.

With the G8, especially its American and continental European members, already sensitized to energy, the new ecological shock of hurricanes Rita and Katrina struck America in September 2005. Katrina was the costliest and one of the five deadliest hurricanes in the history of the US. Then came the unusually cold weather across Russia and Europe, which strained energy supplies in the following months. These shocks were all the more potent in driving summit performance because they connected directly with St. Petersburg's first and defining priority of international energy security. Katrina in particular was a dramatic demonstration of what a climate-induced extreme weather event could do.

Multilateral Organizational Failure
The UN system again failed to respond to these new shocks, with no dedicated organizations for energy or natural disasters. No summits were called to respond. As an organization, the UNFCCC had only 12 senior staff members and an annual budget of $26 million a year (Haas 2008a, 3).

Predominant Equalizing Capability
Equalizing capability also contributed to the substantial success at St. Petersburg, particularly with regard to the priority theme of energy security. Within the G8, the otherwise least powerful partners of Russia and Canada were the G8's and the world's leading full-strength energy superpowers in the core combination of oil, gas, and uranium surpluses and reserves. Germany led in wind and solar. The G5 partner of Brazil stood first in ethanol (with a converted and politically committed US in second place). Mexico along with Canada brought the oil and gas power that America needed from its safe, secure, reliable partners in the North American Free Trade Agreement.

This configuration of overall and specialized capability did much to explain Russia's insistence on security of energy demand, and suggests why its rational G8 partners ultimately agreed. Moreover, Russia's prosperity, important for the G8's combined power and internal equality, was fuelled by its energy exports with a high oil price. Many Russians—and others—feared that a plunge in world energy prices, similar to the collapse in the early 1980s that had bankrupted oil-rich Mexico and many other emerging economies, would swiftly wipe out all the progress that Russia had made since its default in 1998. This time, high and volatile energy demand and prices would be damaging for the G8+5 partners, which America and its allies most needed, and for the war against terrorism in the Middle East and around the world.

Democratic Convergence
Moreover, the spectre of how a deep recession in Russia could destroy its democratic development remained a reality for G8 leaders devoted to this defining principle in their G8 club. Unlike in the Cold War in the 1980s, ensuring energy security, with all its interconnected components, was now a collective G8 and global good (Kirton 2006c).

Political Control, Capital, Continuity, Conviction, and Civil Society Support
Success on the energy security file was not, however, produced by the farsighted, strategically calculating, internationally experienced, politically secure G8 leaders assembling at St. Petersburg. With the important exception of host Putin and past host Blair, the summit leaders were, for the first time in the twentieth century, either unusually new to G8 summitry or politically insecure with a possibly short shelf life at home.

Further contributing to success on the energy security file was the host government's increasing performance on energy issues over time. Russia had, in large measure, contributed to the G8's global governance, especially in the field of energy, since the very start. The Soviet Union was a core member of the G7-centred London Nuclear Suppliers Group formed in 1975 to control the proliferation of nuclear materials after India's explosion of a nuclear device for allegedly peaceful purposes in May 1974. Mikhail Gorbachev's historic move in 1989 to become part of the western world helped the G7 cope well with the third oil shock created by Saddam Hussein's invasion of Kuwait in August 1990 and the liberation of Kuwait by the invading American-led coalition in January 1991. Boris Yeltsin's newly democratic Russia successfully hosted the G8 Nuclear Safety Summit in Moscow in 1996, which helped Russia's democrats produce and win the country's first popular election, and respond on the 10th anniversary to the 1986 Chernobyl nuclear explosion. As host of the first meeting of G8 energy ministers in 1998, Russia pioneered the participation of private sector leaders, through the contribution made at that time by the Business Consultative Council. Russia subsequently participated actively in G8 energy ministerial meetings in 2002 and in environment ministerial meetings in November 2005. Beginning with the 1992 Nuclear Safety Working Group, Russia had contributed to the more than a dozen energy-related official-level working groups established by the G8 (Kirton 2006c).

In the closely related area of climate security, Russia also made a critical contribution. At the G8's 2001 Genoa Summit, it agreed to join Japan and Canada in ratifying the 1997 Kyoto Protocol, which ensured that the protocol would come into formal legal force as the foundation for the world's twenty-first-century climate change control regime.

The Gleneagles inheritance had an impact on St. Petersburg in other ways. One was the important factor of continuity and iteration in the G8's agenda and action from year to year (Bayne 1999). Although energy supply and demand dominated the St. Petersburg agenda, the theme of climate change, featured at Gleneagles, tied in partially with the renewable energy and nuclear energy components of Russia's central energy security theme. Moreover, the concern with African development at Gleneagles flowed into St. Petersburg's concern with energy poverty.

Perhaps the largest and immediate legacy of St. Petersburg would come from deepening democracy in Russia itself. This impact unfolded both in the lead-up to and during the summit. Through the Civil 8 process, several hundred Russian NGOs met their global counterparts face to face for the first time. Through their shared cause of trying to shape the summit's agenda, emphasis and outcomes, they discovered the realities of one another's countries, shared best practices and techniques, and formed mutually supportive relationships and laid the foundations for international coalitions and campaigns. This rapidly emerging civil society movement in Russia received important international reinforcement at a critical stage. It helped promote and advance those issues of core importance to the Civil 8, including sustainability and environmental protectionism.

In particular, at the culminating Civil 8 plenary, Putin promised that should his recently approved NGO bill regulating the activities of civil society organizations in Russia prove more cumbersome than its predecessors, he would remove its restrictive effect.

The G8 summit's news conferences and sessions induced Putin to declare in more detail and with greater directness his support of democratic principles. At his first news conference at the end of the opening dinner on July 15, he indicated his intention to step down as president of Russia, in accordance with the constitution, when his second term came to an end in 2008. Until then, he declared, he was devoted to making democracy and the market economy an irreversible reality in Russian life.

Controlled Club Participation
Controlled club participation also contributed. Despite initial Russian resistance, Blair's insistence on including outside leaders at St. Petersburg helped ensure that the same G5 partners were invited to the summit for the second year in a row.

After the Summit

Nairobi, home of the headquarters of the United Nations Environment Programme, was the site of the COP-12 meeting on November 6–17, 2006. With the Kyoto Protocol already in force for about one year by the time the leaders met at the St. Petersburg Summit, discussions in Nairobi were to continue with future commitments. But there was no compromise as the debate continued over including industrialized developing countries in a future framework. Any discussion of voluntary commitments for developing countries was quickly shut down, resulting in yet another failed round of UN-based climate mitigation discussions (ENB 2006).

These COP/MOP meetings were the first such gathering in sub-Saharan Africa. At the opening session, outgoing UN secretary general Kofi Annan lamented the "frightening" lack of leadership from governments on climate change (UN 2006). He announced the "Nairobi Framework" to help spread the benefits of Kyoto's CDM among the developing countries.

Business issues and those related to the economy took on a more prominent position. The UK government presented a comprehensive economic review that showed that the projected impacts of a changing climate could cost the global economy far more than the steps required to avert them. Despite this evidence, the United States and developing countries still strongly opposed any discussion of binding targets. COP-12 thus made very little measurable progress toward international action beyond 2012, when the current Kyoto commitments were set to expire (Center for Climate and Energy Solutions 2006).

The most contentious issues in Nairobi were the terms of the new Kyoto Protocol review proposed by Russia, in accordance with the all-in principle, to "establish a pathway for developing countries to take on 'voluntary' emission targets" and set their own targets (Center for Climate and Energy Solutions 2006). As these issues concerned only the parties to the protocol, the US opted out and "maintained a relatively low-key posture throughout the conference."

Over the two-week period, only modest steps were taken on climate adaptation. Members debated approaches to reducing deforestation and accelerating technology transfers. Despite the overall lack of substantive progress, much of the debate was marked by a growing sense of urgency about the threats posed by climate change. In private exchanges, several developed country negotiators discussed a "strong new negotiating mandate at next year's conference," tentatively slated for Bali, Indonesia (Center for Climate and Energy Solutions 2006).

Heiligendamm, 2007

The 2007 G8 summit took place from June 6 to 8, when G8 leaders and their now regular G5 partners met in Heiligendamm, Germany, for the G8's 33rd annual summit. Its central challenge was climate change. Leaders sought to get the world's leading carbon-producing powers of the US and China to commit to constrain their greenhouse gas emissions beyond the Kyoto Protocol's pledges before the 2012 deadline. The G8 thus crafted the Heiligendamm Process, which would give the G5 powers more reliable, institutionalized involvement in the G8's climate-related governance than ever before.

As the leaders landed in Germany, there were predictions from many quarters that they would fail. On the defining issue of climate change, German host Angela Merkel herself said it was an open question whether a meaningful consensus could be forged. But the conditions that caused high G8 climate performance in the past—notably high energy vulnerability, institutionalization, iteration, and a capable, committed host—were in place. The Heiligendamm G8 thus went a long way in meeting its critical challenges of getting its Kyoto-unconstrained participants to commit to carbon-controlling actions, and to control their carbon even without firm, binding legally entrenched commitments. To help advance these objectives, G8 leaders doubled their ODA to support clean energy and climate control and reformed the international financial institutions they controlled for this purpose. Two key factors were increasing reliance on Russia and building the G8 system in more institutionalized, inclusive, leaders-driven ways. These features enhanced the G8's recently restored leadership in global climate governance.

Preparations

The German host initially sought to take the summit back to its assumed economic origins, by focusing on the twin themes of growth and responsibility. Discussions on the global economy would include the social dimensions of responsible resource use, alternative energy, emissions reductions, energy efficiency, and climate change. Climate mitigation rather quickly became a central issue. As early as September 2006, Merkel named climate change as her G8 presidency's top priority for the 2007 summit (Kirschbaum 2006). The chancellor stated that she would use Germany's EU presidency to push for reduced energy use and more energy efficiency. Germany was keen on alternative energy sources, leading the world in wind, solar, and biomass. Merkel herself had pledged to augment the energy mix from the current 5 percent from alternative sources to 20 percent by 2020. On Kyoto, she said that "we urgently need agreements for the period after 2012 when the Kyoto Protocol expires. Germany will do all it can within its realm as president of both the G8 and the EU ... We have a great chance next year to have an international impact" (DW Staff 2006).

To further her position on climate change, by early December, Merkel appointed two special advisors—from science Hans Joachim Schnellnhuber, head of the Potsdam Institute for Climate Research, and from industry Lars Goeran Josefsson of the energy company Vattenfall—to develop Germany's climate policy. Stating that "climate change is one of the biggest challenges facing mankind," she emphasized the need for "know-how" on the one hand, and "support from the private sector" on the other (Associated Press 2006).

Although strongly supported by the German hosts, movement on climate change through the early stages of the preparatory process was rather slow. By their second meeting in late February 2007, the sherpas showed little progress toward collective action on climate change. Some sought to go beyond the Gleneagles texts; Blair himself and the British mounted a strong push in this regard. But all recognized that some members were in a state of flux about their own policies. The EU was about to unveil a new policy, the US took a redline defensive position, and the non-Europeans knew that the one way to break EU unity was to suggest stronger text on nuclear energy.

Canada saw the G8 climate discussions as closely tied to its own evolving position. Along with the US, it believed that emerging economies needed to be engaged in making real commitments and that the discussions would benefit from their involvement. Canada recognized that the UNFCCC, with its mandate for conducting the post-2012 negotiations, would likely remain where it last concluded. Canada thus saw a need for more intimate forums for forging agreement on key issues before they were sent to the UNFCCC. A G8-sponsored process could thus prove useful. If the G8 and G5 agreed on critical parameters, a more broadly based body could follow up. This process would not be concluded at Heiligendamm, but could be considered at the Japanese-hosted G8 the following year.

Technology sharing was the other part of the deal for bringing the emerging economies into the climate control club. The G8 needed to frame the issue in terms of making technology available to emerging economies, in an economic framework or regime. For Mexico, the challenge was to enact structural reforms that stimulated growth. Mexican president Felipe Calderón was well positioned for this, particularly having restructured Mexico-based Pemex Petroleum to let the private sector in.

By April, momentum on climate was building. This followed the release of a 21-page draft, based on a 1,400-page UNFCCC assessment of the impact of global warming, to be presented at Heiligendamm (Max 2007). Scientists and diplomats from more than 120 countries had contributed to the UN report. Merkel noted that it "confirmed that climate change is a fact" (Agence France Presse 2007b). She expressed the need to act "rapidly and decisively" to lower the rate at which temperatures were increasing to curb emissions, and promised to raise the issue with her G8 colleagues at Heiligendamm." She intended "to make all nations take responsibility for climate protection." Her attachment to a new inclusive, all-in regime was clear.

A week before the Heiligendamm Summit, on May 31, the US released the New International Climate Change Framework for its post-2012 agenda. Instead of advocating an overarching consensus, the Bush administration's proposal promoted individual approaches to emissions reductions, allowing each country to set its own long-term strategy (US Department of State 2007). The document sought to counter several proposed or inherited G8 pledges, including those to cut greenhouse gas emissions by 50 percent below 1990 levels by 2050, limit global warming to 2°C "this century," and use the UN to tackle climate change (Lovell 2007). According to an anonymous source, the US "rejected any mention of targets and timetables, [did not] want the U.N. to get more involved and [refused] to endorse carbon trading because it must by definition involve targets" (Lovell 2007). In response, another source stated that it remained an open question whether Merkel would accept a "watered-down declaration or break with G8 tradition and declare a failure on climate change." In any case, "the ink would still be wet" when the final declaration was produced.

By the eve of the summit, Merkel's goals were to have the US move incrementally toward accepting that precise limits on greenhouse gas emissions were needed and to sign a UN post-Kyoto agreement in 2012 (Williamson 2007). She stated publicly that she would accept nothing less than the two-degree temperature rise target endorsed in the summit communiqué and that any new climate initiatives must "eventually merge into the UN process" (Benoit 2007). The European Commission declared that climate change discussions would be "difficult," and it would be "unrealistic" to expect the US to make commitments on binding targets at Heiligendamm (Agence Europe 2007). It acknowledged that "the conditions were not in place" for such an advance, even if the Americans had already "made much headway." The objective at this late stage, therefore, was to bring the G8 partners, "as close as possible to the goal and objectives approved by the EU in March." The UK backed Merkel but judged that the German and American positions were not mutually exclusive; although the recent US initiative was an "important step in the right direction," it still needed to be nudged toward accepting the agreed quantifiable climate change goals (Benoit and Williamson 2007).

Canada seemed on board with Germany's overall position. In February, Stephen Harper (2007) had announced that Canada would "regulate greenhouse gas emissions from major industrial sectors" and "the fuel efficiency of motor vehicles, beginning with the 2011 model year," and it would set "enforceable regulatory targets for the short, medium and long term." Moreover, Harper talked about new "eco-energy programs" to support energy efficiency and the production of renewable power, mandate increased ethanol use, and make energy-efficient vehicles more affordable. In his view, the "era of voluntary compliance" was over.

On the eve of the summit, Harper shifted further after an 11th-hour conversation with Merkel on June 2. Canada would now accept the target and timetable of a 50 percent reduction by 2050 to be stated in the communiqué. Two days later in Berlin, Harper told a meeting of the German-Canadian business club that he believed that he and Merkel were "on the same page on this point at least: all countries must embrace ambitious reduction targets, so that the International Panel on Climate Change's goal of cutting emissions in half by 2050 can be met" (Canada, Prime Minister's Office 2007b). The German hosts were most pleased with Canada's move, which had come at a critical time. Indeed, Merkel's last-minute discussions with the leaders of Canada, Japan, and Russia meant that the US would stand alone at the summit, which led Bush to offer an accommodating initiative just before the summit started (Benoit, Ward, et al. 2007).

As Canada declared itself an intermediary between the United States and Europe, success for Canada at the G8 would entail producing a commitment that satisfied European objectives by setting ambitious long-term goals within the post-2012 framework while allowing each member considerable short- and medium-term leeway to develop more individual objectives at the national level.

At the Summit

Just before the summit's opening dinner on June 6, Merkel remarked that Europe's ambitious goals would not be shared immediately by the full global community. Jim Connaughton, Bush's primary environmental advisor, insisted that any targets had to be supported by major developing countries such as China and India, that the US could be a bridge between those now recognizing the need to commit and the reluctant ones that remained, and that action should be taken immediately on the many items already commanding consensus, above all tariff reductions on clean energy technology (Benoit, Eaglesham, et al. 2007).

On the first working day of the summit, the G8 leaders came to consensus on the facts, causes, and consequences of global warming, on how to proceed in a post-Kyoto framework, and above all on the need for the US and G5 members to be bound as carbon-constraining powers there. Further details would be defined through several processes feeding into a new UN agreement by 2009. The next day G8 leaders met with their G5 guests. They all agreed to start real action on a long list of practical measures to reduce emissions through curbing deforestation, sustaining biodiversity, preparing national energy efficiency plans, examining ways to curb fossil fuel consumption, fostering measures to increase energy efficiency in the transport sector, and investing in low carbon technologies, particularly renewables (G8 2007b, 2007a).

Results

Heiligendamm was very successful on climate change. Leaders broke through several deadlocks in discussions on climate at previous summits. The climate section of the "Growth and Responsibility" declaration began by noting that global temperatures were rising and that human activity was largely responsible. Such temperature increases would result in "major changes in ecosystem structure and function with predominantly negative consequences for biodiversity and ecosystems" (G8 2007b). This was a significant strengthening of the principled epistemic consensus on fact and causation encoded at Gleneagles in 2005.

Heiligendamm then consolidated a consensus, for the first time, on how to proceed with a post-Kyoto framework. The US and the G5 would all participate as carbon-constraining powers, even if it would still be centred on the UNFCCC process, based on "common but differentiated responsibilities and respective capabilities" and defined as a "global agreement" by 2009 (G8 2007b). Gone was the old, UN-bred, encoded, and embedded principle that the rich North, responsible for causing the problem, had to cut its carbon while the unconstrained poor South, now suffering from it, was free to develop as it saw fit. Now all countries, including "major emitters," would act together, with the North going furthest and fastest and still transferring resources to the South, and the South constraining its carbon as its responsibility and capacity allowed.

The G8 also agreed to "substantial" reductions in greenhouse gas emissions in the future (G8 2007b). It pledged to "consider seriously" the EU, Canadian, and Japanese approach of reducing emissions by 50 percent by 2050 (from an undefined base) as a long-term goal. The point of convergence was thus clear, even though the G8 avoided setting targets that the developing countries could not accept.

It was a significant step forward for all G8 members to agree collectively to these initiatives, principally as they related to emissions targets, given America's intransigence prior to the summit. On several occasions the leaders implied that the components of the consensus reached were largely political statements, a point iterated by Merkel, Harper, and Putin (German Presidency of the G8 2007; Canada, Prime Minister's Office 2007a; Kremlin 2007). The EU's José Manuel Barroso, however, saw these achievements as remarkable, because they set the stage for further serious discussion on emissions reductions. Ultimately, through this concerted G8 and G5 agreement, the leaders' commitments at Heiligendamm might translate into a post-2012 agreement at the UNFCCC negotiations in Bali later that year. Indeed, the most notable accomplishment of COP-13 in Bali was the agreement on the "Bali Roadmap," which stipulated, in line with the direction set by the G8, that the parties must reach an agreement to finalize a post-2012 climate regime by December 2009 in Copenhagen (ENB 2007).

In the immediate lead-up and during the summit, expectations rose that negotiations would produce an ambitious declaration on climate change. This resulted in a renewed sense of political will to translate these commitments into concrete policy actions. Heiligendamm proceeded to secure full compliance in the two areas critical for the Japan-hosted summit the following year—climate change and outreach (measured by the Heiligendamm Process of a structured, official-level dialogue among the G8 and G5 members) (G8 Research Group 2008).

In particular, Heiligendamm's commitment to at least halve global emissions by 2050 was seen, according to UNFCCC executive secretary Yvo de Boer, as helping reenergise the process in Bali later that year (UN Climate Change Secretariat 2007). Moreover, the G8's agreement to include all major emitters in G8-G5 climate negotiations sent a clear signal to the Bali delegates to immediately launch discussions on a post-Kyoto framework. The effect of the G8 in pushing the UNFCCC toward the new inclusive regime was clear. Indeed, the UN itself acknowledged the G8's leadership. According to a UNFCCC press release at the time of the Bali conference, "the G8 summit in Heiligendamm paved the way for negotiations in Bali and gave climate talks under the auspices of the UN a considerable boost" (UN Climate Change Secretariat 2007). The G8, not the UN, was now again in the global lead on climate change.

Dimensions of Performance

The G8's restoration of G8 leadership on climate change was seen in Heiligendamm's exceptionally, often historically strong performance across all six governance dimensions.

Domestic Political Management
Heiligendamm awarded four communiqué compliments to 44 percent of the G8 members (see Appendix B-1). This was only the second time the G8 had issued such compliments, and they were four times larger than its first summit to do so at Genoa in 2001.

Deliberation
In its deliberation, Heiligendamm issued 4,154 words on climate change, or 12 percent of the final communiqué, about twice as many in both cases as any summit before (see Appendix A-1).

Direction Setting
In Heiligendamm's direction setting, G8 leaders mentioned climate change in the opening sentence of their final communiqué: they met at Heiligendamm "to address key challenges of the world economy, climate change and Africa" (G8 2007a).

Decision Making
In its decision making, Heiligendamm made 44 commitments on climate change, by far the most at any previous summit to date, and the second highest in the G8's 40 years (see Appendices A-1 and D-1).

Delivery
In delivering their Heiligendamm decisions, G8 members complied with the assessed core commitments at a strong level of +0.56, well above St. Petersburg's +0.35 (see Appendices A-1 and E-1). Complete compliance came from the US, Japan, and the UK.

Development of Global Governance
In Heiligendamm's development of global governance, performance was also very strong. Leaders made one reference to their own environmental ministers' body but not in the context of climate change. They also made 16 references to seven outside bodies, a notable increase from the year before (see Appendix F-5). They created the G8-centred Heiligendamm Process, which included energy efficiency as one of its key issues.

Causes of Performance

Causing this strong performance was the G8's recognition that climate change could only be effectively controlled with the participation and partnership of the five outreach partners, both because of shared vulnerabilities to the climate threat and because they commanded the capacity to control carbon emissions for all as well. At Heiligendamm the G8 moved to involve them, outside the UN process, more fully and reliably, with climate change as the first big test. By engaging some of the world's largest emitters, China and India, a hydrocarbon-rich Mexico and a biodiversity-blessed Brazil, the 2007 summit largely succeeded in getting the G5 to join a more inclusive, carbon-controlling club (Kirton 2007c).

Shock-Activated Vulnerability
The first cause of Heiligendamm's very strong performance was shock-activated vulnerability, led by the leaders' first ever communiqué-encoded references to climate vulnerability. The three such references highlighted the vulnerability of the poor and particular regions but embraced all (see Appendix G-1). This public recognition was backed by a spike in world oil prices, which rose more than 38 percent from the start to end of 2007 (see Appendix G-3). While the number of tanker accidents remained small at only four, the number of people affected by and damaged from natural disasters rose (see Appendices G-4 and G-5). Greenhouse gas emissions and China's lead over the US as the primary producer continued to increase, as the G8 and global economies strongly grew (see Appendices G-7 and G-8).

Multilateral Organizational Failure
Multilateral organizational failure also propelled G8 performance. The UN held no climate summits and the Nairobi COP/MOP made little advance. However, the IPCC's Fourth Assessment Report, published in 2007, made a strong epistemic case for action (see Appendix H-1).

Common Democratic Principles
One constraint at Heiligendamm was the reduction of common democratic principles, with Russia's decline in democracy (see Appendix J).

Political Cohesion
A strong cause of Heiligendamm's performance was the high political cohesion of G8 leaders. It was led by host Angela Merkel, a former environment minister attending her second summit and backed by coalition partners and a public that strongly supported German, G8, and global action to control climate change. There was high continuity in the form of Bush, Blair, Putin, Merkel, and even Harper and Italy's Romano Prodi, both attending their second summit (see Appendix K-2).

Compact Club Participation
Another key cause of performance was compact, club participation at Heiligendamm. The G8 leaders invited only their increasingly capable G5 partners, for the third year in a row, and extended this partnership from the Gleneagles Dialogue into the official-level Heiligendamm Process. Indeed, Germany had invited the G5 powers, along with five other guests, because they played a vital role on the summit's most controversial issue of climate change (Williamson 2007).

After the Summit

Half a year later, the 2007 UNFCCC members met in Bali between December 3 and 15, marking the 13th COP and the third MOP to the Kyoto Protocol. With more than 10,000 participants from over 180 countries, negotiations on a successor protocol to Kyoto—set to expire in five years—dominated the discussions.

COP-13 reached important agreements on deforestation, adaptation, and technology transfer. Yet major differences between negotiating parties on emissions reduction targets persisted, as the EU, Britain, and Germany were unable to persuade the US, Russia, and

Japan, among others, to commit to reducing emissions by 25–40 percent by 2020 relative to 1990 levels.

The resulting Bali Roadmap outlined a new negotiating process for a binding agreement to be finalized by 2009 in Copenhagen. It would feed into an international agreement on climate change following Kyoto's expiration in 2012.

The roadmap outlined several pillars, or "building blocks," to allow for the implementation of Kyoto (UN 2008b). They included long-term goals for emissions reductions, enhanced action on mitigation and adaptation strategies, technology development and transfers, forest degradation prevention, and the provision of financial resources for investment in mitigation and adaptation.

But Bali was not without controversy, primarily as the US again pit itself against India and China over their lack of responsibilities under the protocol. Developing countries, on the other hand, countered that the industrialized countries—the primary historic source of toxic emissions—had made little progress in advancing key elements of the convention themselves.

Initial EU proposals called for global emissions to peak in 10 to 15 years and then decline below half of the 2000 level by 2050 for developing countries. For developed countries, on the other hand, the EU proposed to achieve emissions levels 20–40 percent below 1990 levels by 2020 (Pew Center on Global Climate Change 2007). The United States strongly opposed these targets. It was backed by Canada, Japan, Russia, and Australia. In the end, governments could agree only on the most general parameters for the process going forward, leaving virtually all the key issues for future talks.

Conclusion

The Gleneagles Summit launched a three-year period of restoring the G8's global leadership on climate change, after a decade of failure by the UN system and its divided Rio-Kyoto regime. By 2005, it had become clear that the UN's unbalanced, development-first, top-down, convention-to-protocol model of climate change mitigation, with only a small group of developed countries obliged to control their carbon, was failing to rein in the top emitters in both the North and the rapidly rising South and thus to produce adequate solutions to the compounding threat of climate change. The UN's broad consensus strategy, dominated by developing countries, had produced no effective, voluntary commitments on the part of top emitters including China and India. Moreover, nearly all countries failed to meet their targets under the Kyoto Protocol's requirements. The United States, long the world's top emitter of carbon dioxide, had refused to ratify Kyoto.

By bringing all the top emitters and absorbers into a summit-level plurilateral climate dialogue, the G8 and the G5 together achieved what the UN had failed to do: include all key parties excluded from the binding control commitments of the UN regime. Major emitters including Brazil, China, India, South Africa, and Mexico dealt with climate issues very differently within the various international forums. An advantage in working in the G8 was that these outreach countries did not have the support and pressure of the other G77 developing countries behind them, as they did at the UN. This presence was clearly an impediment in breaking logjams and getting movement within the UN's broader climate negotiations. The more intimate forum of the G8 and G8+G5 was conducive to dealing with climate issues apart from and before they were sent to the UNFCCC.

With the release of the G8's first stand-alone document devoted entirely to climate mitigation, the 2005 Gleneagles Summit acknowledged that climate change was a human-induced condition, thereby setting in motion a period that called on both G8 members and other countries with significant energy needs to join in a new climate dialogue, focused on mitigation strategies that would meet everyone's energy needs sustainably. As global climate challenges quickly ascended in prominence during this period, so did the G8's ability to respond in an effective, constructive, more inclusive way.

During this phase, the G8 started to constrain top emitters, including to some degree the United States. This was largely due to the allure of the G8's soft law practices over those of the hard law, legalized UN and, more centrally, to Tony Blair's actions as host of the Gleneagles Summit. With climate change at the top of his agenda, he was able to effectively bring the rising G5 and the US into the deliberations and develop an alternative set of principles and process as the foundation for a more appropriate global climate governance regime.

Gleneagles saw an unprecedented rise in the amount of climate discussions. In direction setting, it was a ground-breaking summit, as its action plan overtly linked the problems of climate, energy, and sustainable development for the first time. Moreover, the dialogue on climate change was unlike anything the G8 had ever done before, marking an independent and ambitious view of the problem. In decision making, Gleneagles saw a sharp spike in the number of climate commitments reached. Delivery was similarly successful, with the compliance scores generating some of the highest in the G8's history at +0.80. This suggests that the new autonomous approach taken by the G8 during this period was an enormous boost for the summit's compliance and accountability.

The momentum of Gleneagles continued into 2006. The St. Petersburg Summit launched an ambitious plan to tackle energy security, with an emphasis on transparency, efficiency, conservation, renewables, and climate change control. It reflected the values of environmental openness to a considerable degree. Noting that protecting the environment and tackling climate change were indeed serious challenges, the leaders linked climate with several other issues, including high and volatile oil prices. They affirmed the need for "environmentally responsible supplies of energy" (G8 2006). The St. Petersburg action plan reaffirmed the principles set down at Gleneagles, particularly those linking climate and energy.

Heiligendamm continued this trajectory in 2007. It launched the new high-level Heiligendamm Process with the G8 and Brazil, China, India, Mexico, and South Africa, with one of the four key themes being climate change (Cooper and Antkiewicz 2008). Specifically, the leaders noted the importance of "sharing knowledge for improving energy efficiency and technology cooperation" in order to contribute to reducing carbon dioxide emissions, consistent with the Gleneagles Dialogue on Climate Change and the St. Petersburg Plan of Action (G8 2007b). This G8+G5 construct—initiated at Gleneagles, but growing in prominence and institutionalization at Heiligendamm—proved useful in producing meaningful discussions among all the world's largest emitting countries. Not only did the G8 and G5 meet to discuss climate-related issues at Heiligendamm, but meetings among sherpas and officials also took place throughout the year, and in the lead-up to the summit, further reinforcing the dialogue and connection between both groups. One such meeting, in April 2007, was specifically dedicated to climate change and the environment. Having those countries at the table was a key step in developing helpful plurilateral approaches on several climate mechanisms, including technology advances and partnerships on renewable energy.

An important cause of the G8's restored leadership on climate change from 2005 to 2007 was the shock-activated, equalizing vulnerability among G8 members and among their G5 partners beyond. This started with spikes in oil prices and extended to extreme weather events, with unrelated terrorist shocks bonding G8 members with their new G5 partners too.

The first shock came from the unprecedented high, sustained oil prices that hit a historic high around the time of the St. Petersburg Summit in July 2006. A second shock was the ecological strikes of hurricanes Rita and Katrina that struck the US in September 2005. Unusually cold weather across Russia and Europe strained energy supplies in the ensuing months. These shocks were all the more potent in driving the G8's performance as they connected directly with St. Petersburg's first and defining priority of international energy security, and the Gleneagles and Heiligendamm priority of climate change. The multilateral UN failed to respond to all.

Equalizing capability also mattered. The otherwise least powerful G8 members of Canada and Russia were the G8's—and indeed the world's—leading energy superpowers, with Germany a global leader in renewable energy. Among the G5 members, Brazil stood first in ethanol production, with Mexico a key player in oil and gas. This combination of overall and specialized capability of the G8 and its partners explains Russia's insistence on energy security as the focus of its summit. With Russia's rising prosperity largely fuelled by energy exports, its democratic development largely depended upon avoiding a plunge into recession. Ensuring energy security, with all the connected components including climate change, thus became a leading G8 preoccupation and propeller of its new regime from 2005 to 2007.

This laid the foundation for the 2008 Japanese-hosted Toyako-Hokkaido Summit, with its long list of environmental objectives, to deliver on its much-anticipated climate platform. This included not only a proposal to be endorsed by the G8 for the post-2012 framework but also an ambitious global target to cut emissions by 50 percent by 2050.

References

Agence Europe (2007). "European Commission Expectations for Heiligendamm Summit." Brussels, June 1.

Agence France Presse (2007). "Merkel Says UN Report Shows Need for Action of Climate Change." April 6.

Asahi (2006). July 18, p. 8.

Associated Press (2006). "Merkel Pledges to Make Climate Protection Key Part of Germany's G8, EU Presidencies." December 1.

Bayne, Nicholas (1999). "Continuity and Leadership in an Age of Globalisation." In *The G8's Role in the New Millennium*, Michael R. Hodges, John J. Kirton, and Joseph P. Daniels, eds. Aldershot: Ashgate, pp. 21–44.

Bayne, Nicholas (2005a). "Overcoming Evil with Good: Impressions of the Gleneagles Summit, 6–8 July 2005." G8 Research Group, July 18. http://www.g8.utoronto.ca/evaluations/2005gleneagles/bayne2005-0718.html (January 2015).

Bayne, Nicholas (2005b). *Staying Together: The G8 Summit Confronts the 21st Century.* Aldershot: Ashgate.

Bayne, Nicholas (2010). *Economic Diplomat.* Durham, UK: Memoir Club.

BBC News (2006). "PM Praises Business Pledge on CO2." June 6. http://news.bbc.co.uk/2/hi/uk/5050774.stm (January 2015).

Benoit, Bertrand (2007). "Bush Faces Isolation on Climate at G8 Talks." *Financial Times*, June 4, p. 1. http://www.ft.com/intl/cms/s/0/96e5bc2c-1237-11dc-b963-000b5df10621. html (January 2015).

Benoit, Bertrand, Jean Eaglesham, Fiona Harvey, et al. (2007). "Merkel Accepts Defeat in Bid to Win US Emissions Pledge." *Financial Times*, June 7, p. 6.

Benoit, Bertrand, Andrew Ward, and Hugh Williamson (2007). "Cheers All Round for 'Winner' Merkel." *Financial Times*, June 9, p. 6. http://www.ft.com/intl/cms/ s/0/7274c3b6-15e7-11dc-a7ce-000b5df10621.html (January 2015).

Benoit, Bertrand and John Williamson (2007). "Merkel to Push Bush on Climate Change." *Financial Times*, June 4, p. 2. http://www.ft.com/intl/cms/s/0/e8f229d8-1237-11dc-b963-000b5df10621.html#axzz3Sm9i1Ukk (January 2015).

Black, Richard (2006). "UK Appoints 'Climate Ambassador'." June 8. http://news.bbc. co.uk/2/hi/science/nature/5057678.stm (January 2015).

Blair, Tony (2010). *A Journey*. London: Random House.

Brown, Paul, Patrick Wintour, and Michael White (2005). "Leaked G8 Draft Angers Green Groups." *Guardian*, May 28. http://www.theguardian.com/society/2005/may/28/ environment.greenpolitics (January 2015).

Bueckert, Dennis (2005). "Martin Blamed for Cool U.S. Response to Climate Talks: Washington Angered by PM's Critical Comments." *Calgary Herald*, December 9, p. A11.

Campbell, Jennifer (2006). "Tamil Tigers Blacklisting Lauded." April 12.

Canada. Prime Minister's Office (2007a). "The 2007 G8 Summit." Heiligendamm, June 8. http://www.pm.gc.ca/eng/news/2007/06/08/2007-g8-summit (January 2015).

Canada. Prime Minister's Office (2007b). "Prime Minister Stephen Harper Calls for International Consensus on Climate Change." June 4, Berlin. http://pm.gc.ca/eng/ news/2007/06/04/prime-minister-stephen-harper-calls-international-consensus-climate-change (January 2015).

Center for Climate and Energy Solutions (2006). "Twelfth Session of the Conference of the Parties to the UN Framework Convention on Climate Change and Second Meeting of the Parties to the Kyoto Protocol." http://www.c2es.org/international/negotiations/cop-12/summary (January 2015).

Cooper, Andrew F. and Agata Antkiewicz, eds. (2008). *Emerging Powers in Global Governance: Lessons from the Heiligendamm Process*. Waterloo: Wilfrid Laurier University Press.

Downing Street (2005). "Prime Minister Blair Concludes Climate Change Conference." London, November 1. http://www.g8.utoronto.ca/environment/env_energy051101-blair.htm (January 2015).

DW Staff (2006). "Merkel to Target Climate Change as G8, EU Leader." September 28. http://dw.de/p/9BHk (January 2015).

Earth Negotiations Bulletin (2005, 12 December). "Summary of the Eleventh Conference of the Parties to the UN Framework Convention on Climate Change and First Conference of the Parties Serving as the Meeting of the Parties to the Kyoto Protocol: 28 November–10 December 2005." 12(291). http://www.iisd.ca/download/ pdf/enb12291e.pdf (January 2015).

Earth Negotiations Bulletin (2006, 20 November). "Summary of the Twelfth Conference of the Parties to the UN Framework Convention on Climate Change and Second Meeting of the Parties to the Kyoto Protocol: 6–17 November 2006." 12(318). http://www.iisd. ca/download/pdf/enb12318e.pdf (September 2014).

Earth Negotiations Bulletin (2007, 18 December). "Summary of the Thirteenth Conference of Parties to the UN Framework Convention on Climate Change and Third Meeting of the Parties to the Kyoto Protocol: 3–15 December." 12(354). http://www.iisd.ca/download/pdf/enb12354e.pdf (January 2015).

Elliott, Larry and Patrick Wintour (2006). "Blair Wants Developing Nations in New G13 to Help Secure Key Deals." *Guardian*, July 13. http://www.theguardian.com/politics/2006/jul/13/uk.topstories3 (January 2015).

G7 (1985). "The Bonn Economic Declaration: Towards Sustained Growth and Higher Employment." Bonn, May 4. http://www.g8.utoronto.ca/summit/1985bonn/communique.html (January 2015).

G7 (1995). "Halifax Summit Communiqué." Halifax, June 16. http://www.g8.utoronto.ca/summit/1995halifax/communique/index.html (January 2015).

G8 (2005a). "Chair's Summary." Gleneagles, July 8. http://www.g8.utoronto.ca/summit/2005gleneagles/summary.html (January 2015).

G8 (2005b). "Climate Change, Clean Energy, and Sustainable Development." Gleneagles, July 8. http://www.g8.utoronto.ca/summit/2005gleneagles/climatechange.html (January 2015).

G8 (2005c). "Gleneagles Plan of Action: Climate Change, Clean Energy, and Sustainable Development." Gleneagles, July 8. http://www.g8.utoronto.ca/summit/2005gleneagles/climatechangeplan.html (January 2015).

G8 (2006). "Global Energy Security." St. Petersburg, July 16. http://www.g8.utoronto.ca/summit/2006stpetersburg/energy.html (January 2015).

G8 (2007a). "Chair's Summary." Heiligendamm, June 8. http://www.g8.utoronto.ca/summit/2007heiligendamm/g8-2007-summary.html (January 2015).

G8 (2007b). "Growth and Responsibility in the World Economy." Heiligendamm, June 7. http://www.g8.utoronto.ca/summit/2007heiligendamm/g8-2007-economy.html (January 2015).

G8 Energy Ministers (2006). "Chair's Statement of G8 Energy Ministerial Meeting." Moscow, March 16. http://www.g8.utoronto.ca/energy/energy060316.html (January 2015).

G8 Research Group (2006a). "2005 Gleneagles Final Compliance Report." June 12. http://www.g8.utoronto.ca/evaluations/2005compliance_final/index.html (January 2015).

G8 Research Group (2006b). "St. Petersburg Performance." July 17. http://www.g8.utoronto.ca/evaluations/2006stpetersburg/performance.html (January 2015).

G8 Research Group (2008). "2007 Heiligendamm G8 Summit Final Compliance Report." Toronto. http://www.g8.utoronto.ca/evaluations/2007compliance_final/ (January 2015).

German Presidency of the G8 (2007). "Pressekonferenz zum G8-Gipfel." Audio available in English, Heiligendamm, June 8. http://www.g8.utoronto.ca/summit/2007heiligendamm/g8-2007-merkel.html (January 2015).

Haas, Peter M. (2008). "Climate Change Governance after Bali." *Global Environmental Politics* 8(3): 1–7.

Hajnal, Peter I. (2008). "Meaningful Relations: The G8 and Civil Society." In *G8: Hokkaido Toyako Summit 2008*, Maurice Fraser, ed. London: Newsdesk, pp. 206–08.

Harper, Stephen (2007). "Prime Minister Harper outlines Agenda for a Stronger, Safer, Better Canada." Ottawa, February 6. http://pm.gc.ca/eng/news/2007/02/06/prime-minister-harper-outlines-agenda-stronger-safer-better-canada (January 2015).

Kirschbaum, Erik (2006). "Germany to Put Global Warming Back on G8 Agenda." *Reuters*, September 26.

Kirton, John J. (2005a). "Gleneagles G8 Boosts Blair at Home." August 1. http://www.g8.utoronto.ca/evaluations/2005gleneagles/coverage.html (January 2015).

Kirton, John J. (2005b). "Gleneagles Performance: Insittutions Created." July 9. http://www.g8.utoronto.ca/evaluations/2005gleneagles/2005institutions.html (January 2015).

Kirton, John J. (2005c). "What to Watch For at Gleneagles." G8 Research Group, July 6. http://www.g8.utoronto.ca/evaluations/2005gleneagles/kirton2005-0706.html (January 2015).

Kirton, John J. (2006a). "The G8 and Global Energy Governance: Past Performance, St. Petersburg Opportunities." Paper presented at a conference on "The World Dimension of Russia's Energy Security," Moscow State Institute of International Relations, April 21, Moscow. http://www.g8.utoronto.ca/scholar/kirton2006/kirton_energy_060623.pdf (January 2015).

Kirton, John J. (2006b). "Shocked into Success: Prospects for the St. Petersburg Summit." Paper prepared for the G8 Pre-Summit Conference on "G8 Performance, St. Petersburg Possibilities," Moscow State Institute on International Relations, Moscow, June 29. http://www.g8.utoronto.ca/conferences/2006/mgimo/kirton_prospects_060717.pdf (January 2015).

Kirton, John J. (2007). "Prospects for the 2007 G8 Summit." In *G8 Summit 2007: Growth and Responsibility*, Maurice Fraser, ed. London: Newsdesk, pp. 18–21. http://www.g8.utoronto.ca/newsdesk/G8-2007.pdf (January 2015).

Kirton, John J. (2013). *G20 Governance for a Globalized World*. Farnham: Ashgate.

Kirton, John J., Courtney Brady, and Janel Smith (2005). "Gleneagles Performance: Money Mobilized." July 8. http://www.g8.utoronto.ca/evaluations/2005gleneagles/2005money.html (January 2015).

Kremlin (2007). "President Vladimir Putin Held a Press Conference Following the End of the G8 Summit." Heiligendamm, June 8. http://eng.kremlin.ru/news/15023 (January 2015).

Louet, Sophie (2006). "France's Chirac Warns Mankind Faces Climate Volcano." *Reuters*, July 16.

Lovell, Jeremy (2007). "G8 Set for Transatlantic Clash on Climate." *Reuters*, May 11. http://in.reuters.com/article/2007/05/11/idINIndia-29779420070511 (January 2015).

Max, Arthur (2007). "Delegates Debate Urgency of Climate Change in Key Policy Report." *Associated Press*, April 4.

Oosthoek, Jan (2005). "The Gleneagles G8 Summit and Climate Change: A Lack of Leadership." *Globalizations* 2(3):443–46. doi:dx.doi.org/10.1080/14747730500409447.

Pew Center on Global Climate Change (2007). "Thirteenth Session of the Conference of the Parties to the UN Framework Convention on Climate Change and Third Session of the Meeting of the Parties to the Kyoto Protocol." Bali. http://www.c2es.org/docUploads/Pew%20Center_COP%2013%20Summary.pdf (January 2015).

Putin, Vladimir (2006). "Speech at Meeting with the G8 Energy Ministers." Moscow, March 16. http://www.g8.utoronto.ca/energy/energy_putin060316.html (January 2015).

Putnam, Robert and Nicholas Bayne (1984). *Hanging Together: Co-operation and Conflict in the Seven-Power Summit*. 1st ed. Cambridge MA: Harvard University Press.

Putnam, Robert and Nicholas Bayne (1987). *Hanging Together: Co-operation and Conflict in the Seven-Power Summit*. 2nd ed. London: Sage Publications.

Russia's G8 Presidency (2006). "Final Press Briefing with President Putin." St. Petersburg, July 17. http://www.g8.utoronto.ca/summit/2006stpetersburg/putin060717.html (January 2015).

Smith, Heather (2008). "Canada and Kyoto: Independence or Indifference?" In *An Independent Foreign Policy for Canada: Challenges and Choices for the Future*, Brian Bow and Patrick Lennox, eds. Toronto: University of Toronto, pp. 207–21.

United Nations (2006). "Citing 'Frightening Lack of Leadership' on Climate Change, Secretary-General Calls Phenomenon an All-Encompassing Threat in Address to Nairobi Talks." Press release, New York NY, November 15. http://www.un.org/press/en/2006/sgsm10739.doc.htm (January 2015).

United Nations (2008). "Report of the Conference of the Parties on Its Thirteenth Session, Held in Bali from 3 to 15 December 2007: Addendum. Part Two: Action Taken by the Conference of the Parties at Its Thirteenth Session." FCCC/CP/2007/6/Add.1, March 14. http://unfccc.int/resource/docs/2007/cop13/eng/06a01.pdf (January 2015).

United Nations Climate Change Secretariat (2007). "UNFCCC Executive Secretary: G8 Document Reenergises Multilateral Climate Change Process under the United Nations." Bonn, June 7. http://unfccc.int/files/press/news_room/press_releases_and_advisories/application/pdf/20070607_g8_press_release_english.pdf (January 2015).

United States Department of State (2007). "A New International Climate Change Framework." Washington DC, May 31. http://2001-2009.state.gov/g/oes/rls/fs/2007/92156.htm (January 2015).

Williamson, Hugh (2007). "G8 Summit: Merkel Keeps Focus Broad." *Financial Times*, June 4, p. 5.

Chapter 9
Reaching Out, 2008–2009

As climate change became one of the world's most pressing issues, global attention increasingly focused on plans for shaping the post-Kyoto regime after 2012. Thus climate leadership centred in the Group of Eight (G8), which had been restored since in 2005, accelerated in 2008–2009. Recent G8 principles, processes, and some promises had provided a starting point for creating a new, inclusive global climate change regime. Yet there remained much room for the G8 to expand its outreach and lead new plurilateral international institutions that had recently taken up the cause of climate change control. As the world searched for solutions from the developed and emerging worlds, the G8's potential was never greater, especially as it constituted the core of the two new summit-level bodies born at the leaders' level in 2008: the Major Economies Meeting (MEM) on Energy Security and Climate Change, which added Australia, Indonesia, and Korea to the G8 and Group of Five (G5), and the Group of Twenty (G20).

December 2009 was set to be the culmination of the Kyoto Protocol process. The 15th Conference of the Parties (COP) in Copenhagen was designed to agree and approve a post-Kyoto climate change regime. During the long preparations, the G8 had two major opportunities—one in Hokkaido, Japan, in 2008 and one in L'Aquila, Italy in 2009—to shape a positive result. It would do so above all by taking the lead on crafting more effective mitigation and adaptation solutions. It would need to facilitate climate governance within the context of the United Nations, an organization with a very different structure from the informal, direct negotiation of the like-minded G8 members. It would also need to turn potential "competitor" institutions into partners, by mobilizing the new summits of the MEM (which became the Major Economies Forum [MEF] on Energy and Climate in 2009), the G20, and the BRIC group of Brazil, Russia, India, and China for supportive global leadership on climate change.

It became increasingly apparent that the UN was not making the much-needed progress in advancing negotiations for the upcoming COP-15 in Copenhagen. Disagreements persisted between the developed and developing worlds as consensus among the 192 UN members became tenuous at best. This prompted leaders to take action through new institutions free from the same constraints. New climate-focused, leaders-level institutions, including the MEF and Copenhagen Accord five countries of Brazil, China, India, South Africa, and the United States, sprung up respectively in 2008 and 2009, while other informal, flexible, restricted-membership, comprehensive summit institutions, led by the G20 and BRICs, became involved in the climate debate. The G8 worked with these new institutions to further the global climate change regime, by turning them from potential competitors into partners. The G8 also devoted more effort to climate change in the lead-up to Copenhagen. There was a strong potential role to galvanize the floundering negotiations at the UN Framework Convention on Climate Change (UNFCCC) by sending signals, giving instructions, and taking strong, commitment-oriented leadership on climate change's most pressing issues.

The G8, working with the MEM/MEF and G20, did succeed in advancing the new G8-led inclusive regime during these two years. It had all the large carbon powers now assembled in the MEM/MEF—starting with the US, Russia, and the G5—affirm the principle that they must all must control their carbon in the years ahead, in ways more shared and politically obligatory than the special and differential principle central to the UNFCCC regime. It had the G8 and thus the US set a long-term reduction target. It introduced a bottom-up, sectoral approach that offered real reductions in key areas. This approach was expanded by the G20 with its historic commitment to phase out fossil fuel subsidies at its Pittsburgh Summit in September 2009.

This increasingly inclusive G8 leadership on climate change did not save the UN from the failures of its old 1992 regime at Copenhagen, nor did it render effective the last-minute, ad hoc alternative proposed by the smaller Copenhagen Accord five. COP-15 and its Copenhagen Accord produced only a weak, non-binding agreement, and a one-time meeting of those five countries (Dimitrov 2010b, 2010a). While some still hoped that the UN Conference on Sustainable Development in Rio de Janeiro in June 2012 might revive the UN's momentum, it was left to the G8, with its MEF and G20 allies, to advance its own inclusive regime from 2008 to 2009 and beyond.

Mobilizing the Major Economies Meeting at Toyako, 2008

This process began at the G8's Hokkaido Toyako Summit, hosted by Japanese prime minister Yasuo Fukuda on July 7–9, 2008. Climate change was declared and became in practice the defining challenge. Hokkaido ultimately met the challenge. It produced 55 climate change commitments, the G8's largest and most robust set ever. Leaders agreed to reduce greenhouse gas emissions by at least 50 percent by 2050 and announced a bottom-up, sectoral strategy that would involve both the developed and developing worlds. The leaders of the major emitting countries, joining G8 leaders as the MEM for the climate change session of the Hokkaido Toyako Summit, agreed in principle to control their carbon in the short and medium terms, and set a long-term reduction goal. They discussed financing and technology for developing countries and liberalizing trade in environmentally enhancing products. The G8 thus successfully incorporated the MEM into its governance process, and encouraged it to adopt many of the new G8 principles.

To be sure, the new G8-led, MEM-approved consensus lacked the comprehensive details and firm obligations necessary for a full-blown new global climate change regime. Notably absent were decisions on short- or medium-term emissions reduction goals, a baseline year from which these reductions would be determined, or new ways to assess or improve compliance with the stated key goals. The MEM remained constrained by the major points of contention blocking progress at the UN—differences between North and South, division of responsibility between them, and how to reduce emissions in the short term.

Preparations

At the end of 2006, Japan began indicating that the central themes for its 2008 G8 summit would be energy, the environment, and intellectual property. The entire Japanese leadership team, including the sherpas, changed by the end of September 2007, when Prime Minister Shinzo Abe resigned and Yasuo Fukuda became prime minister. However, climate change remained on the table, as the new government sought discussions on a "new framework that [would] ensure participation by the United States and China, the world's largest greenhouse-gas emitters" (Agence France Presse 2007a). By this standard, the Hokkaido Toyako Summit would succeed if all major carbon polluters agreed to control their carbon emissions. This did not, however, lead to ambitious goals set by the Japanese host, even if the international community expected more than simply an acknowledgement of the need, as a basic principle, by all MEM members to control carbon emissions.

The summit was due to receive the report, mandated at Gleneagles in 2005, on how to carry forward the sustainable energy dialogue as well as the interim report on the Heiligendamm Process, which included energy efficiency, as specified in 2007.

On January 26, 2008, Fukuda (2008) announced Japan's summit agenda at the World Economic Forum in Davos, Switzerland. He presented a wide-ranging, internally interlinked, and ambitious agenda that added surging oil prices, terrorism, G8 accountability, and multistakeholder participation to the earlier list. It was an unusually specific agenda, accompanied by details about the proposals, goals, and initiatives Japan would propose and unilaterally take. It positioned climate change as one of the "biggest human security issues." It clearly steered Hokkaido Toyako toward using the G8 once again as a global fundraiser. The $10 billion Climate Investment Fund, initiated by the United Kingdom and backed by the US and Japan, would be launched at Toyako, with a request to G8 and outreach partners to help fund it.

At the time, about 95 percent of Japanese people were seriously concerned about the impact of climate change and its effects on rising sea levels, biodiversity, and agriculture. Japan's focus on environmental issues, particularly climate change, was not surprising, given it was the most energy-efficient economy in the developed world by a significant margin. As a proportion of its gross domestic product, Japan spent almost three times as much money on energy technology than any other G8 country.

By mid May, the first priority theme of climate change and the environment still stood as the make-or-break issue for the Hokkaido Toyako Summit. The first task was for all G8 members and their G5 partners to accept the most recent ominous scientific findings of the Intergovernmental Panel on Climate Change (IPCC) and recognize that climate change imperilled both the global environment and the world economy, and to agree that major measures on the part of all major emitters were urgently required. The G8 then needed to define the essential framework of a climate regime beyond Kyoto—one that was effective, inclusive, and based on binding targets accepted by all countries that counted. To do so, G8 leaders would have to conclude their hard bargaining on long- and medium-term targets, timetables, and baselines, and on the contribution that Japan's bottom-up sectoral approach would make. An effective climate change regime would also need to address climate financing, and to account for developing countries undergoing costly setbacks to their growth and poverty reduction as a result of climate change.

The G8's European and Pacific powers had long been divided on these issues and compromise was essential. Part of the solution lay in agreeing on technology development and transfer, forestry, sinks and biodiversity, funding for technology and adaptation, and links to the summit's work on development, Africa, food, and health. Also relevant was the role of various negotiation forums, notably the UN process, the Gleneagles Dialogue due to conclude in 2008, and the MEM, whose first summit was likely to constitute the concluding climate change session of the Hokkaido Toyako Summit.

Japan's sectoral approach fit with the individual priorities and preferences of several G8 leaders. Canada had previously announced support for regulations on specific sectors, which fit much better than the more general reductions targets calculated using a baseline year (CBC News 2007). Russia agreed with its G8 partners, and diverged from all its BRICs ones, on the key principle that there be binding targets applicable to all major emitters in the post-Kyoto regime (Luckhurst 2013). Yet the UK stood by the need for clear reduction targets, even if they were more difficult to achieve under the industry-based, sectoral strategy (Fujioka 2008).

At the end of June, following a special meeting of sherpas and foreign affairs sous-sherpas (FASS) on the margins of the annual ministerial meeting of the Organisation for Economic Co-operation and Development (OECD), climate change was still front and centre. It was the issue with the least consensus, above all on medium- and long-term emissions reduction goals. Acknowledging the importance of the MEM, the US sought to have the G8 endorse a process that would offer medium- and long-term targets on behalf of all MEM members. It also sought a strong G8 statement on clean technology. Europe and Japan resisted prejudging the MEM before its forthcoming ministerial meeting in Seoul on June 21–22, which would be followed by a G8 sherpa meeting the following day. While the issue was contentious, there was much mutual respect in the room, with suggestions from all sides on as well as a good conversation on how the G8 might move forward.

On the summit's eve, climate change remained Hokkaido Toyako's centrepiece. The G8 and its major outreach partners sought agreement on the central architecture of a climate control regime to replace Kyoto, which would expire in 2012. G8 leaders could endorse several principles and initiatives to do so. They could move forward on the sectoral, bottom-up approach that would enable all parties to contribute to carbon control immediately and improve their performance as knowledge, technology, and competitive pressures grew. They could also accept the relevance of carbon sinks, beginning with tackling deforestation that would allow the great biodiversity and forestry powers of Brazil, Indonesia, the US, Canada, and Russia to make an enhanced contribution that would finally be counted as a control measure on the regime. But G8 leaders would also need to tackle the questions that were proving difficult to resolve in all relevant global governance forums, most notably the UN. These included mid-term emissions reduction goals, a baseline for determining these reductions, specific and adequate funding for climate adaptation financing, and, perhaps most contentious, how the developed and developing worlds were going to work together and fairly share the responsibilities necessary to prevent the dire consequences of climate change (UNFCCC 2009b).

The G8 needed to provide a common reference point and strong starting position for the upcoming UNFCCC negotiations. This would give movement and credibility to the looming COP meeting, by already laying the groundwork for some of the most difficult issues it would face.

The commitments put together by Japan in preparation for the summit contained some, but not all, of these items. It detailed commitments on a sectoral approach, heavily favoured by the Japanese, and on carbon sinks. Japan also wanted G8 leaders to endorse and fund the Climate Investment Fund to make available several billion dollars—and at least $10 billion in the near future—to finance the flow of clean technology needed by China, India, and so many other carbon-afflicted countries. It also called for increased free trade in carbon-reducing products and services, so that these, and the technology embedded in them, could flow faster, wider, and less expensively to benefit all. However, there were no commitments on mid-term emissions reductions or the division of responsibilities between North and South. The initiatives proposed were capable of doing much good, but taken together, the package did not suggest that the G8 would produce a robust climate change regime at Toyako of the scale and detail necessary to replace Kyoto and take the world into the next phase of climate governance.

For any such agreement to come, it would be necessary for Fukuda, the G8, and its MEM partners to get America's George Bush to agree to control US carbon emissions.

Bush would do so if all his MEM partners at the leaders' level agreed to control their emissions, as their officials had agreed to in a painstakingly worded draft communiqué negotiated at the MEM ministerial meeting in Seoul in June. In a series of personal pleas to the Americans on the eve of the summit, Fukuda and his sherpa succeeded in getting Bush to adjust just enough for the Japanese package to move ahead.

At the Summit

Strategically selected as an environmental showpiece, the island of Hokkaido offers a pristine natural landscape full of lakes and mountains. The summit facilities highlighted Japanese green technology, with the international press centre produced from 100 percent recycled materials. Hybrid cars and buses were used exclusively for transporting leaders, dignitaries, and the international press corps.

At an on-site bilateral meeting with Bush the day before the summit, Fukuda sought movement from the US toward accepting specific goals on climate change (Tanaka 2008).

Eight African leaders were invited to the first working session of the summit: Algeria's Abdelaziz Bouteflika, Ethiopia's Meles Zenawi, Ghana's John Kufuor, Nigeria's Umaru Yar'Adua, Senegal's Abdoulaye Wade, South Africa's Thabo Mbeki, Tanzania's Jakaya Kikwete, and, as leader of the African Union, Gabon's Jean Ping. There was no discussion of any long-term targets, commitments, or quantified national targets on climate change. The African leaders requested that the G8 support African countries in preventing further desertification and in protecting forests and improving access to renewable energy. There was a reference to Japan's proposed "Cool Earth" partnership (Japan, Ministry of Foreign Affairs 2008). Japan noted it was prepared to provide $10 billion to support developing countries willing to introduce environmentally friendly technology into their economies.

In subsequent sessions, where the G8 leaders met alone, climate change directly entered the discussions. At the same time, G5 leaders, meeting in Sapporo before going to Toyako, pushed for emissions reductions by developed countries of between 80 percent and 95 percent, as opposed to the 50 percent target already identified by the G8.

On the last day of the summit, G8 leaders met with the G5 leaders to discuss the Heiligendamm Process. They were later joined by Australia, Indonesia, and Korea to form the MEM, along with the heads of the UN, the World Bank, the International Monetary Fund (IMF), OECD, and the International Energy Agency (IEA). They discussed global issues and the world economy, which included rising oil prices, rising food prices, and development—as well as both mid-term and long-term climate goals and their determination to operate with and through the UN process.

In the middle of the night of July 7, after the summit's first day, Fukuda had instructed his sherpa, Masaharu Kono, to renegotiate the climate change passages with his US counterpart; Fukuda himself called Bush to solicit further concessions. Early the next morning, on July 8, the sherpas agreed to a draft declaration that included a long-term goal, worded in a way that included the emerging economies and not just the US and other G8 members. Without this inclusiveness, US agreement would not have come.

Results

Hokkaido Toyako did much to advance the new G8-led inclusive climate change control regime. It got all G8 members including the US and Russia to commit to major, specific, long-term reductions and the major emerging country emitters in the MEM to adopt serious control obligations. It did so through a new bottom-up sectoral approach, including serious action on climate sinks.

The first G8-wide achievement was essential for advancing the comprehensive, top-down approach of setting long-term timetables and targets, for including all members, and thus for including the additional MEM emitters, through a bottom-up sectoral approach they were then ready to adopt. The G8 made its strongest statement on comprehensive, long-term targets and timetables to date, indicating that G8 members would "consider and adopt" the UNFCCC's "vision" of reducing global emissions by 50 percent by 2050 (G8 2008a, 2008b). Largely due to US and Russian resistance, previous communiqués had seen the G8 promise only to consider such a goal. The 2007 communiqué had stated that the European Union, Canada, and Japan were the only members committed to this target by 2050, but the 2008 declaration on climate change now included all the G8.

More ambitious G8 action came in the nearer term, under the new bottom-up sectoral approach. G8 leaders identified specific subjects for action, in particular energy efficiency, clean energy, renewable energy, clean coal, and the broad deployment of carbon capture and storage (CCS) technology by 2010. In the short term, it specified the aviation, maritime, and sustainable biofuel sectors, a nuclear energy infrastructure initiative, and the completion of 20 CCS demonstration plants by 2010.

To be sure, there were several shortcomings. Hokkaido Toyako's advances were unable either to save or replace the problematic Kyoto regime. The G8's distant long-term goal of a 50 percent reduction by 2050 was still below the level of action scientists said was required, and it lacked a baseline year from which progress could be measured. Although the G8 recognized that mid-term targets were needed, it did not specify them. In regard to key sectors, there was only a small step to endorse nuclear energy as a critical zero-emission source at the generation stage. No effort was made to end the use of carbon-saturated coal beyond developing the costly, experimental CCS technology, nor were there any specific measures to stimulate renewables including wind, solar, geothermal, and hydro (although second-generation biofuels got a mention). There was only a passing nod to energy conservation and the need to reduce and recycle, and no direct reaffirmation of the 1997 commitment to reduce greenhouse gases by 2010—now only two years away. Still, while the passage on goals did not use the word "agreed," UN secretary general Ban Ki-moon welcomed the G8's long-term goal of reducing emissions by at least 50 per cent by 2050, as "a clear step forward" (UN 2008a).

Developing countries claimed that the action on carbon sinks did not amount to the necessary large-scale undertakings and structural shifts, and that climate financing promises represented no new funding. Even the long-term target of 50 percent by 2050 was seen to fall far short of the G5 call for an 85–90 percent emissions reduction by the G8. The question of how developed and developing countries would fairly divide the responsibilities for controlling and reducing emissions remained.

Dimensions of Performance

Overall, the Hokkaido Toyako Summit was a strong success. On most performance dimensions, it expanded the momentum of previous years toward a new G8-led inclusive climate change control regime (see Appendix A-1).

Domestic Political Management

In its domestic political management, Hokkaido Toyako's performance was small. It issued no communiqué compliments, in contrast to the previous year at Heiligendamm, where the G8 issued four (see Appendix B-1).

Media approval was mixed. Various outlets acknowledged that Bush's agreement to consider and adopt emissions cuts of 50 percent by 2050 marked a major step forward in the US position, even if he would be leaving office half a year hence. Canada's government immediate praise of the G8's progress on climate change was duly reported in the media (Canada, Prime Minister's Office 2008). Yet many news agencies and international bodies noted that developing countries immediately rejected the G8 commitments as weak, vague, and diluted (Wintour and Elliott 2008). They highlighted the G8's failure to meet crucial goals such as defining medium-term emissions reduction targets, choosing a baseline year for the 50 percent cuts by 2050, or agreeing on how adaptation would be financed in developing countries.

Deliberation

In its public deliberation, Hokkaido Toyako's performance was strong. Its 2,568 words on climate, while below Heiligendamm's total, represented 17.5 percent or more of the communiqué, the highest thus far and well above the 12 percent on climate change in 2007 and the 9.3 percent in 2005 (see Appendices A-1 and C-1). Most passages on climate change appeared in a separate document entitled "Environment and Climate Change," showing that climate and environmental issues were considered urgent enough to be treated in their own right.

Direction Setting

Hokkaido Toyako's performance on direction setting was also strong. It made eight references to climate change as a priority in the chair's summary (see Appendix A-1). This was the third highest thus far, if slightly below the ten references in 2007 and 2005. Moreover, led by Japan as host, the G8 affirmed the core principles of the new sectoral approach, while the MEM accepted the need for all members to constrain their carbon emissions.

Decision Making

In its decision making, Hokkaido Toyako's performance was extremely strong. The summit produced 54 climate commitments, the most in G8 history through to 2014. They injected precision and obligation into the new G8-led inclusive regime. While the G8 did speak about the UN in the opening of its declaration on environment and climate change, the rest of the commitments were markedly separate from the UN process. Whereas in previous years the G8 had supported that process, it was now decidedly operating outside and taking a different direction that focused on smaller-scale, nationally directed projects.

The MEM endorsed most of the G8's plans on climate change, but fell short of fully embracing the G8 commitments. It accepted the sectoral approach and, through it, promised to "improve significantly energy efficiency" (MEM 2008). In the medium term, the MEM emphasized how sinks could help stabilize greenhouse gases in the atmosphere and singled out deforestation, forest degradation, forest fires, forest governance, and the effects of diminishing arable land. As such, the MEM and the G8 converged on many basic principles and some promises. The MEM, however, did not accept the G8's only overall numerical long-term target—50 percent reduction by 2050. It also strongly supported the UN process.

Delivery
Hokkaido Toyako's performance was solid. Its five assessed climate commitments averaged a compliance score of +0.53. The leaders' promise to implement "ambitious economy-wide midterm goals," with the specific targets left to each country to decide, was complied with by all but the US (G8 2008a). This commitment only required announcing targets, rather than making concrete progress toward meeting them (G8 Research Group 2009).

On measures to facilitate climate change adaptation in developing countries, most G8 members complied fully with their commitment to provide financial and technical assistance, although efforts and amounts varied by country. Most members also participated in discussions on how to make this assistance and investment more readily available in the future.

The commitment to take new steps toward implementing large-scale CCS projects received less compliance. France, Italy, and Russia failed to implement any new, large-scale projects in the year following the summit.

A related energy commitment "to develop science-based benchmarks and indicators for biofuel production and use" secured compliance of +0.89 (G8 2008c).

Development of Global Governance
In its development of global governance, Hokkaido Toyako's performance was extremely strong (see Appendix A-1). Inside the G8 it mounted the first MEM summit, as an integral part of the annual G8 summit, and made two reference to the G8 environment ministers' work.

Outside the G8, 22 references were made to 11 different international institutions—the highest ever in both respects (see Appendix F-5). As in 2007, it spelled out G8 support for the UNFCCC process, reaffirming the G8's determination to secure global agreement on climate change through it by 2009. But beyond the UN, it invoked the most international institutions in a climate change context in G8 history. In keeping with the bottom-up, sectoral approach, the International Civil Aviation Organization (ICAO), the International Maritime Organization (IMO), the World Trade Organization (WTO), the International Atomic Energy Agency, and the Global Environment Facility were all to play a key role in mitigation. In this way, the G8 kept climate change at the forefront of the global agenda well beyond the UNFCCC process and embedded climate issues in a broad range of diverse, functionally specific multilateral forums.

Causes of Performance

Hokkaido Toyako's strong performance in shaping a new inclusive control regime flowed from the high levels on the six causes that regularly propelled G8 success in climate change control.

Shock-Activated Vulnerability
The first cause was strong shock-activated vulnerability, from oil and gas prices reaching historic highs, from cyclones and floods, and from a food crisis that showed the costs of uncontrolled climate change. G8 leaders themselves recognized the vulnerability to the "adverse impacts of climate change" of the poorest countries, the least developed countries, and small island developing states and their need for adaptation, including through "disaster risk reduction" (G8 2008c). As 2008 and Japan's year as host began, world oil process reached a historic high of $99.81 per barrel, exceeding the previous high reached

in January 1981 of $99.00 (see Appendix G-3). The toll from natural disasters remained high (see Appendix G-5). While greenhouse gas emissions of most G7 members, even including the US, started to decline, those of the G5 countries continued their rapid rise (see Appendices G-7 and G-8). China continued to pull away in an ever stronger global lead.

The absence of severe shocks in the related fields of energy supply or security constrained even stronger climate control action, as did diversionary shocks from newly insolvent banks hurting G8 citizens already suffering from soaring food prices, falling home and stock prices, contracting credit and confidence, and slowing wages and jobs.

Multilateral Organizational Failure

The second cause was the increasing multilateral failure of the UN system. This was led by the COP at Bali on December 3–14, 2007, which made no serious advance in implementing or extending the Kyoto Protocol, even as the 2009 deadline to replace it drew nigh. Indeed, 2009 did not serve as an action-forcing deadline but led many, including some G5 countries, to delay acting on climate change at Toyako in the self-interested hope of a better deal later. The next likely UN summit dedicated to sustainable development was a distant four years away. The Hokkaido Toyako Summit landed in the middle of the five-year gap between the most proximate climate-relevant UN summits, on the Millennium Development Goals in 2005 and 2010. And they paid limited attention to climate change.

In sharp contrast, the arrival of the plurilateral summit of the MEM provided strong support for G8 leadership at Toyako. It assembled all the current major carbon-producing powers including those beyond the G5, made their high-profile summit actions interdependent with the G8 ones, inspired the G8 to reach more ambitious agreements just before its leaders met with their MEM colleagues, and led the G8 to shift to a bottom-up sectoral approach that was easier for MEM's emerging powers to accept, especially without their many developing country colleagues there to confine them to the old G77 line.

Predominant Equalizing Capability

The third cause was globally predominant and internally equalizing capability. The overall global predominance of the G8, while still large, declined, while that of the G5, still small, was on the rise. Internal equality among G8 members grew, due to an economically slowing US with a dropping dollar and an expanding EU with a rising euro. This enhanced both the G8's need for carbon-controlling action from the G5 and thus MEM as well as the G8's ability to agree on new regime, including long-term targets, with which the US and Russia now agreed.

Common Principles and Characteristics

The fourth cause was the incremental increase in the common principles and characteristics of G8 members and their MEM partners. Within the G8, democracy deepened in Russia, with the electoral replacement of Vladimir Putin with Dmitry Medvedev, a leader widely thought to be more committed to market and political openness and the rule of law. The expansion of the G5 to the MEM added three new partners—Australia, Indonesia, Korea, all democratic states—with oil-rich but non-democratic, climate control–resisting Saudi Arabia excluded.

Political Cohesion

The fifth cause, generating some advance but more constraint, was the small domestic political cohesion of the G8 members. Many of the most powerful, including host Japan, sent leaders to the summit who did not firmly control their parties or legislatures, who were deeply unpopular with their voters, who had little summit experience or environmental competence, and who would not be in office long enough to deliver on the promises they made at Toyako. While Bush, at his eighth summit, finally agreed to ambitious long-term targets, his lame-duck status produced a strong temptation to delay achieving any big deal on climate change until 2009, when a new president and congress would arrive to fulfil the hope that they would accept and deliver the G8 and G5 partners' most audacious demands. Yet a push for immediate action did come from G8 citizens, who had increasingly become concerned about climate change. In France, for example, in 2008, 50 percent thought climate change was the predominant risk, compared with only 20 percent in 2002 (Erlanger 2008).

Constricted Club Participation
The sixth cause was significant constricted club participation, particularly as the expansion to the three new participants of the MEM was small, controlled, and grounded in ministerial meetings. Most G8 and G5 leaders met yet again to discuss climate change. For the first time, G5 members now had equal status with their G8 colleagues under the umbrella of the MEM.

The cadence of summit sessions contributed to the clubbiness of the group, in an instrumental sense at least. On the first day, the G8 met with eight invited African leaders, many of whom were familiar participants at G8 summits. On the second day, G8 leaders assembled alone to discuss their broad, pressing agenda as well as come to their own intra-G8 consensus on climate change. On the third day, they came together with their familiar but now equal G5 partners and their three new democratic MEM ones to focus on climate change.

After the Summit

Following the Hokkaido Toyako Summit, a potential reinforcement for G8-MEM leadership came from two meetings before the end of the year—the first G20 summit in Washington on November 14–15 and the UN's COP in Poznan, Poland, on December 1–12. Neither translated potential on climate change into practice.

The G20 summit, hastily arranged in only 24 days in response to the global financial crisis catalyzed by the collapse of Lehman Brothers on September 15, focused fully on crisis response through macroeconomic, financial regulatory, and trade measures (Kirton 2013a). Its declaration made only a passing reference to climate change, stating that G20 leaders were "committed to addressing other critical challenges such as energy security and climate change, food security, the rule of law, and the fight against terrorism, poverty and disease" (G20 2008). Facing the great shock of the fresh global financial crisis, the G20 concentrated on its first core mission of fostering financial stability rather than its second objective of making globalization work for the benefit of all.

At the 14th COP in Poznan in December 2008, much attention focused on an agreement that the first draft of a concrete negotiating text for a global climate change deal would be available at the next UNFCCC gathering in Bonn in June 2009, to be adopted at COP-15 in Copenhagen in December. But several emerging and developing economies described Poznan's outcome as having a "vision gap," with one negotiator accusing industrialized countries of "callousness" . One major issue of contention was the Adaptation Fund, used to protect developing countries against floods, drought, and storms. When the industrialized countries failed to increase the fund's levy from 2 percent to 3 percent, the talks resulted in an "inevitable collapse." Opponents of the increase included Australia, Canada, the EU, Japan, and Russia. Attention now turned to Copenhagen, with renewed hope for producing a clear, political intent to constrain carbon and respond effectively to climate change, in both the short and long term.

Mobilizing the MEF at L'Aquila, 2009

On July 8–10, 2009, G8 leaders assembled at the recently selected, earthquake-scarred site of L'Aquila in central Italy for their 35th annual summit, hosted by Italian prime minister Silvio Berlusconi. Climate change and clean energy were high on the agenda. The latest scientific evidence showed that climate change was a more ominous and urgent problem than predicted by the most recent report of the IPCC. The UN's Kyoto Protocol had failed to control global carbon emissions. The UN system remained deadlocked as its December Copenhagen conference to devise a successor climate control agreement drew near. It was left to the G8 to make the progress needed to secure success at Copenhagen and elsewhere (Kirton and Boyce 2009).

Preparations

While the G8 had achieved significant compliance with its climate change commitments at Hokkaido Toyako in 2008, the challenge now was to boost their strength and give them the direction and force necessary to let the rest of the world know that the G8 members and their new partners in the MEM—now expanded to the Major Economies Forum on Energy and Climate—were serious about controlling climate change. Such signals could propel success at the upcoming COP meeting in Copenhagen.

On the eve of the L'Aquila Summit, the G8 clearly, and all MEF members more vaguely, were on course to accept that they, as the major carbon-producing and -removing powers, must meaningfully control their carbon in a regime that would go beyond Kyoto in both time and space. Led by the G8, they seemed set to agree that warming must be limited to no more than 2°C, and that emissions by 2050 should be reduced by 50 percent globally and by 80 percent in the developed world. This common acknowledgement of the problem and possible solutions at hand would be a minimum starting point for real progress on climate change control.

Such numbers provided a starting point for settling how the developed and developing worlds would work together. The next challenge was to identify how the G8 could help with financing, technology, trade, investment, and adaptability to help get both North and South on track, while ensuring that human lives were protected from the worst impacts of climate change. It would take considerable effort to develop an acceptable set of numerical targets or timetables for both sets of countries, especially given that the G8 had set no concrete mid-term emissions reductions targets at Toyako. Even more difficult would be getting all countries to agree on a baseline for those reductions. However, with the urgency of climate change now peaking, there was reason to believe that it could be done.

While other helpful advances could come on short-term targets, energy efficiency, and carbon sinks, the most important tasks lay starkly before the G8. If the leaders could overcome the differences that were blocking progress, the G8 would be able to lay the foundational normative architecture for a beyond-Kyoto deal at Copenhagen that could appropriately and adequately control the compelling climate challenge facing the global community.

At the Summit

The L'Aquila Summit opened on July 8, when the leaders of the G8 met alone. At their first working lunch, they dealt with the world economy, followed by a working session on global issues focused primarily on climate and energy. On day two, they gathered with leaders of the G5 and Egypt to discuss global issues, development policies, and the Heiligendamm Process. At a working lunch they were joined by the heads of the IEA, the International Labour Organization (ILO), the IMF, the OECD, the UN, the World Bank, and the WTO to discuss future sources of growth. After lunch, the G8 plus G5 and Egypt were joined by Australia, Indonesia, Korea, and COP host Denmark for the MEF session on climate change and energy—co-chaired by Italian host Silvio Berlusconi and recently elected US president Barack Obama. Berlusconi then hosted a large dinner, which included all 17 MEF members, the leaders of Egypt, the Netherlands, Spain, Turkey, Algeria, Angola, Ethiopia, Libya, Nigeria, and Senegal, and the heads of the African Union, the IEA, the ILO, the IMF, the OECD, the UN, the World Bank, the WTO, the Food and Agriculture Organization, the International Fund for Agriculture and Development, and the World Food Programme. The summit concluded on the final day with 40 leaders, under Obama's leadership, producing the well-funded L'Aquila Initiative on Food Security.

Results

The G8 leaders agreed at L'Aquila that even with current and increased mitigation measures, the world would nonetheless experience the seriously adverse effects of climate change. They agreed that all major carbon powers must control their carbon in a beyond-Kyoto regime. They further agreed to a cap of 2°C additional global warming beyond the level of the pre-industrial age. They agreed to follow the science in doing so, but did not explicitly state how they would stay below that limit. They reiterated the Toyako goal of a 50 percent global reduction of greenhouse gas emissions, with a reduction of 80 percent or more for developed countries. The category of "developed" remained undefined, but it could conceivably include all existing and future members of the OECD, which now included Mexico and Korea. They further called for more specific targets in the form of national and medium-term goals.

The talks covered forests and land degradation, energy efficiency for buildings and cars, and a marginally stronger affirmation of the advantage of nuclear power as part of the energy mix. They enhanced the emphasis on carbon capture and storage.

Reaching out to their MEF and developing country partners, G8 leaders stressed that poverty eradication and climate change mitigation were linked and agreed on the importance of green growth plans for developing countries. They promised to work toward a greener economy and agreed that the global recession should not affect efforts to mitigate climate change. They called for continued work on lowering tariffs on goods related to climate change mitigation through the WTO and for letting the free market lead. They further agreed that efforts such as carbon cap and trade systems and taxation were appropriate policies. The leaders also called for accelerated investment in green technology and better diffusion, deployment, and cooperation with developing countries. They reiterated their Toyako initiative to launch 20 CCS projects by 2010 with increased investment and collaboration with developing countries, the IEA, and the Carbon Sequestration Leadership Forum (G8 2009). The G8 offered pro forma strong support for creating a post-Kyoto climate regime at COP-15 in December.

However, G8 leaders did not identify their medium-term goals, nor did they set a baseline year on which targets could be based and progress judged. They did little to demonstrate support for the contentious mitigation measures at the centre of the COP negotiations and largely avoided a major point of disagreement—adaptation financing in developing countries. They agreed to discuss adaptation at Copenhagen, and promised funding in principle, but committed to no specific amount and at no particular time. They urged the MEF to adopt quantifiable goals for emissions reductions by a specified year.

MEF leaders, meeting on the summit's second day, reiterated many of the G8 agreements. They agreed to a cap of 2°C unanimously and pledged to reduce greenhouse gas emissions from a business-as-usual baseline in the medium term. They set no quantifiable targets or timelines for emissions reductions either globally or among themselves. But they agreed to negotiate more concrete targets and timetables at Copenhagen. They also reiterated the importance of adaptation, with a focus on helping the world's poorest. They launched a global partnership for low-carbon, climate-friendly technology, and stressed the need for transparency and predictability of investment in green technology. They agreed, as a soft medium-term target and timetable, to double public sector investment in low carbon technology by 2015, and, in the very short term, to establish national technology roadmaps and action plans by November 15, 2009.

Major emerging and some developing countries, for the first time, thus agreed to reduce emissions as part of a new global climate control regime. The G8 and MEF both referred two crucial items—overall emissions reduction targets and adaptation funding—to the Copenhagen conference.

Dimensions of Performance

The L'Aquila Summit was a strong success on climate change, across most dimensions of summit performance (see Appendix A-1).

Domestic Political Management
L'Aquila's domestic political management, was solid, as all G8 leaders attended. However, within the G5, China's Hu Jintao had to rush home upon arrival in Italy to deal with the domestic shock of a separatist uprising in his country's far western region. His absence, as the leader of one of the world's two top greenhouse gas emitters, constrained what the MEF could do. No communiqué compliments were issued to any G8 member on climate change.

Deliberation
In public deliberation, performance at L'Aquila was very strong. The 5,559 words on climate change were the highest of any G8 summit through 2014 (see Appendices A-1 and C-1). Climate-related passages represented one third of the communiqué, again the highest portion ever and about twice as much as the second-placed summit in 2008. About 16 percent of the communiqué's paragraphs dealt directly with climate change (see Appendix C-1). Three of the five summit documents mentioned climate change.

Direction Setting

In direction setting, L'Aquila's climate performance was also very strong. It gave priority placement to climate change 17 times in its communiqué, by far the highest total of any summit ever (see Appendix A-1). Moreover, for the first time it directly connected climate change with the G8's foundational mission and principles of democracy and human rights, linking climate with democratic principles five times and with human rights once.

The G8 furthered its explicit support for climate science, a considerable success given the US government's former position on climate change. It recognized the "broad scientific view" that global temperatures must not rise more than 2°C, and indicated willingness to take part in the global response necessary to meet that goal (G8 2009). Leaders expressed understanding of the need to see global emissions peak as soon as possible, and then begin to decline.

Decision Making

In its decision making, performance was very strong. G8 leaders made 42 commitments on climate change, the third highest ever and only slightly fewer than the peak of 54 in 2008 and the 44 in 2007 (see Appendices A-1 and D-1).

These commitments ranged widely in their scope, specificity, and connectivity to other issues. Given the great financial crisis, the leaders "committed to promoting economic recovery together with a significant change in investment patterns that will accelerate the transition towards low-carbon, energy efficient growth models" (G8 2009). Other commitments identified specific sectors or international bodies where action would be directed: for example, "participation in ICAO, IMO, and UNFCCC processes" would enable "an agreed outcome for the post-2012 period to rapidly advance towards accelerated emissions reductions for the international aviation and maritime sectors." A handful expressed support for the upcoming UNFCCC conference, but few of them were directed at driving forward the conference in any concrete way. Beyond the call for a 80 percent reduction by 2050, few specific overall reduction targets and timetables were set (G8 2009).

Delivery

In delivering on these climate commitments, L'Aquila's performance was significant. Compliance with the five climate change commitments assessed averaged +0.64, up from +0.53 the year before (see Appendix E-1). All countries complied with their commitment to increase investment in research and development on clean technology. All except Russia complied with the call to implement a post-2012 technology and financing regime. With a score of zero, CCS remained a work in progress, with weaker compliance than the year before. The commitment to reduce deforestation and forest degradation, including by assisting developing countries, averaged +0.67 with three countries—Russia, Italy, and Canada—failing to comply.

Development of Global Governance

L'Aquila's performance was strong in developing global governance. It made one reference to the G8 environment ministers, but made 19 references to 10 different international institutions outside the G8 (see Appendices F-3 and F-5). As in 2008, G8 leaders spelled out explicit and repeated support for the UNFCCC process, invoking the UN and Copenhagen more than a dozen times. This strong support for UN governance of climate negotiations in 2009 as in 2008 underscored the G8's understanding of its inherited responsibility in making the upcoming Copenhagen negotiations a success.

But the G8 went well beyond a singular reliance on the UN's COP/MOP, the UNFCCC secretariat, or the UN itself. Other frequently referenced international organizations included the IEA and the WTO. Calls for action or declarations of support on climate were also made to the IEA, Joint Oil Data Initiative (JODI), MEF, World Meteorological Association, the Global Earth Observation System of Systems, and the UNFCCC's Reducing Emissions from Deforestation and Forest Degradation (REDD) program.

Causes of Performance

Shock-Activated Vulnerability
The first cause of L'Aquila's strong performance was high shock-activated vulnerability. This vulnerability came from the shock of droughts, forest fires and other extreme weather events, as well as soaring world food prices, high world oil prices, and the natural disaster visible to the G8 leaders when they toured earthquake-devastated L'Aquila during their summit.

The leaders noted in their communiqué "the increased threats of natural disasters and extreme weather phenomena caused by climate change, such as increased flooding, storm surges, droughts and forest fires" (G8 2009). They also made several references to the threat and vulnerability of developing countries from climate change. In 2009 oil prices doubled from \$44.81 on January 1 to \$81.82 by January 1 the following year (see Appendix G-3). The number of natural disasters remained high (see Appendix G-5). And overall greenhouse gas emissions of both the G8 and G5 were also high, even with a slight decline due to the great global recession. Emissions from China and India continued to rise steadily, with India's surpassing those of all G8 members save the US and the EU (see Appendices G-7 and G-8).

The global financial crisis, which had begun late in 2008 and was now reaching its peak, did not constitute a diversionary shock. It was left to the G20—now meeting at the summit level—to cope with the crisis at its April and September encounters, leaving the G8 free to concentrate on other issues such as climate change.

Multilateral Organizational Failure
The second cause of L'Aquila's strong performance was substantial multilateral organizational failure, above all by the UN and its dedicated process and organization for controlling climate change. The UN process was making little progress as preparations for its defining Copenhagen conference entered the final five months. Although leaders were scheduled to participate, no other UN summits lent support.

Predominant Equalizing Capability
The third cause of L'Aquila's strong performance was the predominant equalizing capability of the expanded, equal MEF and G20, which had now arrived alongside the G8 to govern climate change. The global financial and economic crisis, starting in the US and Europe, had hit the developed G8 members harder and faster than it hit major emerging ones. This made the expanded MEF and G20 far more globally predominant than the G8 and now more internally equal than before (Kirton 2013a).

Common Principles and Characteristics
The fourth cause was common democratic principles, which caused G8 success but constrained the larger G20. In 2009, G8 democratic like-mindedness increased due to

advances in Russia, while a reduction of rights and liberties was experienced by G20 members Brazil, Indonesia, and Mexico (Kirton 2013a). A further constraint came from Berlusconi's broader group of L'Aquila guests, which included the leaders of Libya and Egypt.

Political Cohesion
The fifth cause of L'Aquila's success was solid political cohesion, which constrained any stronger success. A very experienced host, Silvio Berlusconi was neither committed to nor competent at climate change control. The highly popular Barack Obama and his party, which controlled both chambers of Congress, were far more committed than George Bush had been. But Obama was attending his first G8 summit and chose to invest his formidable political capital on the easier and more urgent food security initiative instead. He also faced growing opposition at home to strong action on climate change, as the US economy continued to falter and unemployment remained high.

Among the G8 veterans, Germany's Angela Merkel was committed to climate change. Canada's Stephen Harper was not, especially given the minority status of his government and his political and economic interests in further developing the oil sands in the province of Alberta. All the leaders of major non-G8 countries present at L'Aquila lacked personal commitment and competence on climate change control.

Constricted Club Participation
The sixth cause, constricted club participation, was very small at L'Aquila and was thus a strong constraint on performance. Going well beyond the controlled, somewhat continuing and socialized leaders of the G8 and MEM-17, Berlusconi kept inviting more leaders to the summit on an ad hoc basis, including those from Egypt and Libya, which were not a member of any club, hub, or network that the G8 valued. Raising more money from willing G8 members for food security on the summit's last day was possible. Having all of the 40 participants take transformational steps to control their carbon was not.

Mobilizing the G20, 2009

The new G20 summit, with far fewer of the same countries that met at L'Aquila, and almost all their leaders, met twice in 2009, following its first gathering in November 2008 in Washington. The G8, extending to include other rising powers in global climate governance, now beyond the G5 and MEF, had its fullest, firmest, and most successful expression in this G20, which was more constricted and controlled than the L'Aquila G8 in taking up the climate cause (Kirton 2013a).

London, April

The second G20 summit, held in London on April 1–2, 2009, had already started actively to address climate change before the G8 met at L'Aquila in July (Kirton 2013a). The determination of host British prime minister Gordon Brown to put climate on the agenda was driven in part by his Commonwealth associates from Africa. When the leaders of several African countries, including Ethiopia's Meles Zenawi and Tanzania's Jakaya Kikwete, visited Brown to discuss their expectations for the summit, they began not by pressing the issue of development, as had been expected by the British, but by expressing the urgent need for climate change control. They said the melting ice cap on the top of Mount Kilimanjaro was a signal of the current climate crisis, which required an immediate response. Committed, competent, influential civil society leaders were also calling for action, as veterans from the Club of Rome met with the British hosts to press them to add climate change to the summit agenda and its communiqué.

Persuaded by the entreaties of Commonwealth colleagues abroad, civil society at home, and key advisors such as Mark Malloch Brown within government, the British tried to advance the climate cause through the G20 process. It was not an easy task. There was a divide between those who sought action and those who resisted expanding the summit agenda in such a way, especially as the financial and economic crisis was cresting. Several G20 members felt that finance and the economy should be the sole focus, arguing that the G20 was the wrong place to negotiate a deal on climate change and that the existing UNFCCC's COP/MOP was the logical forum.

The British first tried to introduce climate change in a G20 sherpa meeting in mid March. Because the meeting was soon after the Obama administration had taken office, prospects for advancing climate change control looked much better than they had at the Washington Summit chaired by George Bush. The European countries were interested in moving the US forward on climate change. Canada, however, took the position that the G20 should deal with the financial crisis first, although it would likely follow if the US committed to something specific on climate change control. Japan, which had emphasized climate change at its 2008 G8 summit, wanted to build on those initiatives. Some non-G8 members of the G20 would also welcome a climate change agenda.

In the end, the G20 leaders came together at London to advance climate change. They endorsed the principles of intergenerational equity and sustainability and pledged to "build an inclusive, green, and sustainable recovery" (G20 2009b). Their specific commitment to climate change, however, was added to the last article in the communiqué under the penultimate section titled "Ensuring a Fair and Sustainable Recovery for All." It promised only that the leaders would "address the threat of irreversible climate change, based on the principle of common but differentiated responsibilities, and to reach agreement at the UN Climate Change conference in Copenhagen in December 2009." These supportive statements were confined to the UN, and contained the first G20 summit commitments on climate change.

The key cause of this G20 start was the visible effect of climate change, seen atop Mount Kilimanjaro and elsewhere. The negative consequences of a changing global atmosphere were arriving in the world's poor countries and a North-South coalition of countries could come together to press for expanded controls. With this chronic climate crisis now underway, it was up to a range of international institutions involving developed and developing countries, including the G20, to act. Another contributing cause was the start of civil society engagement in the intergovernmental G20 summit.

Pittsburgh, September

A strong subsequent step came soon, at the third G20 summit held in Pittsburgh on September 24–25, 2009. The popular Obama sought to build on London's start by creatively linking climate change to the financial and fiscal issues at the G20 agenda's core.

On September 23, G20 sherpas saw the communiqué produced by the American hosts for the first time. At the meeting that immediately followed, the only difficult issue was climate change finance. It had been on the sherpas' agenda all along, but the US had not received much support. It was trying hard to secure an agreement on financing climate change that the finance ministers would address. It was still early enough in the Obama administration to make the Europeans hopeful. Canada was also interested, as long as the result did not eliminate the role of the private sector. Most of the sherpas wanted to drive financing toward the private sector, except for the British sherpa, who emphasized public sources of finance. Almost everyone was able to find something to agree on during the lengthy deliberations that evening.

The exception was China, which had been reluctant from the start. It argued that such negotiations should take place at the UN. During lunch the next day, on September 24, as the G20 leaders were meeting and their sherpas were reviewing the climate change text, China vetoed the draft passage on climate change finance. The US recognized that the discussion would not proceed any further and suggested dropping the longer paragraphs on climate finance. As a result, the communiqué contained only a short passage on climate change, with very little substance. It did, however, put principles in place to guide where the required money would come from, how it would be governed, and to whom it would flow. This would ultimately help developing countries mitigate their carbon pollution and cope with the destructive consequences of atmospheric temperature change already underway.

In another, far more important advance, taken outside the framework of the divided Rio-Kyoto regime, G20 leaders pledged to "phase out and rationalize over the medium term inefficient fossil fuel subsidies" (G20 2009a). They thus simultaneously and synergistically made a major contribution to fiscal consolidation, energy security, climate change control, and much else. This commitment was an American initiative, contained in the preparatory papers that had been distributed among the sherpas in advance. It was initially well received—consumer subsidies harmed the price setting that would provide consumers shelter from high electricity prices driven by fossil fuel–fired plants. The wording spoke of voluntary actions. But the BRICs objected to the American draft and produced an alternative text. At 3 a.m. on September 24, a compromise was reached. The BRICs, with India as their agent, succeeded in including the word "rationalize" in the commitment, in addition to the full "phase out" of subsidies. The sherpas put in place a work plan that pleased everyone, including the oil producers, notably Saudi Arabia.

In a third advance, G20 leaders promised to reduce tariffs on environmentally enhancing and climate friendly goods and services, as part of an agreement to reduce other barriers to trade and investment. However, no concrete measures were specified to achieve this goal.

Continuing the trend from previous G8 and G20 summits, G20 members used closely connected plurilateral summit institutions to advance the G20's work. The leaders of advanced countries, including the US, subsequently met with Mexico and the UN for a breakfast strategy session at the leaders' meeting at the Asia Pacific Economic Cooperation (APEC) forum in Singapore in November. Mexican president Felipe Calderón also held a video conference with several key leaders and the UN's Ban Ki-moon to discuss climate change in the lead-up to the COP summit in Copenhagen that December.

Pittsburgh thus made one important innovative initiative to control climate change: the commitment to eliminate fossil fuel subsidies. The IEA among others estimated that, if fully implemented, such action would cut a full one-tenth of the emissions causing climate change. Much like the G8, the G20 successfully reached out to put a new, inclusive, expanded, innovative regime in place. But it did not revive the failing old UN regime at its next COP, taking place in Copenhagen 10 weeks later.

Creating the Copenhagen Accord Five, 2009

The Copenhagen COP was a major failure for global climate governance as grounded in the principles, promises, and processes of the old, now obsolescent UN Rio-Kyoto regime (Dimitrov 2010b, 2010a). The much-hyped meeting, held at the leaders' level, produced only a non-binding political agreement with voluntary targets and few concrete steps to reduce climate emissions or create effective climate mitigation strategies. All the tough climate questions came to a head, with no more opportunities to delay or postpone an agreement. COP-15 could no longer sidestep the questions of binding mid-term emissions reduction targets, baseline years for gauging emissions reduction, or specific numbers for adaptation financing. The negotiations failed to produce an adequate post-Kyoto climate control architecture. Moreover, its small steps to salvage something came not from the full multilateral COP but from a new, small, ad hoc, one-off plurilateral summit subset of the G8-G5 and G20—the so-called Copenhagen Accord five.

The Copenhagen Accord was thus the product of a last-minute response to the UN's failure, arranged by an ad hoc group of the five big core climate polluting powers, rather than a product of the comprehensive consensus and cooperation that the UN process required. This group of five countries had been whittled down from an earlier group of 20 that had taken over the negotiations from the full COP. The UN process had been unsuccessful due to its own internal, structural flaws, as the prospect of securing a 192-country consensus was too daunting, and the complicated, bureaucratic negotiating procedure too cumbersome and complex.

Preparations

The original purpose of COP-15 had been to negotiate a binding treaty that would replace the Kyoto Protocol, due to expire in 2012. Under the UNFCCC process, COP-15 would be the culmination of two years of dialogue and meetings initiated at COP-13 in Bali in 2007, where the parties had agreed to the Bali Action Plan. It established a subsidiary body to address the main issues in future negotiations: long-term cooperative action, mitigation, adaptation, finance, technology, and capacity building. Parties also agreed to the Bali Roadmap, which outlined the next two years of negotiations to create a post-Kyoto agreement. The Copenhagen conference would be the defining moment when a new, legally binding text would emerge. However, as the deadline and the conference approached, a successor to Kyoto had still not taken shape. Even before the December negotiations began, Copenhagen looked unlikely to be able to deliver what was expected. In the absence of substantial progress in other international forums, traditional divides blocked progress on key unresolved issues, including baseline years, mid-term targets, and the responsibilities of developing countries.

In the lead-up, developing countries, including Brazil, China, India, and South Africa, stated clear goals: they expected all parties to again adopt the old principle of common but differentiated responsibilities, which meant binding, ambitious targets for developed countries and voluntary, less ambitious, individualized plans for developing countries. Uncertainty remained about what China would accept, causing concern over the negotiations from the outset. Without world's largest emitters on board, even a robust agreement would mean little.

Another priority for developing countries was funding for climate adaptation. Before the summit began, there was disagreement over how much funding was necessary. This had been unaddressed in the G8 summits with all their guests. Thus the amount that developed countries would agree to give was still up in the air. The US proposed a $10 billion fast-track fund, but developing countries criticized it as vastly inadequate—some sources estimated that long-term adaptation financing would require up to $800 billion (Goldenberg 2009).

On mitigation, the major concern was the American position. Thus far the Obama administration had proposed only minor reduction targets for the US, lowering emissions by 17 percent, but with 2005 as a baseline year (US Department of Energy 2009). In stark contrast stood the stronger EU commitment of a minimum 20 percent reduction from 1990 levels, with a potential reduction of 30 percent overall (González-Aller Jurado and Falkenberg 2010). With Canada agreeing to commit the same as the United States, and Japan uneasy about making an ambitious commitment before other developed countries did, many predicted a disappointing outcome on mitigation.

At a preparatory meeting in Bangkok in October, after the US, Japan, Canada, and Australia had several proposals, there was a discussion on "general mitigation" that meant requirements applicable to all parties (Rajamani 2009, 13). The Americans proposed a differentiated reporting schedule different from the UNFCCC's, with a new category for "developing countries with greater than [x] per cent of world emissions."

At Copenhagen

At the Copenhagen conference, launched on December 7, several official meetings and negotiation tracks met at once, most notably the annual COPs and MOPs (Dimitrov 2010b, 2010a). For the first week, officials participated in events, talks, and meetings. Environment ministers and negotiators also began laying the groundwork for a draft agreement, to be given to leaders to accept on the conference's concluding days. Bilateral talks, side meetings, and small group negotiations tried to steer the conference toward consensus among the 192 countries in attendance. Many days passed without the appearance of a satisfactory draft text (Dimitrov 2010b).

Prospects for creating a binding treaty diminished quickly. Under UNFCCC rules, such a treaty required unanimous agreement. By December 15, an acceptable draft had still not been circulated. The subsidiary body that had been created at Bali completed its final sessions without producing a working text, despite having worked on drafts for two years. Passages were too "heavily bracketed"—meaning they were still being negotiated—to be realistically considered for adoption (Dimitrov 2010b, 808). Even before the heads had met, failure seemed imminent.

Countries began turning to other methods of negotiation. On December 17, one day before the scheduled close of the conference, COP-15 established a group of "Friends of the Chair" to facilitate the deadlocked discussions. Twenty-two core countries took over the negotiating process. This move met with great disapproval from some developing countries, although the group's membership represented every major region. Resistance to that group squandered more of the conference's remaining hours. Such a conflict symbolized the seemingly irreconcilable tension between "the UN principle of global democracy and the pragmatic need for problem-solving" (Dimitrov 2010b, 809). Nonetheless, the group went ahead, and leaders joined in as they arrived. The real participants were now the United States, Britain, Sweden, Spain, Saudi Arabia, Russia, Norway, Mexico, the Maldives, Lesotho, South Africa, Bangladesh, Algeria, Denmark, Germany, France, India, Ethiopia, Colombia, Korea, China, and Brazil.

This "Friends of the Chair" group did not make the progress necessary for an agreement to be formally adopted. It did manage to produce a draft text, but following a process that diverted sharply from previous UNFCCC negotiations. While the previous process had produced "draft texts of 200-plus pages negotiated over two years," the new group created a 2.5-page draft (Dimitrov 2010b, 810–11). Whittling down the membership and text so drastically represented a move toward a more G8/G20-style process of controlled membership with leaders meeting without scripts, speeches, or stacks of documents, and instead simply putting on paper the things that they felt were most important. It potentially promised to produce the results that the 192-country UN process could not.

Yet even within these few pages, disagreements remained. China refused to permit third-party verification of its emissions reductions, as long as the treaty was legally binding. According to some accounts, a few developing countries, including China, India, and Brazil, resisted any wording that implied a legal obligation, favouring the language of "discussion" and "dialogue" versus the "verification" and "scrutiny" preferred by Obama (Chengappa 2009).

As the final day was ending, many officials began to accept that a deal was out of reach. While the world's media and others began to head to the airport, Obama had one more scheduled meeting with Chinese prime minister Wen Jiabao. Upon arriving, Obama unexpectedly found the prime minister conversing with Brazil's Luiz Inácio Lula da Silva, India's Manmohan Singh, and South Africa's Jacob Zuma, apparently invited by Wen. This "accidental" meeting of the five leaders sparked a breakthrough that would lead to the "Copenhagen Accord." Disagreements over the wording of the document were settled by simply removing the parts that China and others had objected to. The five leaders agreed to drop the reference to legally binding as well as any commitment to monitoring and verification, resulting in a weak political agreement. Nonetheless, the very fact that they had produced an acceptable document meant that the informal, plurilateral Copenhagen Accord five had accomplished what the fully multilateral UNFCCC process could not.

Late Friday night, Obama announced that the five countries had negotiated a non-binding political agreement. Calling it "meaningful and unprecedented," he praised it as the first step toward a concrete, legal text that would hopefully be in place by the end of 2010 (Lee 2009). But the accord would still need formal and unanimous approval and that approval was not immediately forthcoming. Venezuela, Bolivia, Cuba, and Sudan were the main source of this disagreement, claiming they could not accept the accord given its lack of ambitious emissions reduction targets. The conference's conclusion was extended to December 19. In the end, the conference agreed merely to "take note of the Copenhagen Accord," since the consensus required to adopt it officially was not possible (UN 2010). Most countries were expected to accept the accord in the upcoming months, with Ban Ki-moon urging countries to do so as soon as possible. But as Copenhagen ended, it remained unclear how many countries would support the agreement.

Results

In the end, the Copenhagen Accord failed to revive or repair the failing Rio-Kyoto regime (Dimitrov 2010b, 2010a; Rajamani 2009; Dubash 2009). It did acknowledge climate change as "one of the greatest challenges" (UN 2010, 5). It recognized that the global increase in temperature should not exceed 2°C. It stated that deep cuts in global emissions should be pursued through the old concept of "common but differentiated responsibilities" based on "respective capabilities." Countries were asked to submit voluntary emissions reductions targets for 2020 by January 31, 2010. The accord failed to include the goals pledged at the two previous G8 summits, which called for a 50 percent reduction in global emissions by 2050, with an 80 percent reduction in developed countries. It did, however, go further than the G8 by asking countries to submit specific mid-term goals by a certain date.

Most developed countries had already determined their mid-term goals nationally, meaning that in most cases the Copenhagen Accord would require them to do little more than before. It did retain the agreement to limit the rise in temperature to 2°C, but provided no guideline for how this would be done. It briefly noted the need to provide technology transfer, capacity building, and financial resources for adaptation in developing countries, but without many details. Assistance to developing countries was set at $30 billion by 2012 and $100 billion by 2020, with the money to be split between adaptation and mitigation efforts. Finally, it established the Copenhagen Green Climate Fund to support mitigation, adaptation, capacity building, technology transfer, and reforestation projects, although no money was specifically set aside for it. Given that the Copenhagen Accord was not legally binding, and contained little in the way of concrete, coordinated action, it was at best a minor advance.

Yet in one key area the Copenhagen Accord represented a step toward securing the new G8-G20–led inclusive regime. It allowed for one schedule, one containing "economy-wide emissions targets" set by developed Annex 1 countries and the other that would "document mitigation actions by non-Annex 1 countries" (Dubash 2009, 9). This "dual schedule approach" suggested that "a single, if somewhat separated track" was likely.

Causes of Performance

The failure of the UN and its Rio-Kyoto regime at Copenhagen had several causes. First, the UNFCCC process involved difficult criteria for producing an official document, as all parties, without exception, had to agree. Adding to this unit veto predicament were the different goals of developed and developing countries, essentially pitting the North against the South. UN culture can often send negotiations into a polarized developed versus developing world, with North and South separating into camps, a division that may not even reflect the true interests of individual countries (Stavins and Stowe 2010). It was only when the UN process was abandoned in favour of the "Friends of the Chair" format that North and South worked together to produce a draft text. The UN process seemed to perpetuate differences, rather than allowing countries to work as equals, making these differences too difficult to overcome in order to reach a legally binding agreement.

Moreover, the UNFCCC process itself was confusing and complicated. It lacked the efficiency and personal, high-level interaction and comprehensive agenda of a G8 or G20 summit. Eleven days of negotiations involving thousands of participants, multiple decision-making bodies, contact groups, drafting groups, and formal and informal sessions meant an extremely complex challenge for negotiators. The COP and MOP were often sidetracked by the complex negotiations process, leaving limited time for meeting and discussing real agenda items. Even prior to the end of the summit, certain items had been pushed onto the agendas for the following year (Massai 2010). Leaders were present only during the final days of the conference, and little progress was made before their arrival. Without a high-level presence, most of the conference was an unnecessary exercise between negotiators, delegates, and other government representatives, who ultimately did not possess the political capital required to reach an acceptable agreement. The COP-15 summit showed the relative success of the flexible, leaders-only style of negotiations. Constricted participation finally removed the immense coordination problems that had deadlocked the rest of the conference. Although the final agreement was disappointing, this negotiation allowed an agreement to be produced. Had the G8, MEF, and G20 had the political will to do some of this work in the lead-up to Copenhagen, the 2009 Copenhagen conference could have produced much more.

Conclusion

Overall, there were major advances in 2008 and 2009, but there were also many disappointments in the governance of climate change. First, the G8 itself did not meet the growing global challenge of controlling climate change. Although it embarked on several worthwhile initiatives during this time, the cumulative impact was not enough either to put all the pieces in place for a new global climate regime or to facilitate such a regime at the UN. Second, the MEM/MEF also yielded disappointing results. The G8 succeeded at incorporating this new institution into its summit process, but not in mobilizing it to make strong, ambitious commitments. The G20 picked up some of the work on climate change, but devoted nowhere near the attention to it that the G8 had during this time. G8 members did, however, push the G20 to place climate change further up its list of priorities, helping make the transition of the G20 from a financial body to one concerned with the broader issues of global environmental governance—a significant accomplishment indeed.

The G8's greatest failure lay in its lack of action to mobilize success at Copenhagen. Such great potential existed in the two years leading up to COP-15, but the G8, MEM/MEF, and G20 were unsuccessful. The divide at the UN between developed and developing countries was insurmountable in the end.

In the aftershock of Copenhagen's failure, it remained to be seen if the G8 and G20 would pick up the slack and move global climate governance forward. In some ways, the UN's failure renewed the demand and potential for G8 and now G20 leadership on global climate governance. If and how the G8 and G20 would assume this responsibility, however, remained unknown.

References

Agence France Presse (2007). "Japan Aims to Lead Post-Kyoto Climate Change Fight." Tokyo, March 20.

Canada. Prime Minister's Office (2008). "PM Hails Breakthrough on Climate Change at 2008 G8 Summit." Toyako, July 9. http://www.pm.gc.ca/eng/news/2008/07/09/pm-hails-breakthrough-climate-change-2008-g8-summit (January 2015).

CBC News (2007). "Critics Target PM's G8 Climate Change Message." June 5. http://www.cbc.ca/news/canada/critics-target-pm-s-g8-climate-change-message-1.640238 (January 2015).

Chengappa, Raj (2009). "The Earth in ICU." *India Today*, December 24. http://indiatoday.intoday.in/story/The+earth+in+ICU/1/76465.html (January 2015).

Dimitrov, Radoslav S. (2010a). "Inside Copenhagen: The State of Climate Governance." *Global Environmental Politics* 10(2): 18–24. doi: 10.1162/glep.2010.10.2.18.

Dimitrov, Radoslav S. (2010b). "Inside UN Climate Change Negotiations: The Copenhagen Conference." *Review of Policy Research* 27(6): 795–821.

Dubash, Navroz K. (2009). "Copenhagen: Climate of Mistrust." *Economic and Political Weekly* 44(52): 8–11.

Erlanger, Steven (2008). "France Sticks With Nuclear Power Oil Prices and Climate Change, It Says, Vindicate '50s Choice." *International Herald Tribune*, August 18.

Fujioka, Chisa (2008). "Britain Dismisses Japan Climate Change Plan." March 14, Reuters. http://www.reuters.com/article/2008/03/14/us-climate-g-idUSSP6817320080314 (January 2015).

Fukuda, Yasuo (2008). "Special Address on the Occasion of the Annual Meeting of the World Economic Forum." Davos, January 26. http://japan.kantei.go.jp/hukudaspeech/2008/01/26speech_e.html (January 2015).

G8 (2008a). "Chair's Summary." Hokkaido, July 8. http://www.g8.utoronto.ca/summit/2008hokkaido/2008-summary.html (January 2015).

G8 (2008b). "Environment and Climate Change." Hokkaido, July 8. http://www.g8.utoronto.ca/summit/2008hokkaido/2008-climate.html (January 2015).

G8 (2008c). "G8 Hokkaido Toyako Summit Leaders' Declaration." Hokkaido, July 8. http://www.g8.utoronto.ca/summit/2008hokkaido/2008-declaration.html (January 2015).

G8 (2009). "Responsible Leadership for a Sustainable Future." L'Aquila, Italy, July 8. http://www.g8.utoronto.ca/summit/2009laquila/2009-declaration.html (January 2015).

G8 Research Group (2009). "2008 Hokkaido-Toyako G8 Summit Final Compliance Report." June 30. http://www.g8.utoronto.ca/evaluations/2008compliance-final/index.html (January 2015).

G20 (2008). "Declaration of the Summit on Financial Markets and the World Economy." Washington DC, October 15. http://www.g20.utoronto.ca/2008/2008declaration1115.html (January 2015).

G20 (2009a). "G20 Leaders Statement: The Pittsburgh Summit." Pittsburgh, September 25. http://www.g20.utoronto.ca/2009/2009communique0925.html (January 2015).

G20 (2009b). "Global Plan for Recovery and Reform." London, April 2. http://www.g20.utoronto.ca/2009/2009communique0402.html (January 2015).

Goldenberg, Suzanne (2009). "Obama's Arrival Expected to Inject Fresh Momentum into Copenhagen Talks." *Guardian*, December 17. http://www.theguardian.com/environment/2009/dec/17/barack-obama-copenhagen-hillary-clinton (January 2015).

González-Aller Jurado, Cristóbal and Karl Falkenberg (2010). "Letter: Expression of Willingness to Be Associated with the Copenhagen Accord and Submission of the Quantified Economy-Wide Emissions Reduction Targets for 2020." January 28, European Union and European Commission, Brussels. http://unfccc.int/files/meetings/cop_15/copenhagen_accord/application/pdf/europeanunioncphaccord_app1.pdf (January 2015).

Japan. Ministry of Foreign Affairs (2008). "Financial Mechanism for 'Cool Earth Partnership'." Tokyo, November. http://www.mofa.go.jp/policy/economy/wef/2008/mechanism.html (January 2015).

Kirton, John J. (2013). *G20 Governance for a Globalized World.* Farnham: Ashgate.

Kirton, John J. and Madeline Boyce (2009). "The G8's Climate Change Performance." August 12. http://www.g8.utoronto.ca/evaluations/g8climateperformance.pdf (January 2015).

Lee, Jesse (2009). "'A Meaningful and Unprecedented Breakthrough Here in Copenhagen'." December 18. http://www.whitehouse.gov/blog/2009/12/18/a-meaningful-and-unprecedented-breakthrough-here-copenhagen (January 2015).

Luckhurst, Jonathan (2013). "Building Cooperation between the BRICS and Leading Industrialized States." *Latin American Policy* 4(2): 251-68. doi: 10.1111/lamp.12018.

Major Economies Meeting on Energy Security and Climate Change (2008). "Declaration of Leaders Meeting of Major Economies on Energy Security and Climate Change." Hokkaido, Japan, July 9. http://www.g8.utoronto.ca/summit/2008hokkaido/2008-mem.html (January 2015).

Massai, Leonardo (2010). "The Long Way to the Copenhagen Accord: Climate Change Negotiations in 2009." *Review of European Community and International Environmental Law* 19(1): 104–21.

Rajamani, Lavanya (2009). "'Cloud' over Climate Negotiations: From Bangkok to Copenhagen and Beyond." *Economic and Political Weekly* 44(43): 11–15.

Stavins, Robert N. and Robert C. Stowe (2010). "What Hath Copenhagen Wrought? A Preliminary Assessment." *Environment* 52(3): 8–13. http://www.environmentmagazine.org/Archives/Back%20Issues/May-June%202010/what-wrath-full.html (January 2015).

Tanaka, Miya (2008). "Japanese PM to Seek Bush's Cooperation on North Korea, Climate Change." *Kyodo*, July 6. BBC Monitoring Asia Pacific.

United Nations (2008). "G8 Summit Good Start But Further Action Needed to Tackle Global Crisis, Says Ban." July 9. http://www.un.org/apps/news/story.asp?NewsID=27312#.VPC7obPF-hk (January 2015).

United Nations (2010). "Report of the Conference of the Parties on Its Fifteenth Session, Held in Copenhagen from 7 to 19 December 2009. Addendum. Part Two: Action Taken by the Conference of the Parties at Its Fifteenth Session " FCCC/CP/2009/11/Add.1, March 30. http://unfccc.int/resource/docs/2009/cop15/eng/11a01.pdf (January 2015).

United Nations Framework Convention on Climate Change (2009). "Fact Sheet: 10 Frequently Asked Questions about the Copenhagen Deal." November. http://unfccc.int/files/press/fact_sheets/application/pdf/10_faqs_copenhagen_deal.pdf (January 2015).

United States Department of Energy (2009). "President Obama Sets a Target for Cutting U.S. Greenhouse Gas Emissions." Washington DC, December 2. http://apps1.eere.energy.gov/news/news_detail.cfm/news_id=15650 (January 2015).

Wintour, Patrick and Larry Elliott (2008). "Bush Signs G8 Deal to Halve Greenhouse Gas Emissions by 2050." *Guardian*, July 8. http://www.theguardian.com/world/2008/jul/08/g8 (January 2015).

Chapter 10
Realization, 2010–2011

After the surprising disappointment of the 15th Conference of the Parties (COP) meeting in Copenhagen in December 2009, the United Nations process lost much of its momentum. Governing climate change faded from the agendas of most global forums. Still in shock after Copenhagen, the Group of Eight (G8) dropped the subject from its priorities, particularly at the Muskoka Summit in 2010 and the Deauville Summit in 2011. However, the Group of 20 (G20) continued to become more involved in the issue, pulling back only slightly at Toronto in June 2010, then returning in full force with strong performances at Seoul in November 2010 and Cannes in November 2011. Leadership in global climate governance thus passed to the G20, with the emerging powers of Mexico and Korea as well as the established power of France at the head.

Muskoka G8, June 2010

Preparations

Early in the preparations for the 2010 G8 summit, to be held at a resort in Muskoka in Ontario, Canada, on June 25–26, it seemed that climate change could be a central theme for Prime Minister Stephen Harper (Kirton, Kulik, et al. 2014b). The preliminary outline for his summit, drafted at the unusually early date of June 2008, had identified climate change as one of the three core subjects the 2010 summit would address. Even before the G8's L'Aquila Summit in July 2009, and now in the depths of the great recession, Harper (2009) had written about his 2010 summit as follows:

> The international negotiations in the United Nations on climate change will culminate in Copenhagen this December. Canada is working actively and constructively to achieve an ambitious and comprehensive new agreement, one that covers the vast majority of global emissions and includes binding commitments by all major economies. At the same time, a successful agreement in Copenhagen must also support and enable sustainable growth, including through the expansion of secure and affordable global supplies of clean energy.

> Achieving this goal will require leadership from Canada and its G8 partners, as well as from the other countries participating in the US-led Major Economies Forum (MEF) on Energy and Climate. A new partnership will be required among major developed and developing countries if real progress is to be achieved in the coming months. The MEF provides an important new process in this regard, one designed to provide political momentum to the UN climate change negotiations while also deepening global collaboration on the development and commercial deployment of clean energy technologies.

The phrase "binding commitments by all major economies" confirmed that Canada and the G8 intended to lead global climate governance by producing an expanded, inclusive regime. Harper did not list the G20 as a forum for addressing climate change. He instead chose the MEF. He also repeatedly indicated that he preferred climate change to be dealt with primarily at the United Nations Framework Convention on Climate Change (UNFCCC) and the MEF, which were specifically created for that purpose, unlike the G8 (Rennie 2010).

Canada assumed the chair of the G8 in January 2010, a month after announcing it would also host a G20 summit that year. In the immediate aftermath of Copenhagen, Harper quickly dropped climate change as a priority. He outlined his agenda and aspirations for his two summits at the World Economic Forum in Davos on January 28, referring specifically to climate change only once. He noted that "climate change disproportionately threatens the peoples least capable of adapting to it," but, beyond recognizing this threat, said little about how the G8 should proceed (Harper 2010b). By the end of the month, he had stated that the climate change agenda at Muskoka would be limited to "informal discussions" (Harper 2010a). Climate change had been downgraded on the G8's agenda to a very minor place. It would, if anything, be a small aspect of the development pillar, rather than a pillar in its own right as it had been in the years before.

This did not go unnoticed. An array of world leaders and environmentalists criticized Harper's choice, eventually prompting the Canadian government to remark that it did "anticipate" that climate change would "come up" at both the G8 and G20 summits (Rennie 2010). Yet Muskoka would join George Bush's Sea Island Summit in 2004 as the second G8 summit in the 16 years since 1994 where the G8 environment ministers would not meet beforehand to lay any groundwork (G8 Research Group 2010a). Moreover, Harper did not facilitate a discussion between the members of the G8 and the members of the Group of Five (G5), as had become customary in recent years. What support Harper did give on climate change involved clean energy and technology transfer, rather than direct mitigation or adaptation efforts (G8 Research Group 2010b). But it remained unclear how he would reach out to the MEF members, or what the MEF's next steps would be.

Divergence existed among the G8 members themselves over climate mitigation strategies. On the critical issue of the baseline year, the European Union countries favoured 1990, while the North Americans and Japan leaned toward 2005. There were considerable differences in how to gauge progress on reducing emissions. On medium-term reduction targets, Europe generally favoured a cut of 20 to 30 percent cut by 2020 (with the 1990 baseline), while the United States, Japan, and Canada proposed much lower cuts. As both the G8 and G20 summits approached, therefore, key climate change issues were no closer to resolution.

At the Summit

World leaders arrived in Toronto's airport on June 24 to prepare for the official welcome the following day in Muskoka about 200 kilometres away. At Muskoka Harper held official welcomes for G8 leaders alone, followed by a welcome and working session with the invited leaders from Africa: Algeria's Abdelaziz Bouteflika, Ethiopia's Meles Zenawi, Egypt's Hosni Mubarak, Malawi's Bingu wa Mutharika, Nigeria's Goodluck Jonathan, Senegal's Abdoulaye Wade, and South Africa's Jacob Zuma. This was followed by a welcome to the extended outreach leaders from Colombia's Álvaro Uribe, Haiti's René Préval, and Jamaica's Bruce Golding, then a working session with all the partners. These talks left little designated space for climate change to be addressed.

At the G8 working dinner on Friday evening, June 25, climate change was barely discussed. No G8 leadership on mitigation materialized, as the Canadian host had little desire to advance beyond the unspecified voluntary mid-term targets offered at previous G8 summits and now the Copenhagen Accord. Nor did any new commitments on adaptation materialize. Some observers noted how "strangely quiet" outreach leaders were on climate change during both the Friday and Saturday sessions, despite their previous calls for strong climate commitments (Woods 2010).

According to Harper, discussions did occur on the UN climate negotiations, but they were deliberately confined to assessing the state of the UN process and potential strategies for breaking the deadlock, rather than on what the G8 itself could do. Harper succeeded in framing the conversation solely in the context of the UNFCCC process. Throughout the Muskoka Summit, he maintained his position that the G8 did not have a responsibility or mandate to deal with climate change. During a news conference, Harper reiterated that "the real process is the UN process," so the G8 did not need to add to the discussion (Harris 2010).

There was little opposition to this position. This stood in deep contrast to the call of leaders such as the UN's Ban Ki-moon in the immediate lead-up for G8 action on climate change and the outrage of some world leaders and many civil society members who had condemned Canada for not including climate change on the summit's agenda. However, apart from a few minor mentions, most nongovernmental organizations (NGOs) were also silent on climate change. Only the environment-focused organizations, including Greenpeace, tried to keep the issue alive. Other development-based organizations had moved on from discussing the impacts of climate change on the world's poor to focus on the follow-up to the L'Aquila Food Security Initiative and the new Muskoka Initiative on Maternal, Newborn, and Child Health, which had become the summit's signature theme and key result (Kirton, Kulik, et al. 2014b). The aftermath of COP-15 had clearly destroyed momentum across the board, within the G8 process, NGO community, and UN.

Results

Nonetheless, Muskoka made some headway on developing the new inclusive regime on climate change control, notably on identifying baseline years, affirming the all-in principle, and promising fast-start climate finance and improved accountability.

On baselines, for the first time, the G8 communiqué suggested a year and range from which to measure and assess emission trends and reductions. In restating the goal of reducing greenhouse gases in the developed world by 80 percent by 2050, it added "compared to 1990 or more recent years" (G8 2010a). This offered a set of G8-agreed baseline years for the first time.

On the all-in principle, G8 leaders called for a "comprehensive, ambitious, fair, effective, binding, post-2012 agreement" that would include responsibilities for all major economies in the post-Kyoto world (G8 2010a). They thus reaffirmed their inclusive approach, seeking an agreement that required developing and emerging economies to take action. But there was no detail offered on what such an agreement would look like, with the answers presumably left to the COP meeting in Cancun in December that year.

On climate finance, the G8 promised to mobilize members' fast-start finance contributions in the immediate future. The declaration stated that the leaders were "putting in place" the contributions agreed upon at Copenhagen. However, while Copenhagen had called for "new and additional" funding, the G8 did not indicate that this would be the case (UNFCCC 2009a).

On accountability, just prior to the summit, the G8 had released its first full-scale, systematic accountability report, including a wide range of development-related issues (G8 2010b). At the summit itself, in a spirit of self-monitoring, the leaders approvingly assessed their progress on creating projects for demonstrating carbon capture and storage (CCS). Looking outward, in calling for full implementation of all parts of the Copenhagen Accord, they specifically noted "those related to measurement, reporting, and verification thereby promoting transparency and trust" (G8 2010a).

Advances were made on the bottom-up, sectoral approach introduced at the 2008 Hokkaido Toyako Summit. Several countries committed to accelerate the CCS process and achieve full implementation by 2015. The Muskoka Declaration spoke about the importance of clean energy and low carbon technology, as Harper had intended. It expressed support for the Global Bioenergy Partnership and committed to "facilitating swift adoption of voluntary sustainability criteria" in this regard (G8 2010a).

A few innovative principles were affirmed. The leaders endorsed "green recovery" as an aspirational concept (G8 2010a). They also noted the importance of climate-resilient economies, promising to share "national experiences and plans for adaptation" at an upcoming conference in Russia in 2011.

The Muskoka Declaration also reiterated G8 support for the UNFCCC. The G8 specifically pledged support for the new Copenhagen Accord. In a desire for inclusivity, it encouraged other countries to do the same.

Dimensions of Performance

Muskoka had substantial success on climate change (see Appendix A-1). Strong performances in domestic political management, deliberation, and direction setting were offset by only a significant one in decision making, a small one in delivery, and a very small one in the development of global governance.

Domestic Political Management
In its domestic political management, Muskoka's performance was strong. One communiqué compliment related to climate change, making Muskoka one of only four summits to issue such approval. That one compliment was exceeded only by the four issued to four different countries in 2007 (see Appendix B-1).

Deliberation
Muskoka's performance on deliberation was also strong. With 1,282 words on climate, taking 12 percent of all the communiqués, Muskoka was the sixth highest summit ever in the number of words and the third highest in the portion of attention accorded to climate change (see Appendix A-1). This was, however, a sharp drop from the all-time peak of 5,559 words or 33.3 percent of the communiqué at L'Aquila the year before (Guebert et al. 2011). There was no section of the communiqué dedicated to climate change at Muskoka, and it was grouped under the heading of "Environmental Sustainability and Green Recovery" (G8 2010a).

Direction Setting

In its direction setting at Muskoka, performance was strong. One priority placement reference to climate change made it one of only nine summits, even if this was a sharp drop from the peak of 17 the year before (see Appendix A-1). Moreover, it was only the second summit to link climate change to democracy, with two references following the five made at L'Aquila the year before. Muskoka also reaffirmed the all-in principle and highlighted the new green recovery one.

Decision Making

Muskoka's performance on decision making was significant. Its 10 climate change commitments were above the multi-year summit average of eight (see Appendix D-1). The total was the seventh highest ever, although it drop sharply from the 42 and 54 of the previous two summits. A few of these commitments were new. They contained a move to mobilize money, in the form of the commitment on fast-start climate finance.

Delivery

However, in delivering on these decisions, Muskoka's performance was small. On the three commitments assessed, compliance averaged +0.26, a sharp drop from the +0.64 the year before (see Appendix E-1). Only five summits where data for climate compliance are available did worse.

Compliance with the summit's two key commitments—implementing the Copenhagen Accord and setting goals on mid-term emissions reductions—was low, demonstrating a lack of political will to comply with even the modest promises made. Both Canada and the United States failed to take any action on implementing mid-term emissions reductions goals, while most European countries generally fared well. Implementation of the Copenhagen Accord was poor, with no member fully complying (see G8 Research Group 2011).

Development of Global Governance

Muskoka's performance on developing global governance was very small. There were no references to the G8 environment ministers, who did not meet that year, nor to the MEF, which did not assemble at the summit as it had for the previous two years. No new climate-relevant G8 institutions were created.

Outside the G8, there were five references to three bodies, in both cases below the summit's multi-year average on climate change (see Appendix F-5). Only the UNFCCC and its Copenhagen Accord received more than one reference (G8 2010a). The G8 pledged support to the UNFCCC, the Reducing Emissions from Deforestation and Forest Degradation (REDD) initiative, the Global Bioenergy Partnership, and the Intergovernmental Platform in Biodiversity and Ecosystem Services. It also issued brief instructions for the International Energy Agency (IEA).

Toronto G20, June 2010

On the same day that the Muskoka Summit ended, Harper and his G8 colleagues travelled down to Toronto for their expanded G20 summit. Here, too, climate change was not a priority and limited progress was made (Kirton 2013a).

Preparations

Even after announcing in December 2009 that he would host the next G20 summit, Harper had showed even less interest in addressing climate change at the G20 than at the G8. At Davos on January 29, 2010, he stated that the G20 would deal only with finance and economics, with no mention of climate change, and that no expansion of the agenda would be welcomed (Harper 2010b). At the final G20 sherpa meeting on May 24–25, the issue of fossil fuel subsidies arose, but any environment- or climate-related discussions had ended there (Kirton 2009b).

Throughout the preparatory process for Toronto, there was a firm economic focus. The summit would continue to work on the Framework for Strong, Sustainable, and Balanced Growth, but there would be no environmental or climate connection, even under the rubric of a green recovery. If the G20 was to deal with climate change at all, it would be on its previous Pittsburgh commitment to phase out inefficient fossil fuel subsidies. It was also possible that the advanced economies might start to offer their share of fast-start financing for climate change adaptation, according to the Copenhagen commitment, or discuss prospects for encouraging green growth.

In the weeks before the summit, Canada was criticized for neglecting climate change on the G20's Toronto agenda, just as the G8's Muskoka one had. An editorial entitled "Why Is Canada Not Putting Climate Change on the G20 Agenda?" asked the question most directly (Williams 2010). Other media sources noted the sharp remarks from Mexico, the EU, the UN, and an array of environmental groups on Canada's choice to keep the agenda narrowly economic. In response, Harper indicated that climate change would not be entirely excluded, noting, in the House of Commons that "a lot of subjects will be discussed [at the summits], including some issues surrounding climate change" (Rennie 2010).

At the final sherpa meeting in Toronto on the summit's eve, as at the previous sherpa meeting in May, the US tried to get the G20 to act on climate change (Kirton 2013a, 340). The BRIC group of Brazil, Russia, India, and China objected. The Chinese, in particular, did not want to deal with the issue at the G20. They had always taken the position that countries should meet their obligations under the Kyoto Protocol and discussions should be carried out within the existing UN framework. Everything else was unacceptable. The Indians, from all the ministries other than finance, shared this vision. The sherpas of some smaller G20 members had received no instructions about how to deal with climate change at the G20. They felt that the G20 had a three-tier membership, with them in the third tier. Consequently, Ban Ki-moon's public call for the G20 to advance climate change control at Toronto, proclaimed in Canada on the summit's eve, did not succeed.

At the Summit

Mexico insisted on putting climate change on the G20 agenda (Kirton 2013a, 346). China agreed that Mexican president Felipe Calderón speak about the UN climate change conference it would host in Cancun in December. Yet China remained adamant that climate change not be negotiated at the G20, even as a supplement to the UN. Its well-known position was that the G20 had been designed to deal with finance and should not discuss other issues, which were being managed elsewhere.

Climate change was thus scheduled to be discussed at lunch, along with trade. However, the discussion on trade lasted about an hour, leaving little time for other issues. When the topic of climate change finally arose, at Harper's request Calderón led off with an update on plans and prospects for Cancun. Ban, who was present, admitted that Copenhagen had not been a success, but argued that the G20 must be more pragmatic at Cancun by not insisting on a comprehensive deal. China's Hu Jintao did not speak. All the other emerging country leaders who did, save for Calderón, supported China's position on climate change.

In summing up the session, Harper said it was clear that the UN process was a failure and that leaders had to look elsewhere for progress, to forums such as MEF. He concluded by saying, with all due respect to the UN secretary general, that there was no way the UN could make a deal. He thus had the G20 essentially endorse the American-invented and -led MEF. In the Toronto communiqué the G20 leaders merely thanked Mexico for hosting the Cancun climate change conference.

The leaders did, however, reaffirm in the communiqué their Pittsburgh promise to phase out inefficient fossil fuel subsidies. But they did little to extend the definition or move the actual implementation ahead. They also noted their commitment to a green recovery.

At Russia's initiative, the communiqué referred to the accident at the Deepwater Horizon offshore drilling rig in the Gulf of Mexico—one of the worst ecological disasters in US history. Russia proposed the creation of a legally binding agreement on a marine life protection area, by working through a subgroup of the G20's Energy Working Group. Its proposal flowed from a broader Russian view that the G20 should begin with discussions, then move to consensus on principles, and then progress to decisions that could be converted into legally binding agreements and conventions in fully multilateral, legally authorized international organizations.

Results

The Toronto Summit produced very little directly on climate change. It was mentioned only in passing, in connection with other issues, but without any climate-related commitments. The much-noted Framework for Strong, Sustainable, and Balanced Growth, launched at the Pittsburgh Summit, was not linked with environmental sustainability or climate change. Countries reaffirmed their support for the Copenhagen Accord, the UNFCCC, and the upcoming Cancun conference, much as the G8 had done. The G20 briefly mentioned its commitment to green recovery and sustainable global growth, but without any details or implementation plans. It reaffirmed but did not expand its Pittsburgh promise to phase out fossil fuel subsidies, a practical, tangible commitment that could have encouraged movement away from climate-damaging fossil fuels while also freeing up funds that could be devoted to climate adaptation in developing countries.

Dimensions of Performance

In overall performance, however, Toronto was a solid success (see Appendix A-2).

Domestic Political Management

Toronto's performance in domestic political management was substantial. There was one communiqué compliment, equalling the high set at Pittsburgh and the second highest in G20 summitry overall (see Appendix B-2). However, Toronto was the first G20 summit that lacked full attendance as Brazil's Luiz Inácio Lula da Silva and Australia's Julia Gillard did not judge it in their domestic political interest to come.

Deliberation
In deliberation, Toronto's performance was solid. Climate took 838 words or 7.4 percent of the communiqué (see Appendix A-2). This was a fewer words than at Pittsburgh and below the average of 916 words.

Direction Setting
In direction setting, Toronto's performance was also solid. There was no priority placement for climate change, unlike the four such references at Pittsburgh. However, there was the first link to democratic values, starting a trend that endured for the three summits that followed (see Appendix A-2).

Decision Making
Toronto's performance in decision making was solid. It produced three climate commitments, the same as the two summits before, although its one commitment on energy was much fewer than those produced at Pittsburgh and all summits afterward (see Appendix D-2).

Delivery
In delivery, performance was solid. The three assessed climate commitments were complied with at a level of +0.42, below the overall eight-summit average of +0.31. The one assessed commitment on energy was complied with at a level of +0.45.

Development of Global Governance
In the development of global governance, performance was small. Toronto did nothing inside the G20 (see Appendix A-2). Outside it made only three references to three institutions, the lowest of any G20 summit after the first two (see Appendix F-6).

Causes of Performance

The substantial success at the G8's Muskoka Summit and the G20's solid success at Toronto are explained well by the concert equality model of G8 governance, extended into the systemic hub model of G20 governance.

Shock-Activated Vulnerabilities
At the Muskoka G8, the level of recognized shocks and vulnerabilities was strong, indeed the second highest in G8 history, if only half the peak at L'Aquila the year before. At the Toronto G20 there were no such communiqué-recorded recognitions, compared to three at Pittsburgh the year before.

Outside the summits, world oil prices had almost doubled to $81.82 in the six months up to January 2010, but softened to $78.36 by June 2010 (see Appendix G-3). Natural disasters rose somewhat from the year before (see Appendix G-5). The nascent recovery from the great recession caused greenhouse gas emissions to rise from the year before for both the G8 and G5 (see Appendices G-7 and G-8).

A strong diversionary second shock arose. It came from the eruption of a new financial crisis in Europe from Greece in the first half of 2010, which reawakened the vivid memory of the recent recession and the great global financial crisis that had caused it in 2008. Containing this euro crisis, through a firm commitment to fiscal consolidation among the developed members of the G20 became the Toronto Summit's singular focus and leading achievement. This left little space for climate change and, in particular, commitments on climate finance.

Multilateral Organizational Failure

The collapse of the UN's Copenhagen conference just six months earlier also shaped the climate performance of both the G8 and G20, primarily now in a constraining way in the short term. While such a failure might have bred great immediate G8 and G20 ambition, in the immediate aftermath of Copenhagen there was very little global political will for action on climate change. The world was still trying to determine what steps to take next on climate control.

This prompted the G8 and G20 to focus on other issues where success in delivering strong results was more assured. The major choice was development, where the UN had failed to meet its Millennium Development Goals (MDGs) and was planning for its second review conference in New York in September 2010. This UN failure and summit landing spot induced Harper to focus at the G8 on the MDGs 2 and 4 on maternal, newborn, and child health, which were two strong, specific goals, but furthest behind in meeting their 2015 deadline (Kirton, Kulik, et al. 2014a). In contrast, climate change had no MDG, being covered only generally by the weak MDG 7 on the environment. Such factors also led Harper's G20 to focus on a new approach to development, pushed by Korea and pulled by Korea's priority for the G20 summit it would host in November 2010.

Moreover, the leaders of the Copenhagen Accord five, who appeared at the Toronto G20, including Barack Obama at the Muskoka G8, had an interest in giving Copenhagen a chance to work.

Harper's shift as host from his planned G8 priority for climate change prior to Copenhagen to his later dismissal confirms this Copenhagen failure effect. So does the end of 2010 MEF summitry in 2010 for good.

Indeed, the MEF did not meet alongside the G8 in Muskoka. Instead, officials met in Rome on June 30–July 1, 2010. Not only did the MEF fail to meet alongside the G8, but it also failed to meet in the same country, suggesting a disconnect from the G-summit process. Moreover, the MEF did not meet at the leaders' level, but instead held a meeting by government officials and leaders' representative meeting, a move that represented a possible backslide in the G8's authority on global climate governance.

Predominant Equalizing Capability

Changes in relative capability offered another, much smaller constraint and a spur to a solid performance, too. In terms of gross domestic product (GDP), the global predominance of the G8 declined from 55 percent in 2009 to 53 percent in 2010, while that of the G20 remained the same at 77 percent. This helped lead the G8 to do less and look to broader plurilateral forums to do more. Internal inequality among the G8 members declined, while that among G20 members rose (Kirton 2013a, 451).

Common Principles

A convergence on common democratic principles contributed a little. All G8 and G20 members retained their same generally high levels of democracy at home.

Political Cohesion

Modest political cohesion contributed a little to the substantial or solid success of the twin summits. Canada's Stephen Harper, hosting both, had four years of G8 and three years of

G20 summit experience, but led a minority government, with a political base in the oil-rich province of Alberta and no personal commitment to climate change control or the UN. At the G8, however, he faced the more politically secure veteran German chancellor Angela Merkel, and US president Barack Obama at his second G8 and third G20 summit. Both were committed to the climate cause. They were joined at the G20 by the veteran leaders of China's Hu Jintao, India's Manmohan Singh, and Saudi Arabia's King Abdullah bin Abdul Aziz Al Saud, all of whom strongly thought that climate change should be left to the UN and its Rio-Kyoto regime.

Constricted Club Participation

The modest levels of constricted controlled club participation also limited the climate change advances of both summits. At the Muskoka G8 there was no third annual MEF summit devoted to or experienced in climate change control, nor was there a G5 substitute. There were many leaders developing countries, three from countries participating at the G8 for the first time, and choosing to emphasize specific subjects such as drugs and development in general, rather than climate change. At Toronto, the number of members expanded to 20, meeting for the fourth time, plus the usual five invited countries, and a growing list of multilateral organizations. This made the G20 more of a hub of a global network, but not at all a club where climate change was concerned, especially as none of the multilateral organizations invited were dedicated to climate or environmental protection. Moreover, the two back-to-back summits both suffered on the climate front from Harper's insistence on keeping climate change out of the G-summit process and, in the case of the G20, keeping the agenda firmly focused on economic and financial issues (Kirton 2010).

Seoul G20, November 2010

After the Toronto Summit, attention shifted to the next G20 summit, which took place in Seoul, Korea, on November 11–12, 2010. It was expected to return climate change to the G20 agenda. This it did, producing a very strong performance that advanced the new all-in regime, now led by the new G20 rather than the old G8.

Preparations

At the end of January 2010, Korea's G20 summit had not yet discovered the climate cause in any robust way. When President Lee Myun-bak, speaking on the same day as Harper at the World Economic Forum in Davos, outlined his agenda and objectives for the G20 summit he would host, climate change was not on his list. Lee's signature theme as host was development, focused on greater balance between the rich industrialized countries and those of the developing world. Yet green growth soon became part of the agenda, providing greater promise that climate change would be included.

Green growth grew increasingly prominent as the Seoul Summit drew closer. In August Young Soo-gil, chair of Korea's Presidential Committee on Green Growth, said that the summit's preparatory committee was intending to mainstream green growth in the agenda, to allow Lee to bring the issue to greater attention at the leaders' level (Kim Jae-kyoung 2010). Korea's goal was to drive the global green growth agenda forward, using its experience and closer connection to developing countries to broker solutions. Underscoring these ambitious plans were a series of other actions by the G20 host, demonstrating Korea's commitment to the green growth cause: it made green growth a key component of its official development assistance in order to set an example, it launched the East Asia Climate Partnership, and it initiated the Global Green Growth Institute hoping to develop it into a treaty-based institution by 2012. Thus, expectations for the success of green growth—and climate change control by proxy—were relatively high.

At the Summit

However, climate change did not arise until the second day of the Seoul Summit, when Ban Ki-moon opened the session on development shortly after 11:30 a.m. He called for continued focus on the MDGs as the universally recognized blueprint for, inter alia, clean water and a clean environment, and promised that the UN's lead on the global environment would continue to be the G20 leaders' essential partner. Hu Jintao called for technology transfer to create conditions for developing countries to achieve green and sustainable development at an early date. Lula referenced Brazil's wind power law.

Soon after, in the same session, climate change arose directly. Vietnamese prime minister Nguyen Tan Dung, representing the Association of Southeast Asian Nations, said that Vietnam endorsed the inclusion of environmental protection and climate change on the G20 agenda. He called for a new legally binding international agreement at COP-16 in Cancun and for the G20 to work with international organizations to create a fund to assist countries vulnerable to climate change through a preferential, accessible financing mechanism. He proposed that G20 members consider marine economic development projects to protect the marine ecosystem and suggested creating a forum for coastal countries to exchange experience and coordinate actions to deal with climate change. He offered to co-host such a forum with a G20 member in 2011 to work on green growth, and assist developing and low-income countries with generating technology transfer, promoting low carbon economies, and implementing green growth solutions.

In similar fashion, Malawi's President Bingu wa Mutharika, representing the African Union, asked the G20 to assist Africa and implement viable strategies to develop infrastructure in Africa, including energy and climate change mitigation and management. He said Africa was concerned that all efforts toward recovery might otherwise be negated by the adverse impacts of global warming and related environmental degradation.

It was notable that at this session on development, no G8 leader or the Organisation for Economic Co-operation and Development (OECD) made any reference to the natural environment or climate change. The demand for climate control, based on growing and massive vulnerability, came from the UN, China, Brazil, and, above all, the two leaders representing poor countries outside the G20, each representing Asia and Africa as a whole. Also noteworthy was the sponsorship of the cause for green growth and clean energy by BRIC members China and Brazil.

In the luncheon session on trade, which followed immediately, Japanese prime minister Naoto Kan connected climate change with the Doha negotiations at the World Trade Organization in a way that affirmed the all-in principles of climate change control. He argued that the world would not be able to reduce global emissions by setting the second commitment period of the current regime. At Cancun, countries should thus build on the Copenhagen Accord and adopt decisions as a balanced and comprehensive package to build a fair international framework with all major countries participating.

At the next session, also held over lunch and devoted specifically to climate change, several leaders spoke in turn, beginning with Korea's Lee as chair, followed by Mexico's Calderón, the EU's José Maria Barroso, Ethiopia's Zenawi as co-chair of the UN secretary general's high-level advisory group on climate change, the World Bank's Robert Zoellick, and the UN's Ban Ki-moon scheduled to conclude, followed by France's Nicolas Sarkozy with a spontaneous addition. Lee led off by calling on all G20 leaders to lend their support to the success of the Cancun COP, which would start in just a few weeks on December 9. Calderón referred to the difficult experience and disappointment in Copenhagen and his decision to moderate expectations by focusing on concrete decisions for Cancun. He reported recent progress, not in reaching a legally binding agreement but on adaptation in developing countries, technology transfer, fast-start financing of $30 billion from 2010 to 2012, the creation of the Green Climate Fund, and the reduction of emissions, soil degradation, and forestry. He called for progress on a long-term target, convergence on the 2°C goal, and action on transparency, measurement, reporting, and verification. He invited all the G20 leaders to attend Cancun, emphasizing that their personal involvement was indispensable. Throughout his intervention, he underscored the need to proceed based on the UN regime, specifically its UNFCCC, Kyoto Protocol, and principle of common but differentiated responsibilities and respective capability.

Barroso then noted that since Copenhagen the very credibility and relevance of the UN process was at stake. It could not afford a new setback at Cancun. With G20 members accounting for 80 percent of global emissions, the group around the table had the economic and political power to tackle climate change. He referred to the EU's goal of an ambitious, comprehensive, legally binding global framework, and said it could only accept a second Kyoto period if key conditions were met, where the gap was still wide. He noted the paradox that most countries, including China, India, Brazil, and those in Europe, were doing more domestically than they were prepared to agree to internationally. He confirmed the EU's commitment to mobilize €0.2 billion for fast-start financing to implement the Copenhagen Accord.

Zenawi reported that his panel had concluded that $100 billion a year was needed, that the price on carbon should be at least $20–$25 per ton, and that the Green Climate Fund was needed to spend the money effectively. He noted that African countries had not yet seen the announced short-term funding arrive.

Zoellick warned the G20 leaders that if they "let the perfect be the enemy of the good," as everyone had in Copenhagen, Cancun would fail again. He said that agreements were possible on measuring, reporting, and verification based on discussions between the US and China, and on REDD, agriculture and soil carbon, fast-start financing monitoring, and a renewable energy initiative for small island developing states. He also called for action on the adaptation framework and funding, technology centres, and the role of the private sector.

Ban said immediate results were needed on forests, finance, adaptation, and technology, even as negotiations would continue on the thorny political issues of mitigation and of monitoring, reporting, and verification. But he asked G8 members, outside these formal negotiations, to act on fast-start funding, the commitment of $100 billion a year by 2020, and research and development in green technology and energy efficiency. He specifically called for accelerated efforts to phase out fossil fuel subsidies, to improve climate efficiency, and to generate sizable revenues for climate action and green growth. Sarkozy spontaneously called for all to provide the money promised at Copenhagen, to keep faith with Africa in particular.

At the summit's concluding session on energy and corruption, Russia's Dmitry Medvedev and Italy's Silvio Berlusconi spoke on energy, with Medvedev emphasizing marine environmental protection. But neither made a direct link to climate change.

Results

Despite Korea's ambitious domestic actions on green growth, Seoul's final communiqué did not deliver much in meaningful plans for its implementation. The G20 final documents spoke about "strong, sustainable, balanced growth" only in a non-environmental context. At their working lunch, the G20 leaders had discussed climate change and green growth promotion alongside trade protectionism, but the communiqué did not reflect much progress on these fronts. The G20 publicly recognized the potential benefits of green growth policies that employed energy efficiency and clean technology, promising to create "enabling environments that are conducive to [their] development and deployment," but details on how this would be done were limited to a single statement (G20 2010). The leaders pledged support for a "Clean Energy Ministerial" and tasked their Energy Experts Group to report back on research and development and on regulatory measures at the 2011 G20 summit in France.

However, little else was agreed. In the view of civil society, the lack of financing plan or goals for green growth-driven emissions cuts limited the force of the green growth statements (CAFOD 2010). Seoul missed out on an important opportunity to make green growth a major centrepiece of the summit, although it did manage to keep the issue on the agenda and keep interest in environmental and climate-friendly policies alive.

Dimensions of Performance

Nonetheless, in overall climate change performance, the Seoul Summit was a very strong success. It produced a strong performance on domestic political management and direction setting, very strong ones on deliberation, decision making, and the development of global governance and a solid one on delivery (see Appendix A-2).

Domestic Political Management

In its domestic political management, performance was strong. Seoul issued two communiqué compliments related to climate change to two members, which is the highest among the first eight G20 summits (see Appendix B-2). And attendance improved, as only Saudi Arabia's King Abdullah chose not to come and France's Nicolas Sarkozy arrived late. Brazil's Lula also brought his successor, Dilma Rousseff, along to observe.

Deliberation

In deliberation, performance was very strong. Climate was given 2,018 words, taking 12.7 percent of the communiqué. In both cases this was by far the highest in G20 history (see Appendix A-2).

Direction Setting

In direction setting, performance was strong. Climate change was given priority placement twice, and it was linked to democratic principles for the second summit in a row (see Appendix A-2).

Decision Making

In decision making, performance was very strong. There were eight climate change commitments, more than double the three produced at each of the three summits before (see Appendix D-2). In addition, 14 energy commitments were made. The total of 22 climate and energy commitments exceeded the previous peak of 19 at Pittsburgh. Seoul picked up the commitment on phasing out fossil fuel subsidies that Toronto had abandoned, representing the return of a promising opportunity for the G20 to make an important, concrete contribution.

Delivery

In the delivery of these decisions, performance was solid. The two commitments on climate change achieved a compliance score of +0.36. The two relevant assessed commitments on energy were complied with at an average score of +0.51 (see Appendix A-2).

Development of Global Governance

In the development of global governance in the context of climate change, performance was very strong. Inside, the G20 leaders made an all-time high of eight references to G20-centric institutions, five to ministerial and three to official-level ones (see Appendix F-4). Outside, the G20 they made 20 references to other international institutions, double the number at any other summit (see Appendix F-6). They referred to 11 different institutions, led by the Global Marine Environment Protection initiative with four, followed by the UNFCCC with three. They were clearly looking beyond the UN process to advance climate change control.

After the Summit: COP-16 Cancun

At the COP-16 conference in Cancun from November 29 to December 10, 2010, the UNFCCC process regained some of its momentum on climate change control, due in part to the impetus given by the G20 leaders at Seoul. Cancun made some headway on adaptation, financing, and technology transfer, even as it struggled on mitigation and emission targets. With the memory of Copenhagen still alive, the agreements reached at Cancun made the future of climate governance seem slightly brighter. G8 members participated in the conference in different ways, ranging from strong leadership from the EU to resistance from Canada, Japan, and the United States.

In the months preceding Cancun, it was clear that the EU intended to play a strong role in driving the negotiations forward. A few weeks before the meeting, Connie Hedegaard (2010), the EU commissioner for climate action, repeatedly stated the EU's commitment to its own emissions reductions, as well as to a robust, fair global agreement. She expressed Europe's readiness to reach a legally binding accord in Cancun and criticized other parties for not sharing the same goal. She singled out the US for its lack of action, which she saw as affecting the success of UNFCCC process as a whole. These statements made it clear that the EU would not allow itself to be sidelined in these negotiations, as it had been in Copenhagen, and that it would be ready to challenge its fellow G8 members for their lack of commitment to climate change control.

Despite this strong EU leadership, global expectations—including those of the EU—remained low. While 120 leaders had attended the Copenhagen conference, only about 20 were expected at Cancun. There was a keen awareness of how difficult it would be to get all relevant players—the US and China, most notably—to forge a meaningful agreement. Most countries had associated themselves with the 2009 Copenhagen Accord and sent in their voluntary emissions targets. But the sum of these targets, even if they were to be fully met, would not keep global temperatures below the 2°C mark.

One major issue that arose in the lead-up to Cancun was extending the duration of the Kyoto Protocol. Developing countries pushed for an extension into a second commitment period. The EU, consistent with the new all-in principle pioneered in the G8, refused to consider an extension unless the US and other major developing economies also committed to significant reductions. Even Kyoto proponents were growing weary of the effectiveness of Kyoto's two-track process, where the US, China, and other large emitters, were parties to the UNFCCC but not to the Kyoto control commitments. Much had changed since 1992 and 1997, and a climate regime that reflected the new global reality was badly needed.

Mexico as host was deeply dedicated to making its summit a success. Nonetheless, it appeared to lack the needed high political capital. Despite a direct appeal from Felipe Calderón, British prime minister David Cameron refused to attend (Vidal 2010a). Mexico could not be entirely blamed, but the absence of leaders left doubts that Mexico would be able to give the proceedings much guidance. On the optimistic side, however, with fewer than half the delegates and media members expected at Cancun than had at Copenhagen, there was a chance that the conference might be less confusing and disjointed than the year before.

There were other early signs that COP-16 would encounter difficulties. The US continued to push for the weak, voluntary Copenhagen Accord to form the basis of Cancun's talks. Such disagreements on the legal framework threatened to overshadow the wider goals of the conference. As a potential strategy to prevent US disengagement once again, Mexico proposed a package of agreements that would avoid the debate on long-term legal frameworks altogether (Tollefson 2010). Discussion again arose over whether the 192-country UN process was capable of producing any agreement or whether a smaller group, preferably led by leaders, might be able to accomplish more in less time. Both formats had failed at Copenhagen. Would Cancun be able to make one work? Should the smaller, leaders-led forum prove more useful, it could lead to new discussions of a bigger role for the G8 and G20 in climate governance.

COP-16 opened with little excitement or fanfare, unlike the year before. Predictably, gridlock set in quickly, with little ground conceded on any side. All countries seemed determined to hold their negotiating positions without compromise. Then, unexpectedly, G8 member Japan announced that it would be the first Kyoto party to refuse to extend the protocol. It was criticized for this move, although it brought out the harsh truth about Kyoto: it could not go on in its existing form, especially without the world's two largest emitters, China and the US, tied to some sort of commitment.

The deadlock on emissions reductions continued into the final day of negotiations. After two long weeks, packages on softer areas of climate policy were almost concluded, but holding those agreements hostage was a deep and seemingly irreconcilable divide (Davenport 2010). The Kyoto problem could be ignored no longer. On one side, a developing country bloc led by India and China insisted that any new climate treaty include continuing the existing Kyoto regime beyond 2012. Conversely, a developed country bloc, led by G8 members Canada, Japan, Russia, and the US, refused to continue Kyoto as long as developing countries were excluded from its control provisions. An acceptable compromise would require careful negotiation. But it was unclear whether developing countries led by China would accept any emission control commitments, even alongside a Kyoto continuation.

When the conference finally closed on December 10, several agreements were made. However, parties managed to conclude them not by breaking the deadlock, but by decoupling the soft agreements from the tougher decisions on a legal framework for emissions reductions. The Green Climate Fund came into being, to support mitigation and adaptation in developing countries. A new technology mechanism would be fully operational by 2012. The Cancun Adaptation Framework was the centrepiece of the agreements, placing adaptation on the same priority level as mitigation. Outside of the official UNFCCC process, fast-start finance also advanced, although only under voluntary commitments from developed countries.

Thus, while the conference was a considerable success for adaptation, Cancun did not overcome the continued difficulties with a legal framework on emissions reductions and on a future for Kyoto. Media reaction generally highlighted this failure, pointing to the absence of numbers, targets, or legally binding language. Yet there remained a sense that Cancun had done enough to salvage the credibility of the UN process, which had been so badly damaged the year before, by moving beyond the still deadlocked and increasingly abandoned UNFCCC-Kyoto regime.

Deauville G8, May 2011

At the G8 summit in Deauville, France, on May 26–27, 2011, climate change received a little less attention and action than it had at Muskoka the year before. Leadership continued to shift to the G20, which met at Cannes later that year on November 3–4.

Preparations

When the presidency of both the G8 and the G20 moved to France in 2011, the French team was widely expected to feature climate change prominently at both summits. Indeed, the G8 Deauville Summit's preliminary agenda gave significant space for climate discussions by listing green growth as a major priority, alongside the internet and innovation. Progress could be made on shifting to low carbon economies through clean technology and sustainable, green jobs. A financing commitment could give green growth momentum, too. G8 members could also pledge to incorporate the principles of green growth into their development strategies, as Korea had suggested at its G20 summit. More could be done on mapping out mid-term mitigation strategies. As a country strongly on track to meet its 2012 Kyoto commitments, France was well positioned to move forward here.

The G8-Africa Partnership, a second major summit theme, also promised progress on climate adaptation and climate finance, given that such issues were high priorities of African and outreach countries. The status of fast-start financing contributions remained uncertain, in terms of determining whether all financing was new and additional, as the Copenhagen Accord had called for.

The G8 was also slated to address COP-16 and pledge its support for the UNFCCC process once again. On the whole, the outlook was positive, amid confidence that France's G8 would contribute meaningfully to G8 climate governance.

Initial preparations also included with peace and security as a third major theme. Planning continued when sherpas first met on January 24 and their sous-sherpas met on February 3–4. However, international events soon shifted the G8's focus away from climate change.

In mid December 2010, a series of North African and Middle Eastern countries had become swept up in a revolutionary wave of protests against well-established, authoritarian governments. Cast as a democratic awakening in the Arab world, many countries found their largely peaceful, civil resistance efforts met with violence from authoritarian regimes. Egypt, Tunisia, and Libya moved to the forefront of these movements, as regimes were toppled in the former two and a bloody war raged in the latter. This quickly became the defining issue of the Deauville Summit, with a G8 assistance plan for the region assuming a priority for the French hosts.

A second key international event that altered the agenda was the combined natural and nuclear disaster that struck Japan on March 11 (Kirton 2012). The double ecological shock of a massive earthquake followed by a devastating tsunami set off a nuclear disaster at the nuclear power complex in Fukushima. With more than 15,000 people dead, an estimated $35 billion in damages, and the catastrophic potential of nuclear power laid bare, the French hosts quickly shuffled their G8 agenda to include the devastating impact of natural disasters as a key summit theme.

In the end, the final G8 priorities for Deauville were the mounting crisis in the Middle East, progress with democratic development in Africa, the internet, nuclear safety, and climate change. Green growth and innovation moved to a tenth place sub-theme. But climate change remained in some capacity, and thus so did the potential for progress.

At the Summit

The schedule of the Deauville Summit, which began on May 26, reflected these revised priorities. The first working lunch was dedicated to solidarity with Japan (and the global economy), coupling nuclear safety with climate change, and cooperation with emerging economies.

The working dinner that evening focused on the Arab Spring. So did the working session the next morning, with Egypt's Essam Sharaf, Tunisia's Beji Caid Sebsi, and Amr Mohammed Moussa, secretary general of the League of Arab States participating. The Deauville Partnership emerged as the named, centrepiece achievement of the summit. G8 members pledged to help the Arab countries make the transition to free, democratic societies. Other working sessions focused on political issues, African peace and security, African development, and the internet. As a result, climate change failed to secure its initially intended place as a priority for a summit preoccupied with so many other new pressing global concerns.

Results

G8 leaders discussed the role of green growth and innovation as ways to drive sustainable development and control climate change. The communiqué listed the many benefits provided by green growth. On mitigation, leaders' reaffirmed Muskoka's baseline year of "1990 or more recent years" and an 80 percent reduction in the developed world by 2050 (G8 2011). Such a result had been expected at Muskoka, given the Canadian government's position on climate change, but more had been expected from the French.

Action on renewable energy was promised through ongoing support of G8 initiatives, including the International Partnership for Energy Efficiency Cooperation, the Global Bioenergy Partnership, the International Renewable Energy Agency, and various others. G8 leaders pledged to "intensify [their] efforts to contribute to progress for the next steps," highlighting the June 2012 UN conference on sustainable development in Rio de Janeiro and the next COP in Durban in December 2011 as important to achieving the "comprehensive, ambitious, fair, effective and binding agreement" that the G8 so often spoke of (G8 2011). The G8 reiterated its commitment to fulfilling the provisions of the Copenhagen Accord. The G8 also pledged to preserve biodiversity and support the upcoming UN-centred meeting on the Convention on Biological Diversity in Nagoya.

Dimensions of Performance

Deauville produced a solid success on climate change. It performed a little less strongly than Muskoka did on most dimensions, but better on delivery and the development of global governance. This inversion of the Muskoka pattern suggested that Deauville emphasized putting the new advances from the previous G8 summit into practical effect.

Domestic Political Management
In its domestic political management, Deauville's performance was very small. While all G8 leaders attended, the summit issued no communiqué compliments to any members (see Appendix B-1).

Deliberation
In deliberation, performance was solid (Guebert et al. 2011). It devoted 1,086 words and 5.9 percent of the communiqué to climate change (see Appendix A-1). This was a drop from Muskoka, but close to the overall G8 average of 4.16 percent. The treatment of green growth was expanded, but the rest of the text was largely repeated from previous years.

Direction Setting
In direction setting, performance was significant. Deauville made one priority reference to climate change and linked it to democratic principles once (see Appendix A-1).

Decision Making
In decision making, performance was solid. Deauville's seven climate commitments were a drop from the ten at Muskoka, and just below the overall G8 average of eight (see Appendices A-1 and D-1).

Delivery
In the delivery of these decisions, however, performance was significant. The one assessed commitment's compliance score of +0.67 exceed that of the +0.26 at Muskoka (see Appendix E-1).

Development of Global Governance
In its development of global governance, performance was solid. Inside the G8 Deauville's communiqué made no reference to its environment ministers' work. But outside the G8 it made seven references to six international institutions, an advance from Muskoka and above the G8's multi-year average on climate change (see Appendix F-1).

The G8 mentioned its own initiatives, including the International Partnership for Energy Efficiency Cooperation and the Economics of Ecosystems and Biodiversity study. The communiqué frequently noted the UNFCCC—whether through the Copenhagen Accord, the Cancun conference, or the upcoming Durban conference. Various other UN bodies also got a mention, including the UN conference in Rio de Janeiro on sustainable development and United Nations Environment Programme. The G8 also invoked the IEA and the International Renewable Energy Agency, along with tasking the OECD with issues including continued reporting on the Kobe 3R Action Plan, which had been put forward by the G8 environment ministers during Japan's year as host in 2008.

Cannes G20, November 2011

The G20's Cannes Summit, on November 3–4, 2011, saw a continuation of the G20's new leadership on climate change.

Preparations

In sharp contrast to the Deauville G8, climate change was not on the list when France had first released its G20 priorities in early 2010 (French Presidency of the G20 and G8 2011). This was despite the EU's desire to lead negotiations at the UNFCCC. But with the EU backing the UN as the only legitimate forum for advancing a climate change treaty, and with deep disagreements in the G20, France saw the issue as a lost cause.

Perhaps as a result, Sarkozy's biggest climate-related initiative was promoting a financial transaction tax (FTT). He suggested that such a tax might be used to fund development projects broadly, and climate adaptation projects more specifically (Oxfam 2011). This focus on adaptation and finance was in line with the accomplishments of the Cancun COP, as well as the work to be continued at Durban in December 2011. Moreover, NGOs built up much enthusiasm for the FTT in the months preceding the summit. Sarkozy himself was deeply committed to advancing his summit goals and to using his political capital to push the FTT. There was talk of a potential coalition of pioneering countries, as G20 members Germany, Brazil, South Africa, and Argentina, as well as the African Union, had expressed support for the tax (World Wildlife Fund 2011).

At the Summit

As the summit opened on November 3, the FTT received significant attention. Representing the Bill and Melinda Gates Foundation, Bill Gates attended the summit to directly promote the tax to G20 leaders. He (2011) released a report entitled "Innovation with Impact: Financing 21st Century Development," which underscored Sarkozy's efforts to make the tax a success. There were high hopes that the Cannes Summit would deliver.

But international events derailed this summit's agenda too. Cannes was consumed by a deepening of the Greek debt crisis, panic over Greek prime minister George Papandreou's decision to put the EU debt rescue package to a referendum, and speculation that he might resign by the end of the day. EU leaders negotiated overtime as they worked behind closed doors to quell the crisis. Many of Sarkozy's engagements, including a press conference with Gates about the FTT, were cancelled as a result. Once a major priority, the FTT now moved to the back burner.

Results

However, on November 4, as the summit wrapped up and the final declaration surfaced, there were considerable advances made on climate change control. The G20 paid significant attention to energy markets, promoting efficiency and sustainability. The communiqué also advanced the dialogue between oil-producing and oil-consuming countries. G20 leaders again promised to rationalize and phase out inefficient fossil fuel subsidies for all G20 members, welcomed country progress reports on fossil fuel subsidies, and gave instructions to their finance ministers to "press ahead with reforms and report back next year" (G20 2011). On energy markets, G20 performance was reasonably strong.

The G20 also addressed climate adaptation and finance, lending support to the Green Climate Fund that was to be fully operationalized in Durban. The leaders called for the implementation of Cancun's agreements, which also primarily focused on climate adaptation, as well as for further progress at COP-17. The FTT received only an indirect reference. The G20 merely welcomed the Gates presentation and recognized the need for diverse sources of financing.

Dimensions of Performance

The Cannes Summit was a strong success on climate change, led very strong performances on domestic political management and decision making, and a strong performance on deliberation (see Appendix A-2).

Domestic Political Management
In domestic political management, performance was very strong. The Cannes communiqué gave two compliments to two members, the highest, along with Seoul, that any G20 summit ever made (see Appendix B-2). Attendance was almost complete, rising after the two previous summits that some leaders skipped.

Deliberation
In deliberation, performance was strong at Cannes. Climate change secured 1,167 words or 8.2 percent of the communiqué (see Appendix A-2). While this was a drop from the all-time highs at Seoul, it was above the eight-summit average of 916 words and 6.6 percent of the communiqué.

Direction Setting
In direction setting, Cannes's performance was solid. While climate change was not given priority placement, it was linked once to democratic principles, for the third summit in a row (see Appendix A-2).

Decision Making
Cannes's decision-making performance was very strong. It made eight commitments on climate change, tied with Seoul as the all-time high (see Appendix D-2). It also made 18 commitments on energy, becoming the peak on its own.

Delivery

In delivering on these decisions, Cannes's performance was substantial. The average compliance score for the three relevant commitments assessed, all on energy, was +0.61, higher the one commitment assessed from Seoul the year before (see Appendix E-2).

Development of Global Governance

In the development of global governance on climate change, Cannes's performance was significant. Inside, the G20 leaders made two references to G20-centric institutions, both to ministerial ones (see Appendix F-4). Outside, the G20 made 11 references to seven other international institutions, well above the multi-summit average in both respects (see Appendix F-6).

Assessment

Despite these many advances, the Cannes communiqué lacked any agreements on emissions reduction or the future of the Kyoto Protocol, which were the two most pressing topics on the agenda for the COP meeting at Durban. While G20 members pledged their commitment to Durban's success, the lack of any other details suggested that there was no consensus on what this success might be. The outcomes of the Copenhagen and Cancun conferences had exposed the growing divide among G8 and other G20 members. Some advanced countries had revoked their support for Kyoto, while others still pushed hard; some developing countries wanted robust commitments for everyone, while some emerging economies steadfastly refused to accept binding emissions reduction targets. With such irreconcilably different views, G20 leaders opted simply to leave out any mention of these issues altogether. Without any groundwork laid for agreement at Durban, it would soon make for a difficult two weeks of negotiations.

Causes of Performance

The G20's very strong success at Seoul in November 2010, the G8's solid success at Deauville in May 2011 and the G20's strong success at Cannes in November 2011 were again largely caused by the forces contained in the concert equality and systemic hub models.

Shock-Activated Vulnerability

Shock-activated vulnerability was a consequential cause. The very strongly successful Seoul Summit recognized two shocks and vulnerabilities in its communiqué, while solidly successful Deauville and Cannes recognized only one each. World oil prices rose by 17.45 percent from $78.36 in June 2010 to $92.04 January 2011, but then rose by only a further 9.6 percent to $100.90 in May 2011 and then decreased slightly to $100.39 by January 2012 (see Appendix G-3). The number of natural disasters and people affected by them remained high and the damaged caused by them had spiked in 2010 (see Appendix G-5). Moreover, greenhouse gas emissions and the related cumulating vulnerability declined from G8 members in 2011, but rose again from the G5 countries that year (see Appendices G-7 and G-8).

The force exerted by such climate shocks and vulnerabilities varied as a result of particular diversionary shocks. The European financial crisis receded and went unrecognized in its Irish incarnation at the strongly successful Seoul Summit. The Deauville Summit, which was only solidly successful, was diverted by the Arab Spring. The climate-related shock of Fukushima, and its immediate effect of inspiring Germany and Japan to phase out and suspend nuclear power as a climate-friendly energy source, had a constraining effect on Deauville. The return of the Greek financial crisis to consume the Cannes Summit reduced its level of climate performance from that of Seoul, but did not prevent a strong success.

Multilateral Organizational Failure
The cadence of multilateral organizational failure also played a part. The fresh memory of Copenhagen's collapse spurred success at Seoul one year later once its immobilizing impact had worn off, but exerted less force a year after through to 2011. Ban Ki-moon at Seoul and the COP itself at Cancun in 2010 had begun to focus on making progress outside the legalized confines of the UNFCCC and its Kyoto Protocol. This allowed the G20, with the G5 and other participating major emerging economies, to lead on the new equal, all-in regime. And after the MEF's failure to move forward in April 2011, the EU's Connie Hedegaard admitted that a binding international agreement would not arrive that year.

Predominant Equalizing Capability
Predominant equalizing capability also contributed to the G20's superior performance over the G8 (see Appendix I). The global predominance of the G20 in combined GDP was much superior and increasing. Internally, G20 equalization increased, as from 2010 to 2011 China's GDP increased 23 percent while that of the US grew by only 3.9 percent.

Common Principles
Common principles also contributed to the superior performance of the G20 over the G8 (see Appendix J). From 2010 to 2011, Turkey's domestic democratic character jumped two points from seven to nine, very close to the perfect score of 10.

Political Cohesion
Political cohesion also helped. As host of the Seoul Summit, Lee Myung-bak had firm control of his government and a proven commitment to green growth.

At Deauville, G8 leaders suffered from limited political control, capital, and conviction. France's Nicolas Sarkozy had a modest standing in the polls. Barack Obama lacked control of the US legislature, and his previous attempts at climate change legislation had brought disappointing results. Japan's Naoto Kan was on his way to becoming the fifth of six Japanese prime ministers since 2007. Germany's Angela Merkel faced weaker support in the face of voter anger over euro bailouts. Britain's David Cameron governed with a coalition. Italy's Silvio Berlusconi also operated in a coalition government with waning support. And visible disagreements between President Dmitry Medvedev and Prime Minister Vladimir Putin were on the rise in Russia. Only Canada's Stephen Harper came to Deauville with a new majority government but with his old lack of commitment to climate change control through the UNFCCC.

Most leaders faced low popularity at home. Obama continued to receive criticism for his handling of the economy, and any attempts at environmental protection or climate change mitigation would have quickly been perceived as threatening the economy further. The Japanese, German, and Russian leaders also faced declining popularity.

Civil society support and pressure were low. Climate change received little media coverage at the Deauville Summit and civil society seemed to accept that the G8 (and much of the world) had moved on to other issues. Development groups focused again on G8 initiatives including the L'Aquila Food Security Initiative and maternal, newborn, and child health. Most controversy focused on accountability reporting and discrepancies in funding disbursements for development. Many of these groups also commented on the Deauville Partnership on the Arab Spring, but few said anything about climate change.

Constricted Club Participation

Constricted club participation mattered too. At Seoul, the G20 met for the first time without the Netherlands as a guest, for its first summit where its outreach partners had been fixed at five and defined by constituencies. Thus Spain came for the fifth straight time. The Cannes Summit followed this formula, but with less constriction due to the appearance of Bill Gates and Greece's George Papandreou.

At the G8 Deauville Summit, the overall number of participants rose from the 20 at Muskoka to 22 (see Appendix L-1). The Arab Spring helped generate invitations for Algeria, Egypt, Ethiopia, Equatorial Guinea, Senegal, and Tunisia as well as the Arab League, African Union, the New Partnership for Africa's Development, and the UN.

After the Summit: COP-17 Durban

As the Durban COP meeting approached, pessimism was high that the UN could or would accomplish much. Calls continued for the UN to switch to a different style of negotiation that would cut out bureaucracy, move more efficiently toward goals, and prevent a handful of countries from blocking progress for all. Such critics, largely from a few developed countries, typically suggested smaller negotiating forums and a bottom-up approach that would supposedly produce more robust results (Victor 2011).

Developing countries, and at times the EU, still pushed for the UN to remain the first and only forum for official agreements to be negotiated. But after the MEF meeting in April 2011, the EU's Hedegaard admitted that no binding international agreement would be forthcoming. US climate change envoy Todd Stern stated that the US did not envision the negotiations moving forward until all important emitters were ready to accept a binding agreement citing China, India, Brazil, Russia, and South Africa as the "major players" a deal would need (BusinessGreen 2011). As an alternative to concluding a binding universal agreement, Hedegaard suggested that Durban focus on emissions from the shipping and aviation sectors, which was a priority for the EU internally as well as internationally. The shift to the new, G8-pioneered bottom-up, sectoral approach was gathering force.

Lying just beneath these low expectations was the knowledge that Durban would be the last chance to extend the Kyoto Protocol before it expired in 2012. The world's only binding climate change agreement was about to evaporate, yet the alarm was not being sounded very loudly. Everyone from Hedegaard to Ban Ki-moon played down expectations for Durban (Max 2011). Most expected that COP-17 would be little more than the moment where the world put the final nails into the coffin of Kyoto. After Japan's condemnatory comments the previous year, the media mused about which countries would be next to put a "dagger" into Kyoto (Vidal 2010b).

The answer quickly came. One day after the conference opened on November 29, Canada announced its intention to formally withdraw from the Kyoto Accord. With its full transition from Kyoto champion to Kyoto repudiator now complete, other countries could more easily withdraw (Vidal 2010b). Before long, Japan and Russia confirmed that they too would not take on Kyoto emissions targets after 2012. This left four of the G8 members, including the US and Japan as the two most powerful, now officially outside the Kyoto framework.

The EU did agree to a second Kyoto commitment period, but only on condition that other countries agree to its negotiations roadmap. The EU's proposal stipulated that all major emitters from both the developing and developed world work together to achieve a new binding agreement by 2015, to begin cutting emissions in 2020. Thus, once Copenhagen's voluntary emissions reductions targets ran out, the new agreement would pick up and carry forward with binding targets. The sticking point was that the treaty be legally binding. Even with the EU's promise of a second Kyoto period, it was unclear if developing countries would accept this new constraint.

As the conference wore on, several key parties remained reluctant to accept the EU-backed roadmap. Perhaps the three most important countries in the negotiations—China, India, and the US—had not openly backed it. At one point, Stern was quoted as "speaking approvingly of the roadmap," leading to claims that the US would jump on board, but he quickly clarified that the it still did not support the proposal, and instead preferred an open "'process' of negotiations without strict time limits" (Harvey and Vidal 2011). But with more than 120 other countries already pledging support, including Brazil, South Africa, and Argentina, the pressure was on.

On the final night of the conference, negotiations were on the brink of collapse when China officially rejected the EU proposal. After previously supporting it, an African representative expressed dissatisfaction with the details of the final text, which appeared to spell the end of the Kyoto Protocol, a criticism also made by China. These criticisms came as the EU claimed the proposals were nearing successful completion (Vidal 2011). Amid confusion and conflicting reports, the South African hosts extended the conference closing time in hopes of salvaging an agreement.

Thirty-six hours later, all 194 countries agreed to a roadmap for moving forward. A major breakthrough, it included commitments for developing countries for the first time. The G8-pioneered all-in principle had finally prevailed at the UN. The deal also included an extension of Kyoto for a five-year minimum. Without this the deal would have never secured the support of developing countries. The Green Climate Fund also formally came into being, to serve as a $100 billion adaptation financing mechanism for poor countries. The process for another legal agreement was finally underway, although the details would still need to be negotiated at future COP conferences.

The results of Durban showed the potential impacts of diplomatic pressure when more than 100 countries, including all G20 members, joined together, as well as the role that just one or two countries could play in making or breaking a deal. India was reportedly the last country to agree to the roadmap, but with such a large coalition of countries already on board, it did not want to be labelled as the one to kill Durban's progress (Kuipers 2011). This pressure to avoid being labelled as obstructionist was what drove the climate deal to tentative success. India did, however, secure one concession before agreeing: it insisted that the post-2020 Kyoto replacement be referred to not as a "legal instrument," but as a much weaker "legal outcome" (Shi Jiangtao 2011). As the EU roadmap marked the end of Kyoto's defining principle of common but differentiated responsibility, considerable resistance remained in the developing world, demonstrated by India's desire to keep the new agreement weak.

Thus, Durban produced some possibilities for the years to come. But there was still no guarantee, or even any likely prospect, that the proposed roadmap would keep global warming below the 2°C post-industrial increase threshold. COP-17 demonstrated a deep split among the G8 members, with the Europeans on one side, pushing for robust climate action under the purview of the UN, and the remaining four resisting such progress. The

G20 was similarly split but across the old North-South or established-emerging country divide, with emerging countries such as South Africa and Brazil playing a supportive role on the EU's roadmap, and others such as China and India deeply reluctant to accept a new framework that committed them to carbon control.

Conclusion

By the end of 2011, it was clear that the old UN-Kyoto regime was largely dead, and that the fundamental principle of the G8-pioneered new equal, inclusive regime had prevailed, even at the UN and its COPs at Cancun and Durban. It was also clear that global leadership in advancing the new regime had passed from its G8 originator to the G20, where all the major carbon powers were included as equals, and which, hosted for the first time by an emerging country in November 2010, began to govern climate change in a broad way.

References

BusinessGreen (2011). "Durban Climate Deal Impossible, say US and EU Envoys." *Guardian*, April 28. http://www.guardian.co.uk/environment/2011/apr/28/durban-climate-deal-impossible (January 2015).

CAFOD (2010). "CAFOD Responses to Seoul G20 Communiqué." Notice distributed at the 2010 Seoul Summit, Seoul, November 12. http://www.bond.org.uk/data/files/CAFOD_response_to_Seoul_G20_communique.doc (January 2015).

Davenport, Coral (2010). "Cancun Climate Talks Reach Crucial Stage." *National Journal*, December 9.

French Presidency of the G20 and G8 (2011). "Dossier de Presse." Paris, January 24. http://www.g8.utoronto.ca/summit/2011deauville/2011-g20-g8_dossier_presse.pdf (January 2015).

G8 (2010a). "G8 Muskoka Declaration: Recovery and New Beginnings." Huntsville, Canada, June 26. http://www.g8.utoronto.ca/summit/2010muskoka/communique.html (January 2015).

G8 (2010b). "Muskoka Accountability Report." June 20. http://www.g8.utoronto.ca/summit/2010muskoka/accountability (January 2015).

G8 (2011). "G8 Declaration: Renewed Commitment for Freedom and Democracy." Deauville, May 27. http://www.g8.utoronto.ca/summit/2011deauville/2011-declaration-en.html (January 2015).

G8 Research Group (2010a). "2010 Canada Summit Expanded Dialogue Country Assessment Report." June 27. http://www.g8.utoronto.ca/evaluations/2010muskoka/2010-muskoka-ed.pdf (January 2015).

G8 Research Group (2010b). "Climate Change and the Environment at the G8 and G20." June 11. http://www.g8.utoronto.ca/briefs/ccenv-100611.pdf (January 2015).

G8 Research Group (2011). "2010 Muskoka G8 Summit Final Compliance Report." May 24. http://www.g8.utoronto.ca/evaluations/2010compliance-final (January 2015).

G20 (2010). "The G20 Seoul Summit Document." Seoul, November 12. http://www.g20.utoronto.ca/2010/g20seoul-doc.html (January 2015).

G20 (2011). "Cannes Summit Final Declaration — Building Our Common Future: Renewed Collective Action for the Benefit of All." Cannes, November 4. http://www.g20.utoronto.ca/2011/2011-cannes-declaration-111104-en.html (January 2015).

Gates, Bill (2011). "Innovation with Impact: Financing 21st Century Development." Report to G20 leaders at the Cannes Summit, November. http://www.gatesfoundation.org/~/media/GFO/Documents/2011%20G20%20Report%20PDFs/Full%20Report/g20reportenglish.pdf (January 2015).

Guebert, Jenilee, Zaria Shaw, and Sarah Jane Vassallo (2011). "G8 Conclusions on Climate Change, 1975–2011." Toronto, June 20. http://www.g8.utoronto.ca/conclusions/climatechange.pdf (January 2015).

Harper, Stephen (2009). "The 2010 Muskoka Summit." In *The 2009 G8 Summit: From La Maddalena to L'Aquila*, John J. Kirton and Madeline Koch, eds. London: Newsdesk, pp. 18–19. http://www.g8.utoronto.ca/newsdesk/harper-2009.html (January 2015).

Harper, Stephen (2010a). "Canada's G8 Priorities." Ottawa, January 26. http://www.g8.utoronto.ca/summit/2010muskoka/harper-priorities.html (January 2015).

Harper, Stephen (2010b). "Statement by the Prime Minister of Canada at the 2010 World Economic Forum." Davos, January 28. http://www.g8.utoronto.ca/summit/2010muskoka/harper-davos.html (January 2015).

Harris, Kathleen (2010). "G8 Leaders Take Heat for Failing to Act on Global Warming." *Toronto Sun*, June 26. http://www.torontosun.com/news/g20/2010/06/26/14527901.html (January 2015).

Harvey, Fiona and John Vidal (2011). "Durban COP17: Connie Hedegaard Puts Pressure on China, US, and India." *Guardian*, December 11. http://www.guardian.co.uk/environment/2011/dec/09/durban-climate-change-connie-hedegaard (January 2015).

Hedegaard, Connie (2010). "Cancún Must Take Us Towards a Global Climate Deal." *European View* 9(2): 175–79.

Kim Jae-kyoung (2010). "Seoul to Bring G20 Leaders' Attention to Green Growth." *Korea times*, August 22. http://www.koreatimes.co.kr/www/news/biz/2011/03/301_71799.html (January 2015).

Kirton, John J. (2009). "Prospects for the 2010 Muskoka G8 Summit." Paper prepared for a conference on "The 2009 G8's Sustainable Development Challenge: Initiative and Implementation," Aspen Institute Italia, July 1, Rome. http://www.g8.utoronto.ca/evaluations/2010muskoka/2010prospects090702.html (January 2015).

Kirton, John J. (2010). "Working Together for G8-G20 Partnership: The Muskoka-Toronto Twin Summits, June 2010." *International Organisations Research Journal* 5(5): 14–20. http://iorj.hse.ru/data/2011/03/15/1211462316/4.pdf (January 2015).

Kirton, John J. (2012). "Energy Security amidst Disaster: The Global Governance Response to March 11, 2011." Paper prepared for the Japan Futures Initiative Spring Symposium 2012, Balsillie School of International Affairs, University of Waterloo, March 14, Waterloo ON.

Kirton, John J. (2013). *G20 Governance for a Globalized World*. Farnham: Ashgate.

Kirton, John J., Julia Kulik, and Caroline Bracht (2014, 21 July). "The Political Process in Global Health and Nutrition Governance: The G8's 2010 Muskoka Initiative on Maternal, Child, and Newborn Health." *Annals of the New York Academy of Sciences*. doi: 10.1111/nyas.12494.

Kuipers, Dean (2011). "Progress at End of Durban COP17 Climate talks." *Los Angeles Times*, December 12. http://articles.latimes.com/2011/dec/12/local/la-me-gs-progress-at-end-of-durban-cop17-climate-talks-20111212 (January 2015).

Max, Arthur (2011). "Ban Ki-Moon, UN Chief, Doubts Climate Deal." *Associated Press*, December 6.

Oxfam (2011). "Sarkozy Renews Calls for FTT, Signals Ambitious G20 Agenda on Food Price Volatility." Ottawa, January 24. http://www.oxfam.ca/news-and-publications/news/sarkozy-renews-calls-ftt-signals-ambitious-g20-agenda-food-price-volatili (January 2015).

Rennie, Steve (2010). "Tories Put Climate Change on G8 Agenda after Pressure from World Leaders." *Globe and Mail*, June 14. http://www.theglobeandmail.com/news/politics/tories-put-climate-change-on-g8-agenda-after-pressure-from-world-leaders/article1211785/ (January 2015).

Shi Jiangtao (2011). "Durban Delivers Road Map on Climate Change Treaty." *South China Morning Post*, December 12.

Tollefson, Jeff (2010, 3 December). "Cancún Week One: A Climate of Confusion." *Nature*. doi:10.1038/news.2010.653

United Nations Framework Convention on Climate Change (2009). "Decisions Adopted by the Conference of the Parties." Report of the Conference of the Parties on its 15th session, Copenhagen. http://unfccc.int/resource/docs/2009/cop15/eng/11a01.pdf (January 2015).

Victor, David G. (2011). "Why the UN Can Never Stop Climate Change." *Guardian*, April 4. http://www.guardian.co.uk/environment/2011/apr/04/un-climate-change (January 2015).

Vidal, John (2010a). "Cameron Refuses to Attend UN Climate Change Talks." *Guardian*, November 29. http://www.theguardian.com/environment/2010/nov/29/cameron-cancun-climate-change-summit?INTCMP=ILCNETTXT3487 (January 2015).

Vidal, John (2010b). "Kyoto Protocol May Suffer Fate of Julius Caesar at Durban Climate Talks." *Guardian*, November 29. http://www.guardian.co.uk/environment/blog/2011/nov/29/kyoto-protocol-julius-caesar-durban (January 2015).

Vidal, John (2011). "Climate Change Conference in Trouble as China Rejects Proposals for New Treaty." *Guardian*, December 9. http://www.theguardian.com/environment/2011/dec/09/climate-change-conference-durban-treaty (January 2015).

Williams, Jody (2010). "Why Is Canada Not Putting Climate Change on the G20 Agenda?" *Globe and Mail*, June 11. http://www.theglobeandmail.com/globe-debate/why-is-canada-not-putting-climate-change-on-the-g20-agenda/article1372784/ (January 2015).

Woods, Allan (2010). "Climate Change Gets Short Shrift from G8 Leaders." *Toronto Star*, June 26. http://www.thestar.com/news/world/g8/article/829116—climate-change-gets-short-shrift-from-g8-leaders (January 2015).

World Wildlife Fund (2011). "Small Steps on Climate Finance in Cannes Will Require Giant Leaps in Durban." Cannes, November 4. http://wwf.ca/newsroom/?uNewsID=10082 (January 2015).

Chapter 11
Replacement, 2012–2014

The period from 2012 to 2014 saw the replacement of the old, divided, United Nations regime by the new, inclusive, expanded one with the Group of Eight (G8) and the Group of 20 (G20) as the dominant centre of global governance on climate change. This replacement first saw the G20 emerge alongside and at times ahead of the G8 as the lead plurilateral summit institution on climate change control. Second, it saw the G8 replaced by the Group of Seven (G7) alone at its summit in 2014, as Russia's invasion and annexation of the Crimean region of Ukraine disrupted but did not delay the G7's revival of climate leadership. As the landmark Conference of the Parties (COP) conference to replace the Kyoto regime in Paris in December 2015 drew nigh, the inherited UN regime was being replaced in principle, practice, and, prospectively, in formal promise, by the inclusive, expanded, equal one led by the G8 and G20. As Peter Haas (2008, 6) noted, "the very thinness of the social constructions of the issue developed through the UN mean they can be replaced with other procedural and substantive norms, such as coordinating UN negotiations with discussions elsewhere, or going beyond consensus voting rules."

Camp David G8, May 2012

Preparations

The United States hosted the 37th annual G8 summit on May 18–19, 2012, at the presidential retreat at Camp David, Maryland, just outside of Washington DC. Taking place in the lead-up to the US presidential election in November, in which Barack Obama was seeking a second term in office, the Camp David Summit was designed to return the G8 to its roots as a small, intimate, face-to-face gathering of world leaders in which decisions on several tough global issues could be taken at the highest political level in an informal, relaxed setting.

The summit's agenda was specifically designed to be selective, producing a short, action-oriented communiqué. Over a 24-hour period, beginning on the evening of May 18 and continuing until the late afternoon of May 19, G8 leaders were scheduled to address the still recovering world economy, balanced growth and job creation, trade and investment, the means to deliver on the Millennium Development Goals (MDGs) scheduled for realization by 2015, food security, accountability, and peace and security, particularly in hotspots such as Syria, North Africa, Afghanistan, and the Korean Peninsula. Energy was also considered to be a high priority as disgruntled Americans were paying record prices at the pumps and blaming the incumbent president for high fuel costs. Other energy issues became part of the leaders' agenda, including the energy supply mix, oil emergencies, energy efficiency, nuclear energy, and climate change.

German chancellor Angela Merkel identified the importance of climate change and that 2012 marked the United Nations Conference on Sustainable Development (Rio+20) as well as the 18th COP in Doha. Merkel declared that G8 leaders must "renew our political commitment to sustainable development and advance the Durban Platform in order to ensure that we can conclude a new international climate change agreement by 2015 at the latest" (Merkel 2012). In noting the close policy linkages between the Camp David agenda and the G20's Los Cabos Summit just one month later, she further noted that environmental issues must be considered in shaping the framework for global economic growth, for the "G20 have set ourselves the goal of fostering strong, sustainable and balanced growth." In this context, Merkel further drew the link between the environment the economy, noting that "economic growth and environmental protection go hand in hand and create new potential," with developments in environmental technologies slated to grow to over \$5.7 billion by 2025—a threefold increase compared to 2007. Noting that "investments in energy efficiency … have been producing impressive growth rates for many years," Merkel stressed the importance of "promoting this dynamic growth."

As summit host, Obama had remarked on the importance of preserving the planet through concrete action on climate change when he addressed the UN General Assembly in September 2011. He also noted the need to continue down the path of accountability on climate action by stressing that the G8 "continue our work to build on the progress made in Copenhagen and Cancún, so that all the major economies … follow through on the commitments that were made" (Obama 2012). Reaffirming the need for partnership on this issue, he said that "together, we must work to transform the energy that powers our economies, and support others as they move down that path."

This need to focus on climate change despite the ongoing slow growth in the US and Europe was echoed by those outside policy circles. They asked the G8 to "demonstrate leadership by example by surmounting national interests and providing the necessary funding and technological know-how to enable developing countries to cope with climate change and transit to low-carbon economies" (Agidee 2012). There was a growing sentiment that if G8 leaders "step through the door first, other countries will surely follow."

At the Summit

On their second day of the summit, the G8 held a session on energy and climate change. The expectation was that climate change would be given sufficient time, particularly on issues involving energy efficiency and security, the impact of climate change on food security, low carbon energy targets, fast-start finance mechanisms for climate control, green growth job creation strategies, and a 2015 deadline for a post-Kyoto climate regime.

Results

In the end, the leaders at Camp David took several substantial steps on these issues to produce a sound success. Five important advances stood out in the paragraphs on climate change. The first was the explicit affirmation of an all-in approach to a post-Kyoto climate change control regime, with the delivery of "an agreed outcome with legal force applicable to all Parties, developed and developing alike" by 2015 (G8 2012).

The second was the agreement that all G8 members would join the Climate and Clean Air Coalition to Reduce Short-Lived Climate Pollutants (CCAC) to reduce the impact "on near-term climate change, agricultural productivity, and human health." The CCAC (2014b) defined short-lived climate pollutants as "agents with a relatively short lifetime in the atmosphere—a few days to a few decades—and a warming influence on the climate." Examples of such pollutants are black carbon, methane, hydrofluorocarbons, and tropospheric ozone. According to the G8, comprehensive action to reduce these pollutants was urgently needed as they accounted for more than 30 percent of near-term global warming, as well as 2 million premature deaths a year (G8 2012).

Third was the call, in synergistic support of the G20, for "efforts to rationalize and phase-out over the medium term inefficient fossil fuel subsidies that encourage wasteful consumption" (G8 2012). Fourth was a focus on increased energy efficiency and reliance on renewables and clean energy technologies to address climate change and promote innovation and sustainable economic growth. The leaders committed to advancing appliance and equipment efficiency and promoting industrial and building efficiency through enhanced energy management systems.

Fifth, in producing these statements, the leaders reaffirmed their prior commitment to do their part in effectively limiting the increase in global temperatures below 2°C above pre-industrial levels. Overall, the now more inclusive nature of the G8 on global climate governance once again stood out in its final declaration. The leaders agreed to continue to "work together in the UNFCCC and other fora, including through the Major Economies Forum, toward a positive outcome at Doha" (G8 2012).

Dimensions of Performance

In its climate change governance, Camp David was a substantial success (see Appendix A-1). Relatively robust performances in deliberation, direction setting, and decision making were offset by modest successes in delivery and the development of global governance, with few notable results in domestic political management.

Domestic Political Management
Camp David's domestic political management was small. Russian president Vladimir Putin did not come, breaking the G8's perfect attendance record since its start in 1975. No communiqué compliments related to climate change were issued to any members, as at Deauville the year before (see Appendix B-1).

Deliberation
Deliberation in its public form was strong. Camp David's 789 words on climate change, taking 7.1 percent of the communiqué's text, ranked the seventh highest in number and the fourth highest in portion in summit history (see Appendices A-1 and C-1). This marked a rise from Deauville's 5.9 percent. More specifically, climate change arose in 11 paragraphs.

Direction Setting
Camp David's performance on direction setting was strong. With a one-sentence preamble in the communiqué, the leaders jumped directly into sections on the summit's core thematic issues. Energy and climate change had a stand-alone section. It was placed second in the sequence, preceded only by the section on the world economy.

Connections to climate change abounded, if not those to democracy and human rights. Paragraphs on energy were placed in the context of promoting low carbon policies to tackle climate change. The strategy endorsed to enhance energy production was similarly conditioned by the need to proceed in an "environmentally safe, sustainable, secure, and affordable manner" (G8 2012). The summit further emphasized the need for "high levels of nuclear safety," referencing the extreme weather event of the tsunami and consequent nuclear accident in March 2011 in Japan. Simultaneously, the G8 called to "reduce barriers and refrain from discriminatory measures that impede market access" for energy supplies. This offered a boost for the G8's big energy exporters of Canada and Russia (Kirton and Kulik 2012b).

Decision Making

Decision making was solid. Camp David's five climate change commitments were a little below Deauville's seven and below the average of eight (see Appendix D-1). Yet the number of climate commitments was the ninth highest ever. There were new commitments on establishing best practices in the exploration of "frontier areas" with the use of new technologies, including "deep water drilling and hydraulic fracturing" for the "safe development of energy sources" (G8 2012). In light of the shock of the 2011 Japanese natural and nuclear disaster, new commitments were made for "comprehensive risk and safety assessments of existing nuclear installations."

Delivery

Camp David's delivery of these decisions, however, was small. Compliance on the one assessed commitment on reducing pollution averaged +0.11, a sharp drop from the +0.67 at Deauville (see Appendix E-1). However, compliance with the commitment on joining the CCAC and implementing broad actions to reduce the use of short-lived climate pollutants saw all G8 members save Italy and Russia join the coalition in the following year. By joining the CCAC, the G8 committed to "mitigating short-lived climate pollutants in their own countries, helping others take similar actions, and to actively participating in the work of the Coalition" (CCAC 2014a).

Canada, Germany, and the European Union took the most robust steps in implementing specific initiatives to limit the use of short-lived climate pollutants. Canada became the first country in the G8—and the world—to implement new standards banning the creation of new coal plants using traditional technology key to producing black carbon (G8 Research Group 2013). To support this initiative, the Canadian government offered CA$10 billion in April 2013 in support of the CCAC, with an additional CA$7 billion for projects aimed at mitigating short-lived climate pollutants in developing countries (Environment Canada 2013).

Development of Global Governance

The development of global governance at Camp David was quite small. There were no references to the G8 environment ministers, who did not meet during that year (see Appendix F-1). There was a reference to the Major Economies Forum (MEF), with the G8 agreeing to continue to work together through the MEF process. No new climate-relevant G8 institutions were created. Outside the G8, four references were made to three external bodies related to climate change—the United Nations Framework Convention on Climate Change (UNFCCC), and the United Nations Environment Programme—and one to the forthcoming Doha COP meeting (see Appendix F-5).

Los Cabos G20, June 2012

Just one month after Camp David, the G20 leaders met for the seventh time, assembling in Los Cabos, Mexico, on June 18–19, 2012.

Preparations

According to the Mexican presidency in the lead-up to Los Cabos, the main themes for the summit would be strengthening the financial system and fostering financial inclusion to promote economic growth, improving the international financial architecture, enhancing food security, controlling commodity price volatility, and promoting sustainable development, green growth, and the fight against climate change (Mexican Presidency of the G20 2011).

However, many G20 leaders came to Los Cabos still seeking a way out of the 2008 financial crisis. After four years of high unemployment and continued economic uncertainty, the industrialized countries in particular were divided over what strategies to try next (Hochstetler 2012). To this end, Mexican president Felipe Calderón noted that economic stabilization, structural reform, and unemployment remained key summit priorities, with the goal of correcting macroeconomic imbalances, resisting protectionism, and stimulating global growth. However, he also noted that the severe consequences of a changing climate in the coming decades meant that "many of our current concerns about the world economic crisis will seem superfluous in comparison" (Calderón 2012). With Mexico having just endured its worst drought in 70 years, Calderón (2011) had already declared that there was a "terrible climate change problem … that relates to the viability of our civilization's whole model as we look ahead to the end of this century." Climate change was "affecting the whole planet" and "each country must do its part" (Calderón 2012). He intended to make green growth policies, the phase-out of inefficient fossil fuel subsidies, and contributions to the Green Climate Fund priorities at Los Cabos. In recognizing the link between climate and development, Calderón further noted his aim of "designing tools that enable developing countries to boost their technical and institutional capacities, in order to implement policies that achieve sustainable development" (Calderón 2012).

Mexico's sustainable development plan was to develop green growth strategies based on technological innovations that would improve the environment as well as foreign investments. In a joint article published in May, Calderón and Korean president Lee Myung-bak called for green growth as a key solution to the global financial crisis and pledged to work together with other G20 members to this end (Chang 2012). They said that the current growth paradigm of production and consumption had to become greener so that both economic growth and environmental protection could be achieved. Mexico asked Korea to lead the discussions at Los Cabos on energy efficiency and clean resources, with the intent on developing a concrete set of policy recommendations.

At the Summit

On June 17, just before the Los Cabos Summit opened, the G7 finance ministers and central bank governors issued a statement on the situation in Greece, and the leaders of Brazil, Russia, India, China, and now South Africa (BRICS) held an informal summit on the morning of June 18. The G20 leaders' first working session started at 15:30 on June 18, followed by the evening session over dinner. A subsequent meeting between Barack Obama and the European leaders was cancelled.

The summit got off to a strong start on its central challenges of controlling the escalating euro crisis, producing a credible plan to boost global growth and jobs, and raising resources for any rescues the International Monetary Fund (IMF) might be called on to take. On the second issue of generating global growth and jobs, G20 leaders prepared an action plan that emphasized growth, including through greater government spending by countries with relatively low debt and deficit levels. On the third issue of IMF resources, by the time the leaders went to sleep after their first day, Calderón had announced that the $430 billion total assembled by finance ministers in Washington in April had now risen to $456 billion. A big part of the boost brought by new pledges from 12 countries came from China's announcement at the G20's evening dinner session that it would contribute $43 billion.

Given the preoccupation with the European financial crisis, and concern over the escalating civil war in Syria, there was little time for a full-scale discussion on climate change. Australian prime minister Julia Gillard (2014) resolved that climate change would be one of the summit's three priority subjects when Australia hosted in 2014.

Results

In many ways, Los Cabos was a significant success overall. It finally controlled the urgently needed escalating euro crisis, promising to support Greece's path to reform by taking "all necessary policy measures to safeguard the integrity and stability of the area" (G20 2012b). To generate global growth and jobs, Los Cabos set a credible strategy that stressed immediate stimulus and fiscal consolidation, and, to this end, mandated a broad array of structural reforms.

Under the leadership of Mexico, an important emerging economy and energy producer, Los Cabos made important advances on climate change control. Of the 95 commitments it made, 11 were dedicated to green growth strategies or climate change—the G20's highest to date. With commitments addressing adapting agriculture to climate change and promoting long-term prosperity through inclusive green growth, the G20 (2012a) recognized that "climate change will continue to have a significant impact on the world economy, and costs will be higher to the extent we delay additional action." Leaders welcomed the creation of the newly developed G20 study group on climate finance to consider ways to mobilize resources effectively. They asked the study group to provide a progress report to the G20 finance ministers later that year. They further noted their intent to self-report again in 2013, on a voluntary basis, on efforts and progress on incorporating green growth policies within national structural reform agendas that promoted sustainable development.

On fossil fuel subsidies, the leaders reaffirmed the promise and mandate made at their 2009 Pittsburgh Summit to reduce and phase out inefficient ones in the medium term. They took some steps to encourage implementation, asking the G20 finance ministers to report on progress at the next summit and to "explore options for a voluntary peer review process for G20 members by their next meeting," while welcoming "a dialogue on fossil fuel subsidies with other groups already engaged in this work" (G20 2012a). However, the leaders specified neither a deadline nor interim benchmarks for the phase-out, nor what subsidies should be included—consumer, producer or regulatory (Kirton and Kulik 2012c). The cost was considerable, as such subsidies caused an estimated 10 percent of carbon emissions to be released into the atmosphere each year.

There was significant progress on developing a legally binding protocol that would apply to all parties in ways defined by the new regime. In accordance with the UN's draft declaration for the forthcoming COP, all parties were requested to submit by February 28, 2012, their plans on "options and ways for further increasing the level of ambition and possible further actions" that would enable the effective and sustained implementation of the UNFCCC (UN 2012). These plans were intended to integrate climate change adaptation into new and existing policies and programs within all relevant sectors, based on nationally defined priorities. Furthermore, the member's actions would support the launch of the Green Climate Fund, designated as an operating entity of the financial mechanism of the convention. All parties at Durban would thus be invited to make financial contributions to the fund's start-up. With 16 of 20 G20 members declaring their intent to adhere to the COP-17 decisions, clear signals of the G20's influence in moving the climate agenda forward were sent.

Dimensions of Performance

The Los Cabos Summit delivered a significant success on climate change, led by strong performances on deliberation, decision making, and the development of global governance, with more modest success on direction setting and delivery, and little overall success in terms of domestic political management (see Appendix A-2).

Domestic Political Management
In domestic political management, Los Cabos offered no communiqué compliments to any of its permanent members.

Deliberation
Deliberation in its public form was strong. Climate change secured 1,160 words or 9.1 percent of the communiqué, placing Los Cabos third behind Seoul and Pittsburgh (see Appendix A-2). Los Cabos also stood above the average of 916 words or 6.6 percent of the communiqué on climate change.

Direction Setting
Direction setting was solid at Los Cabos. While climate change per se was not given priority placement, the environment was. Climate change was also linked to democratic principles, for the fourth summit in a row. And although it did not secure its own stand-alone portion of the communiqué, it captured four of the eight paragraphs in the section on inclusive, green growth. Los Cabos also strongly forged links between climate and the economy and between climate and energy.

Decision Making
Decision making was strong. Los Cabos made six commitments on climate change, second only to Seoul and Cannes (see Appendix D-2). It also made 10 commitments in the separate issue area of energy, continuing to emphasize the climate-energy link (see Appendix A-2).

Delivery
The delivery of these decisions was solid on climate and substantial on energy, with a score of +0.38 on the two assessed climate commitments and +0.58 on the one assessed energy commitment (see Appendix E-2). This placed Los Cabos below the Cannes average of +0.61 on its three assessed energy commitments, but just above the G20's eight-summit average of +0.52. Compliance with the core Los Cabos commitment on consenting to the adoption of a legally binding agreement on climate change at Durban received a score of +0.70, with full compliance by most G20 members, Saudi Arabia and Indonesia scoring in the negative range, and Italy and Turkey indicating ongoing work.

Development of Global Governance
The development of global governance was significant. Inside the G20, leaders asked their finance ministers to report on phasing out inefficient fuel subsidies and acknowledged the relevance of accountability and transparency in their reporting. They also welcomed the creation of the G20 study group on climate finance to consider ways to mobilize financial resources, particularly through the operationalization of the Green Climate Fund. The study group was mandated to provide a progress report to G20 finance ministers by November. Outside the G20, the leaders made six references to five other international institutions related to energy and climate change, placing Los Cabos close to the overall summit average in both respects (see Appendix F-6).

Causes

The sound and significant climate performance of the 2012 G8 and G20 summits respectively was driven by several forces.

Shock-Activated Vulnerability
The first was shock-activated vulnerability, which was led by the shocks and vulnerabilities recognized by the leaders and recorded in the climate passages in their communiqués (see Appendices G-1 and G-2). There were none at the G8's Camp David, but one at the G20's Los Cabos, consistent with its superior performance. World oil prices also served as both a driver and somewhat of a constraint. They reached a record high of $100.39 a barrel at the start of 2012 but slipped to a still high $89.52 by May (see Appendix G-3). The drop meant that there was less immediate pressure at Los Cabos to reduce the strain on government budgets brought by soaring oil prices and subsidy costs through ambitious new action on fossil fuel subsidies. The 2011 Fukushima disaster in Japan still had a negative effect on the treatment of climate-friendly nuclear power at the G8, but not at the more multilateral G20 summit. However, the environmental shock of a severe drought in Mexico helped drive the Los Cabos Summit to a greater climate performance than the US-hosted G8 summit in 2012. At Los Cabos, "the sequence of severe and spreading shocks erupting throughout Europe" in finance served as somewhat of a diversion but also a driver of green growth, as a committed Calderón was able constructively to convert and connect the financial shock in this way (Kirton and Kulik 2012c).

Multilateral Organizational Failure
Multilateral organizational failure was a second cause. To be sure, the UN's Durban COP in December 2011 had put an end to the special but differentiated formula. Leaders, including all G20 leaders, had consented to the adoption of a legally binding agreement on climate change, to be prepared "as soon as possible, and no later than 2015" (UNFCCC 2011). A management framework for a Green Climate Fund was also adopted, in addition to the creation of specifically focused working groups to operationalize decisions taken.

However, Durban left much to be done. There was no new assessment report from the Intergovernmental Panel on Climate Change (IPCC) to provide an epistemic boost, five years after the last, still somewhat cautious one. The International Energy Agency was not present at the Los Cabos Summit table to push for the fossil fuel subsidy phase-out or some other climate-friendly energy initiative. And neither G8 nor G20 environment or energy ministers met, even though many other G20 portfolio ministers did for the first time during the Mexican presidency.

At the leaders' level, the Rio+20 summit in Rio de Janeiro on June 20–22, 2012, held right after the G20 Los Cabos Summit, had a dual effect. It galvanized G20 action to help make the more multilateral UN summit a success. But it also was a major constraint, as many G20 members convincingly argued that climate change should be left to the UN summit two days hence. More broadly, a possible precedent for direct G20 summit climate leadership came from the G20's financing for the new IMF firewall fund. It showed that the G20 leaders had to—and did—act to prevent a looming crisis when the IMF alone was unable to.

Capability

Relative capability, with stable ratios, had little effect. The G20 remained far more globally predominant than the G8, but did not increase its lead, because growth in China, India, and above all Brazil slowed (Kirton and Kulik 2012a). Within the G8, internal inequality was maintained, amid the "flight to safety" rise of the US dollar, the historic high value of the Japanese yen against the US dollar, the rise of the Canadian dollar against the US one, and the stability of the British pound, even as the euro and ruble declined a bit. G8 growth was led by the US but also by Canada and Russia.

Democratic Convergence

Democratic convergence had little direct impact. Contextually, the G8 benefited from a direct connection between its foundational core mission of promoting open democracy and individual liberty and Camp David's priorities of democracy for the Deauville Partnership for North Africa and the Middle East and Afghanistan and protecting human rights in Syria. But the absence of Putin, newly elected as Russia's president, raised doubts about the continuing commitment to democracy there. At Los Cabos, democracy was a clear concern in economically afflicted Europe, especially as far right-wing parties increased their electoral strength in France and Greece and many remembered that Greece and Spain had only moved from dictatorship to democracy around the time of the G7's birth.

Political Cohesion

Political cohesion helped cause the superior climate performance of the G20. While it was helped at Camp David by the four new G8 leaders coming with fresh democratic mandates from popular elections just held, it was lessened by Putin's choice to stay at home and send Dmitry Medvedev. At Los Cabos, among the many new leaders, there remained a strong core of committed G20 summit founders and veterans, led by Mexico's climate-committed Calderón as host, who was able and willing to take the lead.

Constricted Club Participation

Constricted club participation boosted climate performance at Camp David, due to the isolated, intimate, informal setting of a rural retreat. At Los Cabos it was enhanced by the intense interaction within and overlapping the G20 members that had made the G20, as an institution, and its leaders, as individuals, the emerging club at the centre of a dense and ever-expansive network of global governance (Kirton and Kulik 2012a).

After the Summits

Rio+20, Brazil, June 2012

Marking the 20th anniversary of the first ever sustainable development summit, the Rio+20 summit took place in Rio in June 2012 immediately following the Los Cabos Summit. It promised to address the mounting international crisis brought on by a combination of volatile oil prices, accelerated ecosystem degradation, global financial instability, and climate-induced extreme weather events. According to participants, these "multiple and inter-related crises" called into question the ability of a growing human population to live sustainably, thus demanding "the urgent attention of governments and citizens around the world" (Earth Summit 2012 Stakeholder Forum 2012).

Although a broad range of sustainable development issues were addressed, Rio+20 leaders offered little in the way of new commitments on climate change. Instead, they reiterated their broad pledges to accelerate the reduction of global greenhouse gas emissions in accordance with the old principle of "common but differentiated responsibilities and respective capabilities" (UN Conference on Sustainable Development 2012). They noted with "grave concern" the significant gap between current emissions targets and maintaining average global temperatures below 2°C above pre-industrial levels, but failed to outline specific targets aimed at mid- to long-term emissions reduction strategies. Although they recognized the importance of funding from a variety of public, private, bilateral, and multilateral sources to support mitigation and adaptation measures, they failed to mobilize money and only "welcomed" the launch of the Green Climate Fund, calling for its "prompt operationalization" and an "adequate replenishment process." The climate section of the final declaration recognized the importance of the progress on climate mitigation already achieved through the COP process, thereby deferring further efforts and commitments to the upcoming meeting scheduled for Doha later that year.

COP-18, Doha, November–December 2012

The 18th COP took place in Doha from November 26 to December 7, 2012. It was the last UNFCCC meeting prior to the scheduled expiration of the Kyoto Protocol on December 31. Pressure mounted over the 12 days, as European countries in particular were keen on marking Doha as the transition from the old climate regime—where only the developed countries were legally obligated to reduce emissions—to a new regime in which all major countries would make legally binding agreements for the first time. Intense consultations resulted in a final 36-hour session in which the EU successfully negotiated an eight-year extension of the protocol. The new deal would require both developed and developing countries to sign a global agreement to cut greenhouse gas emissions by 2015, to come into force by 2020 (Harvey 2012). Reorganizing the protocol negotiations into a single unified set of talks and setting out a work plan up to 2015 were heralded as conference successes.

But with Russia having joined Canada and Japan in announcing its intention to discontinue participating in the protocol, questions about Doha's effectiveness mounted. Russia's support for climate cooperation had long been contingent on participation by the world's largest emitters—including the US, China, and India. If these countries refused to commit to further reductions, Russia—the world's fourth largest emitter (after China, the US, and India)—did not intend to assume the emissions limitations or reduction commitments required of Kyoto's second commitment period (Low 2013).

With an economy rich in fossil fuels, Russia was producing 5.4 percent of the world's greenhouse gas emissions that would no longer fall under a legally binding agreement.

Moreover, pulling out of Kyoto meant that Russia, and its fellow countries that had withdrawn, could avoid penalties for missed targets, setting a dangerous precedent for other industrialized countries to follow.

Overall, Doha managed to establish a new timetable for a climate change agreement that would cover all major countries, to be adopted by 2015. It could not, however, agree on how to scale up efforts beyond existing pledges to curb emissions but remain below the 2°C temperature rise. Doha may have been able to reach an agreement but "at the same time had also fallen short" (Earth Negotiations Bulletin [ENB] 2012).

Lough Erne G8, June 2013

Preparations

In preparing for the 39th G8 summit, hosted in Lough Erne, Northern Ireland, British prime minster David Cameron designed a summit that would draw on the United Kingdom's "experience to prepare a summit that [could] make the most of the G8's current potential" (Bayne 2013, 41). This included an isolated site, a simple format with few lead-up ministerial meetings, and a short and focused agenda, all designed to make the summit a personal instrument of the leaders. Moreover, Lough Erne's location and apparent absence of invited outsiders would allow the leaders to pull together in a way that would allow them to address the most difficult and demanding global problems. In this regard, Lough Erne would see the G8 return its traditional roots with its highly focused agenda, its secluded, informal setting, and its lack of outside invited guests. Cameron's intent was to have only "one table and one conversation, with G8 leaders holding each other to account" and transforming good intentions into action (Cameron 2012).

With its central themes of trade, tax, and transparency, the summit was also expected to return to its assumed traditional economic roots, and in this way, forge a closer, more cooperative relationship with the newer G20. On development, Lough Erne was expected to revert to its historic accomplishments by mobilizing new money for nutrition and food security. And on the political-security agenda, the G8 was expected to deepen democracy and human rights by calling on Syria to end the slaughter of its civil war, reinforcing the political and economic transformation in the Middle East after the Arab Spring, and preventing the proliferation of nuclear weapons in Iran and North Korea.

Important advances were made on several key issues in the lead-up to Lough Erne, including on trade, tax, transparency, nutrition, Syria, and terrorism. Substantial progress had been made in advancing the EU-US trade negotiations, in moving several closed jurisdictions toward exchanging more tax information, and in having the US and Russia set aside their differences on Syria in support of a peace conference that could succeed where the UN's efforts had failed. Together, this progress would demonstrate that the G8 was an effective and unique summit that could provide "badly needed global public goods across a wide front" (Kirton 2013a).

At the Summit

The G8 Lough Erne Summit yielded several significant successes in its macroeconomic policy performance. Leaders made strong commitments on trade, tax systems, and transparency, noting their collective responsibility to "support prosperity worldwide"

(G8 2013a). The G8 appropriately recognized the presence of important vulnerabilities, particularly with Europe still in recession, and called on Europe to move toward a banking union and strengthened bank balance sheets. Here as elsewhere, it emphasized the importance of fiscal sustainability through growth-oriented structural reforms. At the microeconomic level, it appropriately highlighted the acute problem of long-term and youth unemployment, and the importance of well-functioning credit channels for investment through small and medium-sized enterprises (SMEs) (Kirton 2013a).

On development, the core commitments came on food security and nutrition, with progress outlined on commitments made in catalyzing private sector investment to improve access to new technology, manage risk, and improve nutrition.

A substantial amount of time was dedicated to foreign policy and security issues. Terrorism, nuclear safety, the Middle East, North Africa, Syria, and Libya occupied more than one half of the final declaration.

In the end, the G8 was successful in recognizing the necessity of being more inclusive on climate governance issues and decisions. The leaders noted they would pursue "ambitious and transparent action" on climate change through various international forums, including the MEF, the International Civil Aviation Organization (ICAO), and the International Maritime Organization (IMO) (G8 2013a). Moreover, in accepting membership in the CCAC, the G8 formally recognized that climate change is "a contributing factor in increased economic and security risks globally." The leaders acknowledged that "climate change is one of the foremost challenges for our future economic growth and well-being," and that they remained "strongly committed" to addressing the need to reduce greenhouse gas emissions by 2020 and to pursue low carbon policies with a view to "doing our part to limit effectively the increase in global temperature below 2°C above pre-industrial levels, consistent with science." The commitment of hard funds by the largest industrialized countries for climate-related programs and initiatives was also noted, with the G8 reiterating its commitment to jointly mobilize $100 billion of climate finance per year by 2020 through a "wide variety of sources" and in the context of "meaningful mitigation actions." Finally, the leaders committed to working through the UNFCCC "to ensure that a new protocol, another legal instrument or an agreed outcome with legal force under the Convention applicable to all parties is adopted by 2015, to come into effect and be implemented from 2020."

Results

The G8 released its own "Lough Erne Accountability Report: Keeping Our Promises," reinforcing Cameron's emphasis in the preparatory process on transparency and accountability. It covered 61 commitments made over the previous 11 years, with useful country-specific reporting. It devoted a stand-alone section of the 175-page report specifically to commitments on the environment and energy. Most reflected the G8's assistance to the developing world in adapting to climate change, notably managing forests sustainably, facilitating energy access and support, managing shocks related to energy prices, and developing clean, efficient energy sources. According to the report, "these commitments have been complemented by the G8's focus on green growth, which is an essential element for ensuring sustainable global growth" (G8 2013b). It credited the G8 with significant progress in climate finance for the developing world, with more than $10 billion allocated over the prior two years to climate adaptation in more than 100 developing countries. The report noted the continued delivery of funds by the G8 through Climate Investment Funds, supporting increased energy efficiency, renewable energy programs, and low carbon projects in developing countries. Reinforcing the G8's move to become more inclusive, the report concluded that environment and development challenges are best tackled when the G8 closely cooperates with developing countries and other international forums and organizations, reinforcing the need and importance for continued partnerships.

Dimensions of Performance

Across its dimensions of performance, Lough Erne showed solid success on climate change (see Appendix A-1). The most robust performances were in its decision making and development of global governance, followed by domestic political management, deliberation, and direction setting. More modest success came in delivery.

Domestic Political Management

In its domestic political management, Lough Erne's performance was fairly solid. There was one communiqué compliment, issued to France as host of the 2015 COP meeting, where G8 leaders would mobilize political will to generate a "successful global agreement" on climate change (see Appendix B-1). Such a direct reference had only been made at three summits prior to Lough Erne.

Deliberation

Lough Erne's performance on deliberation was small. It generated only 525 words on climate change, or 3.9 percent of the final communiqué (see Appendices A-1 and C-1). This placed Lough Erne as the least productive summit in terms of deliberation since Sea Island in 2004.

Direction Setting

Lough Erne's direction-setting performance was also solid. There was no priority placement of climate change in the opening chapeau of the final communiqué, which focused on trade, tax, and transparency. There was, however, a direct reference to democracy in relation to climate change with summit leaders noting the importance of improving the transparency of international climate finance flows (see Appendix A-1).

Decision Making

Decision making was where success was most evident at Lough Erne. The leaders produced 12 discrete commitments on climate change, more than the previous three summits, and well above the average of eight (see Appendix D-1). The total number of climate commitments was the fifth highest ever. New commitments were generated in market-based measures to address rising aviation emissions.

Delivery

In the delivery of its decisions, performance was small. With a compliance average of +0.22, Lough Erne generated significantly higher compliance than the year prior at Camp David, but remained well below the overall compliance average (see Appendix E-1). Compliance with the summit's core commitment of pursuing action in the UNFCCC by, among other things, joining the CCAC in order to reduce the use of short-lived climate pollutants fared significantly better, at +0.56, than with the commitment on climate finance, at −0.11.

Development of Global Governance

In its development of global governance, Lough Erne showed considerable success. Although there were no direct references to the G8 environment ministers, there were references to the MEF, with the leaders committing to "overcome differences on the road to the global deal in 2015" (G8 2013a). Outside the G8, five references linked to climate change were made to four external bodies (see Appendix F-5).

St. Petersburg G20, September 2013

Preparations

The eighth G20 summit—and the first hosted by Russia—took place in St. Petersburg in September 5–6, 2013. It came amid broad economic challenges from rising interest rates, ongoing fiscal deficits, continued slow US growth, recessions in Europe, and financial fragility in once vibrant China, India, and Brazil.

Given this immediate uncertainty, however, the Russian presidency sought outcomes in 2013 that would have a lasting impact by focusing its agenda on creating balanced growth, jobs and employment, fighting corruption, strengthening financial regulation, enhancing multilateral trade, and increasing energy sustainability.

Russia also had a well-defined plan for the summit preparatory process. It included numerous meetings of G20 sherpas and finance and central bank deputies, as well as labour and employment ministers. Throughout, the G20 would "continue to enhance its efficiency by delivering on the decisions made" and "ensure legitimacy through engagement with non-G20 countries, international organisations, business and trade unions, think tanks and academia, civil society and youth" (Yudaeva 2013, 36).

Energy security was viewed early on as a G20 priority for Russia, as it had been when Putin hosted the G8 summit in St. Petersburg in 2006. On climate change, Russia traditionally considered energy efficiency as part of the energy security agenda. However, not all its G20 partners shared this view. All members had to participate in the solution, including the BRICS countries, the US, Canada, and Japan. Without their participation, Russia would not lead the G20 to move on this issue at St. Petersburg (Kirton 2013b).

By December 2012, the Russian presidency had released its official plan for its summit, detailing Russia's strategic agenda and outlining the priority issues. It opened with a list of eight priority policy areas, including energy sustainability. Russia continued to endorse the connection between energy and growth, calling energy a "crucial growth engine for G20 members" and "key to unlocking world sustainable development" (Russia's G20 Presidency 2012, 16). Over the ensuing months, the Energy Sustainability Working Group (ESWG)—with experts from G20 members and representatives from selected international organizations—worked toward four objectives: making energy and commodity markets transparent and more predictable, promoting energy efficiency and green growth, proposing sound regulation for energy infrastructure, and ensuring protection of the marine environment (Rooney 2013). It would pursue an ambitious strategy with an explicit remit to deliver, by the end of 2013, a progress report on the G20's contribution to enhancing the transparency and functioning of international commodity and energy markets. It would draft principles for efficient energy markets that integrate green growth and sustainable development priorities into structural policies. It would make recommendations on the voluntary peer review process for phasing out fossil fuel subsidies, and draft a plan to develop a database of best practices for green energy and energy efficiency policies used by G20 members (Russia's G20 Presidency 2012).

The ESWG met in February and July to establish the coordination among G20 members that would produce the text of the clauses to be included in the leaders' final declaration at St. Petersburg.

At the Summit

At St. Petersburg, the G20 leaders made several advances on a broad economic and security agenda. On stimulating growth for a slowing world economy, they moved toward coherent growth strategies backed by credible medium-term fiscal consolidation that re-emphasized jobs and the contribution of SMEs. They also proposed key financial regulatory reforms, resisted protectionism, and began innovative work on financing for investment with more attention on private sector involvement. On security, the summit took a substantive step toward a solution to the escalating conflict in Syria, as all 20 leaders agreed that chemical weapons had been used and international law had thus been breached (Kirton 2013c).

Results

St. Petersburg was a substantial success on climate change. Roughly 10 percent of the G20's final declaration was devoted to energy sustainability and climate change (13 of 114 paragraphs). Leaders emphasized reliable and transparent energy markets in boosting economic growth, job creation, and sustainable development through inclusive green growth strategies and clean energy technologies.

Recognizing that it was in "their common interest" to identify opportunities for investment in low carbon, clean, sustainable energy infrastructure, G20 leaders committed to a closer, more inclusive engagement of the ESWG with private sector and multilateral development banks that would bring together the public sector, market players, and international organizations to discuss the measures needed to promote a sustainable, affordable, efficient and secure energy supply. On climate change more specifically, the G20 noted the urgency of the impact on the world economy by remarking that the "cost will be higher to the extent we delay additional actions" (G20 2013a). To this extent, the leaders agreed to support the "full implementation" of the UNFCCC's outcomes, reiterating their political will in successfully adopting a post-Kyoto Protocol by 2015. They further committed to operationalize the Green Climate Fund, referencing the work of the G20 Climate Finance Study Group in finding ways to mobilize additional climate finance. The leaders asked their finance ministers produce a report on approaches to climate finance within one year.

In addition to the commitments made in the final declaration, the G20 also released its second G20 accountability report just before the summit. It covered 67 commitments on food security, financial inclusion, trade and investment, and inclusive green growth. The report noted the efforts of the Development Working Group to deliver outcomes that foster economic growth consistent with the sustainable use of natural resources that address environmental concerns. It emphasized the Los Cabos commitment to scale up private sector finance for inclusive green growth strategies. To this end, the Development Working Group developed a "toolkit," with practical information on 15 specific tools—including one on energy and one on climate change—for crafting strategies that support country efforts to "green their economies" sustainably. Workshops were to be held after the summit to examine the toolkit's practical application, with proposals for further implementation (G20 2013b, 69).

Overall, the Development Working Group's prospective success after the summit was largely attributed to the G20's inclusive efforts to enhance its engagement and consultations with both private sector organizations and civil society, in addition to developing "more coherent and targeted cooperation with international organizations" (G20 2013b, 7).

Dimensions of Performance

The St. Petersburg Summit delivered a solid success in its performance on climate change. Strong outcomes on domestic political management, delivery, direction setting, decision making, and the development of global governance came with modest delivery of the decisions on climate change, but a more robust delivery of its energy ones (see Appendix A-2).

Domestic Political Management
In its domestic political management, St. Petersburg offered one communiqué compliment (see Appendix B-2). It noted its support for France as host of the COP-21 in Paris in 2015.

Deliberation
In deliberation, St. Petersburg's performance was strong. Climate change secured 1,697 words in the final declaration (see Appendices A-2 and C-2). This placed St. Petersburg second behind Seoul in the number of words on climate change, well above the average of 916 words.

Direction Setting
St. Petersburg's performance on direction setting was solid. Climate change was referenced in the opening preamble of the final declaration, in a reference to the G20's commitment "to work together to address climate change and environment protection, which is a global problem that requires a global solution" (G20 2013a). Climate change was also the topic of a stand-alone section, and another section was dedicated to sustainable energy policy. Combined, these two sections produced 13 of the 113 paragraphs of the G20's final declaration.

Decision Making
In St. Petersburg's decision making, performance was also very strong. The summit produced 11 commitments on climate change, making it the best performer to date (see Appendix D-2). An additional 14 energy commitments were produced, well above the 9.8 average (see Appendix A-2).

Delivery
In delivering St. Petersburg's decisions, performance was mixed. Compliance with the assessed energy commitment was at least substantial, at +0.55 (see Appendix A-2). Delivery of St. Petersburg's core climate change commitment, however, was weak at −0.02.

Development of Global Governance
In the development of global governance, St. Petersburg's performance was significant. Inside the G20, leaders made one reference to G20 finance ministers, asking them to report

back to the G20 in one year's time on the work of the Climate Finance Study Group. Two additional references were made to official bodies inside the G20, including the Green Climate Funds and the Climate Finance Study Group. Outside the G20, 10 references were made to seven international bodies related to energy and climate change (see Appendix F-6). This put St. Petersburg second only behind Seoul and well above average.

Causes

Shock-Activated Vulnerability
The substantial climate performance at Lough Erne and St. Petersburg, especially in decision making, was caused first by increasing shock-activated vulnerability. While neither the G8 nor G20 recognized such shocks and vulnerabilities in their communiqués, deadly storms throughout the US and central Europe in 2012–2013 pushed members to react to the deadly consequences of climate change. "Superstorm Sandy," assaulting the northeast US in October 2012, cost the United States 150 lives and $67 billion in damages, followed by deadly tornados in Oklahoma in May 2013 that cost another 27 lives and $3 billion (National Climatic Data Center 2015). Extreme flooding throughout southern Germany and the western Czech Republic in June 2013 resulted in 25 fatalities and more than €12 billion in damages, with research suggesting that rising temperatures were likely to cause more frequent and heavier rains throughout central Europe in the coming years (*Economist* 2013). World oil prices rose from $89.52 a barrel in May 2012 to $96.40 at the start of 2013, and stayed at a high $93.54 by June and $94.53 by January 2014 (see Appendix G-3).

By May 2013, carbon dioxide emissions had surpassed the critical threshold of more than 400 parts per million (ppm) per day, or more than 50 ppm above what scientists considered safe for keeping the earth's temperature rise below 2°C over pre-industrial levels (Rafferty 2013). As Lough Erne approached, diversionary shocks arose from the regular use of chemical weapons in the civil war in Syria by the Assad government and terrorist attacks in Boston in April 2013. The situation in Syria led the leaders at the last minute to devote the entire, extended opening dinner discussion at the St. Petersburg Summit to the issue, leaving less time and energy for advances on climate change (Carr 2014).

Multilateral Organizational Failure
Another cause was the continuing failure of the major multilateral organizations to control climate change. The UN disappointed yet again, both at Rio+20 in June 2012 and at COP-18 in Doha in December 2012. And it was now six years since the last IPCC assessment had been produced.

Predominant Equalizing Capability
Changing relative capability ratios were consistent with the rising climate performance, especially of the G20 (see Appendix I). From 2012 to 2013, China's gross domestic product rose 9.8 percent and that of the US 3.9 percent, fuelling the global predominance and internal equality of the G20. Within the G8, while the US growth rate exceeded that of many partners, while that of the UK rose a higher 5 percent.

Political Cohesion
The democratic character of G8 and G20 members remained constant from 2012 to 2013 (see Appendix J). However, political cohesion was enhanced in several ways. In the US

Barack Obama was re-elected president in November 2012 for a second four-year term. In Russia, a still very popular Vladimir Putin had also returned as president in May 2012, to host the G20's St. Petersburg Summit in September 2013.

Compact Club Participation

Compact club participation contributed as well. There were only six invited guests to the G8's Lough Erne Summit (see Appendix L-1). There were few new additions to the G20's St. Petersburg one (see Appendix L-2).

After the Summits: COP-19 Warsaw

The COP-19 met in Warsaw from November 11 to 23, 2013, was another UN failure. With Supertyphoon Haiyan having just struck the Philippines, and with the Australian government announcing it would discontinue contributing to the Green Climate Fund, talks between the developing and developed countries at Warsaw were strained. Developing countries were increasingly mistrustful that Copenhagen's promise of $100 billion per year by 2020 would be fulfilled. Work on a revised framework entailing various approaches and new market mechanisms got caught up in procedures. With finance issues remaining the most contentious in the UN's climate talks, "relatively timid decisions" on long-term finance and loss and damage were reached, but parties were unable to achieve consensus on other key issues (ENB 2013). COP-19 "did not even meet its modest expectations." Attention turned to 2015, where the 21st COP would convene in Paris, and its efforts toward finalizing a comprehensive and universal climate action agreement to scale up efforts beyond existing pledges.

Brussels G7, June 2014

Preparations

For its scheduled G8 summit in Sochi on June 5-6, 2014, Russia planned a dense preparatory schedule of more than 100 meetings. It commissioned experts' papers and circulated them among the sherpas to discuss at their first meeting in February in Moscow. However, on February 21, Russia's move to invade and annex Ukraine's Crimea region prompted a conference call among the G7 sherpas, who decided to suspend their participation in the preparatory process. An easily negotiated leaders' statement issued on March 1 confirmed this move. A Canadian suggestion that G7 leaders meet to discuss Ukraine on the margins of the scheduled Nuclear Safety Summit in the Netherlands was eventually accepted.

Thus an ad hoc summit was held—without Russia—at the Hague on March 24, 2014. G7 leaders, including the presidents of the European Council and European Commission, engaged in a good discussion about sanctions and decided to hold the regular summit without Russia on the initially intended date in Brussels. They also called for a meeting of G7 energy ministers.

To prepare for this G7 Brussels Summit, at the first sherpa meeting, in Berlin on April 29, Germany and the UK took the lead. Germany was scheduled to host the summit after Russia in 2015, and the UK had hosted the 2013 Lough Erne Summit. The summit agenda was to include energy and climate change, in response to the Europeans' strong desire as France would be hosting COP-21 the following year, as well as the situation in Russia and Ukraine and the global economy.

At their second meeting, at Lancaster House in London, the sherpas focused on drafting the communiqué. Indeed, they completed most of it. Canada wanted greater clarity on climate change. The summit was designed to show solidarity and continuity rather than new initiatives.

Discussions on energy and climate change were scheduled for the second day of the Brussels Summit. Leaders would focus specifically on energy security in light of the Ukrainian crisis, as well efforts to limit global temperature increases through a post-Kyoto global climate agreement. Climate change would thus be linked with energy security to chart a path toward a new global control regime due in 2015. Leaders would be "identifying and implementing concrete domestic policies separately and together to build a more competitive, diversified, resilient and low-carbon energy system," with particular attention on upgrading energy efficiency and clean technology while reducing greenhouse gas emissions (European Council 2014).

To develop this work, G7 energy ministers met in Rome on May 6, a month before the Brussels Summit. They noted, in reference to the escalating situation in Ukraine, that "energy should not be used as a means of political coercion or as a threat to security" (G7 Energy Ministers 2014). They articulated their united determination in providing "various types of assistance that Ukraine needs to strengthen its energy security."

Reducing greenhouse gas emissions was also a key consideration. Discussions centred on the transition to a low carbon economy to support ongoing energy security. To do so, continued investment in research and innovation, diversification in the energy mix, and the deployment of clean and sustainable energy technologies were all seen as important. Energy ministers concluded by proposing to G7 leaders that a working group be established to develop initiatives adopted in Rome and to report back to the ministers within six months.

At the Summit

On June 5–6, 2014, the G7 summit was hosted, for the first time, in Brussels and co-chaired, also for the first time, by the European Council and the European Commission. The priority was the ongoing crisis in Ukraine following its presidential elections on May 25 and escalating tensions with Russia. Attention focused on the continued work to support Ukraine's economic and political reforms as well as the G7's "continued readiness to intensify targeted sanctions and to impose further costs on Russia should events so require" (European Council 2014).

The opening dinner session was opened by European Council president Herman Van Rompuy and quickly handed off to Germany's Angela Merkel, followed by Barack Obama. The discussion was almost entirely about Ukraine. The next day, at lunch, during the development session, the leaders discussed how to deal with Vladimir Putin at the 70th anniversary of the landing at Normandy immediately after Brussels.

On the morning of the second day, G7 leaders first discussed global growth and how to tackle high unemployment, followed by trade, and then the financial reforms undertaken in response to the global financial crisis.

Japan's Shinzo Abe led on the economy, which was a routine discussion. The second working session specifically addressed climate change and energy security, with particular focus on diversification of energy routes and sources. Italy's Matteo Renzi, making his summit debut, spoke about the energy ministerial that had taken place in Rome. On climate change directly, France's François Hollande spoke at length about how important the issue was. Obama agreed. But with the UN dates and deadlines already in place, the discussion was perfunctory. There was little room for initiative at a summit overwhelmingly preoccupied with the situation in Ukraine.

Results

In its public results, the summit endorsed the principles and actions of the Rome G7 Energy Initiative that would, as European Commission president José Manuel Barroso (2014) said, "ensure that our citizens and businesses benefit from energy that is cleaner, safer and more secure than in the past."

The G7 also recognized the link between energy security and climate change, noting that reductions in greenhouse gas emissions and a move to a low carbon economy were both necessary for energy security. In doing so, the leaders said they would do their part to limit the increase in global temperatures below 2°C above pre-industrial levels. They noted their commitment to a new protocol in 2015, a legal instrument, or an agreed outcome with legal force under the convention applicable to all parties. The new regime's core principle of "applying to all" was thus affirmed.

Although no new money was pledged for energy and climate change in Brussels, the G7 leaders reaffirmed their prior Copenhagen Accord commitment of mobilizing an additional $100 billion per year by 2020 from both private and public sources to meet mitigation and adaptation needs in developing countries.

On fossil fuel subsidies, leaders indicated their commitment to eliminating subsidies and to improving the measurement, reporting, verification, and accounting of carbon emissions, consistent with the UNFCCC. They asked their energy ministers to report on progress made in all these areas by the next G7 summit in 2015 (G7 2014).

Dimensions of Performance

On climate change, Brussels was a significant success (see Appendix A-1). The most robust performance came in the G7's deliberation, decision making, and development of global governance. Results in domestic political management and direction setting were less concrete.

Domestic Political Management
In its domestic political management, Brussels made no explicit complimentary references to any member on climate change.

Deliberation
In its deliberation, performance at Brussels was solid. It generated 747 words on climate change, or 14.6 percent of the communiqué (see Appendices A-1 and C-1). This made Brussels the highest performing summit on climate change in the past five years.

Direction Setting
Direction setting was weak: Brussels failed to mention climate change in its preamble or to link climate to democracy and human rights. Although there was a stand-alone section dedicated to climate change issues, it consisted of only two of the 43 paragraphs. An additional four paragraphs were on energy.

Decision Making

Brussels's performance on decision making was strong. The leaders produced 16 discrete commitments on climate change or energy, more than any other summit since 2009 and double the average of eight (see Appendix D-1). The total number of climate commitments was the sixth highest ever. Notably, leaders affirmed their "strong determination to adopt in 2015 a global agreement ... with legal force ... applicable to all parties" (G7 2014).

Development of Global Governance

In developing global governance, Brussels was a substantial success. There were direct references to the G7 energy ministers and their meeting in Rome a month before Brussels. Here, the energy-climate link remained strong, repeating the energy ministers' statements about urgently reducing greenhouse gas emissions and "accelerating the transition to a low carbon economy" as key elements to sustainable energy security (G7 2014).

Outside the G7, seven explicit references were made to six external multilateral organizations linked to climate change (see Appendix F-5). Brussels tied with Deauville as the highest performers since 2009.

Brisbane G20, November 2014

The ninth G20 summit took place in Brisbane on November 15–16, 2014, under the chair of Australia's new conservative prime minister Tony Abbott. It advanced the new regime for climate change, despite Abbott's strong opposition. A strong supporter of Australia's coal industry, he had made it clear from the start that he wanted the summit to focus exclusively on private sector–led growth and to produce a three-page communiqué, and that there would be no scheduled discussion of climate change. That was an issue he claimed should be left to other forums. In the months leading up to the summit, private and public demands to put climate change on the agenda came from the US, France as host of the Paris COP in December 2015, Europe, Mexico, Japan, and Korea.

The week before the summit, on the road to Brisbane, in Beijing while attending the leaders' meeting of the Asia Pacific Economic Cooperation (APEC) forum, US president Barack Obama and Chinese president Xi Jinping made a historic joint announcement that China intended "to achieve the peaking of [carbon dioxide] emissions around 2030" and the US intended to "[reduce] its emissions by 26%–28% below its 2005 level in 2025 and to make best efforts to reduce its emissions by 28%" (White House 2014). This was the first time that China had agreed to control its carbon, thus accepting the fundamental first principle of the new regime pioneered by the G8 and the G20. The more stringent US commitment brought the US closer to the even more rigorous ones announced by the EU.

At the Summit

In Brisbane, the day before the summit started, Obama publicly called for climate change control to protect Australia's iconic Great Barrier Reef and pledged a contribution of $3 billion to the Green Climate Fund. Japan immediately followed with a pledge of $1.5 billion. Abbott, who had publicly pledged with Canada's Stephen Harper a year earlier that they would not contribute, continued to refuse, even after Harper announced at his news conference at the end of the summit that Canada would now give. This it did at the formal pledging conference in Berlin a few days later, where almost all the targeted $10 billion was raised. Only on December 10 did Abbott, now at historic lows in approval in his domestic polls, announce that Australia would contribute AU$200 million (Siegel 2014).

Abbott was forced to allow a robust discussion of climate change at the summit. The first session, scheduled to start at 3:15 and end at 6 p.m., dealt with improving global economic resilience, including tax, trade, and IMF reform. It also covered energy, as declining oil prices were expected to have a major impact for the next two or three years. The discussion on climate change was inspired by the China-US announcement a few days earlier, which was seen as a good start. The G20 emphasized getting prices and incentives right for greener growth, even with oil prices in decline. IMF managing director Christine Lagarde sought to get the IMF involved in climate change control. Turkey's recently elected Ahmet Davutoğlu said climate change would be on the agenda item when Turkey hosted in 2015.

Results

Brisbane thus produced a single paragraph directly on the subject in the concluding communiqué plus one on energy where climate change was also involved. The text on energy efficiency forwarded the bottom-up sectoral approach with the Action Plan for Voluntary Collaboration on Energy Efficiency, covering work on financing for energy efficiency and on emissions performance of vehicle emissions, networked devices, building, industrial processes, and electricity generation. It also reaffirmed the commitment to phase out inefficient fossil fuel subsidies.

The paragraph on climate change went further. All G20 members affirmed the inclusive all-in principle, with their commitment to adopt an agreed outcome at Paris in 2015 that would apply "to all parties" (G20 2014). It encouraged those countries ready to do so to present their "intended nationally determined contributions well in advance of COP 21" by the first quarter of 2015." It also reaffirmed support for climate finance, "such as the Green Climate Fund."

This stance was strengthened by the US and EU shortly after the summit. On December 3 the US-EU Energy Council declared that "coordinated action by the EU, the United States and all major and emerging economies will be essential to tackling the threat of global climate change" (US Department of State 2014). It repeated the call for an agreed outcome applicable to all and added that "all major economies must demonstrate similar leadership and submit their intended contributions by the same time in a manner that is transparent, quantifiable and comparable." It also said that "to ensure a successful outcome in Paris it will be essential to work also in fora outside the United Nations Framework Convention on Climate Change, such as the Major Economies Forum, the G20 and the G7" (see also Busby 2010).

Dimensions of Performance

Across the major dimensions of performance, the Brisbane Summit was a solid success on climate change, above all in its decision making (see Appendix A-2).

Domestic Political Management
Brisbane's domestic political management produced no communiqué compliments (see Appendix B-2). The leader of Saudi Arabia did not attend, and those of Russia and Indonesia left the summit early. Following the summit, host Tony Abbott's approval ratings sunk to historical lows (Kirton 2014).

Deliberation

In its deliberation, Brisbane devoted 323 words to climate change in its communiqué, or 3.5 percent of the total (see Appendix C-2). This was the lowest since Washington in 2008 and London in 2009. It was well below the average of 916 words and 6.6 percent.

Direction Setting

Brisbane made no reference to democracy or human rights in direction setting, nor did it give priority placement to climate change (see Appendix A-2). It failed to recognize any causal connection between climate change and economic growth, financial stability, or globalization working for all.

Decision Making

In its decision making, Brisbane produced seven commitments on climate change, the fourth highest ever and above the overall summit average of 5.4 (see Appendix D-2). It also made 16 commitments on energy, tied for the second highest ever and well above the summit average of 9.8 (see Appendix A-2).

Development of Global Governance

Brisbane made no references in a climate change context to institutions within the G20. However, it did ask G20 energy ministers to meet. It made four references to two institutions outside the G20: the UNFCCC and the Green Climate Fund (see Appendix F-6).

After the Summit: COP-20 Lima

The new, inclusive regime pioneered by the G7/8 and the G20 made a substantial advance at the UN's COP-20, held in an extended session in Lima, Peru, from December 1 to 14, 2014. After difficult and long deadlocked negotiations, the 193 countries produced the first broadly multilateral, UN-incubated deal in which all countries agreed to control their carbon (Davenport 2014; McGrath 2014). All countries committed to submit national plans to reduce emissions by March 31, 2015, or if not by June, indicating how much and how they will cut after 2020. The UN would publish these plans publicly. Countries would set targets beyond their "current undertaking." The UNFCCC secretariat would report on these "Intended Nationally Determined Contributions" in November, and they would form the foundation for the beyond-Kyoto regime to be determined at the Paris COP in December 2015 (Davenport 2014). The UN had thus formally agreed to the G8-G20's all-in principle and bottom-up approach.

To be sure, much was missing from the agreement. With no obligations for any country to make significant cuts to produce the proclaimed ambitious agreement in Paris, the combined national contributions were expected to fall far short of the cutbacks needed to hit the accepted 2°C target. Indeed, countries were free to submit no plans and face no sanctions. China, India, and their G77 allies secured a reaffirmation of the principle of common but differentiated responsibilities from the UNFCCC. They eliminated a proposal requiring quantifiable data on how they planned to meet their emissions targets. They also vetoed any review mechanisms.

The deal thus relied heavily on public and peer pressure. Nonetheless, there were signs that the new principles would have an impact. India intended to submit a plan by June to show how it would lower the rate of its carbon emissions. Russia, the fifth largest carbon polluter, stated it was working on its carbon-reduction plan. Australia announced it would submit its plan in the spring.

The process and outcome of the Lima Accord showed the independent effect of the international institutional context on state behaviour. At the start of the UN's Lima negotiations, China retreated from the flexible position it had taken on the road to the G20's Brisbane Summit, to revert to its traditional, hard-line G77 stance at the UN. But, in the end, both China and India adjusted at the UN to allow the recently approved G20 principles to prevail.

Conclusion

By the end of the Brisbane Summit in 2014, the process of replacement was firmly in place. The old, divided UN regime on climate change was being replaced by the expanded and inclusive G8-centred process. With Kyoto's future fragile at best, this phase of replacement from 2012 to 2014 first saw the G20 emerge alongside and often move ahead of the G8 and restored G7 as the leader of global governance on climate change control. A group of equals, representing the largest carbon emitters in the world, the G20 increasingly placed climate change at the centre of its deliberations, generating commitments and delivering on them in significant ways.

This phase also saw the G8 replaced by the G7, which met without Russia for the first time in more than 15 years. With energy security now topping the leaders' agenda in the wake of Russia's invasion and annexation of Crimea, the G7's revival of climate leadership in Brussels marked the need for "urgent and concrete action" by world leaders in the run-up to the Paris COP in 2015 (G7 2014). This action aimed at replacing the existing Kyoto Protocol with "ambitious contributions" and "actions to reduce emissions and strengthen resilience" that would apply to all.

References

Agidee, Yinka (2012). "Climate Change: Has the Momentum Slowed." In *The G8 Camp David Summit 2012: The Road to Recovery*, John J. Kirton and Madeline Koch, eds. London: Newsdesk, p. 185. http://www.g8.utoronto.ca/newsdesk/g8campdavid2012. pdf (January 2015).

Barroso, José Manuel (2014). "Remarks by President Barroso on the Results of the G7 Summit in Brussels." Closing press conference, Brussels, June 5. http://www. g8.utoronto.ca/summit/2014brussels/barroso.html (January 2015).

Bayne, Nicholas (2013). "The UK's Role in the G7 and G8." In *The UK Summit: The G8 at Lough Erne 2013*, John J. Kirton and Madeline Koch, eds. London: Newsdesk, pp. 40–41. http://www.g8.utoronto.ca/newsdesk/g8lougherne2013.pdf (January 2015).

Busby, Joshua W. (2010). "After Copenhagen: Climate Governance and the Road Ahead." Working paper, August, Council on Foreign Relations, New York. http://www.cfr.org/ climate-change/after-copenhagen/p22726 (January 2015).

Calderón, Felipe (2011). "The Current Challenges for Global Economic Growth." December 13. http://www.g20.utoronto.ca/2012/2012-111213-calderon-en.html (January 2015).

Calderón, Felipe (2012). "Mexico's Vision of Innovation and Cooperation at the Los Cabos Summit." In *The G20 Mexico Summit 2012: The Quest for Growth and Stability*, John J. Kirton and Madeline Koch, eds. London: Newsdesk, pp. 14–15. http://www.g8.utoronto.ca/newsdesk/loscabos/calderon.html (January 2015).

Cameron, David (2012). "In Fight for Open World, G8 Still Matters." *Globe and Mail*, November 20. http://www.theglobeandmail.com/report-on-business/economy/david-cameron-in-fight-for-open-world-g8-still-matters/article5508595/ (January 2015).

Carr, Bob (2014). *Diary of a Foreign Minister*. Sydney: NewSouth Publishing.

Chang, Jae-soon (2012). "Lee, Mexican President Cite 'Green Growth' as Solution to Global Crisis." *Yonhap English News*, May 22. http://english.yonhapnews.co.kr/nation al/2012/05/22/91/0301000000AEN20120522000700315F.HTML (January 2015).

Climate and Clean Air Coalition to Reduce Short-Lived Climate Pollutants (2014a). "Country Partners." Paris. http://www.unep.org/ccac/Partners/CountryPartners/tabid/130289/Default.aspx (January 2015).

Climate and Clean Air Coalition to Reduce Short-Lived Climate Pollutants (2014b). "Definitions: Short-Lived Climate Pollutants." Paris. http://www.unep.org/ccac/Short-LivedClimatePollutants/Definitions/tabid/130285/Default.aspx (January 2015).

Davenport, Coral (2014). "A Climate Accord Based on Global Peer Pressure." *New York Times*, December 14. http://www.nytimes.com/2014/12/15/world/americas/lima-climate-deal.html (January 2015).

Earth Negotiations Bulletin (2012, 11 December). "Summary of the Doha Climate Change Conference." 12(567). http://www.iisd.ca/vol12/enb12567e.html (January 2015).

Earth Negotiations Bulletin (2013, 26 November). "Summary of the Warsaw Climate Change Conference." 12(594). http://www.iisd.ca/vol12/enb12594e.html (January 2015).

Earth Summit 2012 Stakeholder Forum (2012). "Earth Summit 2012: Background." http://www.earthsummit2012.org/about-us (January 2015).

Economist (2013). "Central European Floods: A Hard Lesson Learned." Eastern Approaches: Ex-communist Europe (Blog). June 6. http://www.economist.com/blogs/easternapproaches/2013/06/central-european-floods (January 2015).

Environment Canada (2013). "Canada Invests in Global Climate and Clean Air Solutions." Ottawa, April 10. http://www.ec.gc.ca/default.asp?lang=En&n=714D9AAE-1&news=47727622-982C-4AE8-91CF-F9F50F81FF91 (January 2015).

European Council (2014). "Background Note and Facts about the EU's Role and Action." Brussels, June 3. http://www.g8.utoronto.ca/summit/2014brussels/background_140603.html (January 2015).

G7 (2014). "G7 Brussels Summit Declaration." Brussels, June 5. http://www.g8.utoronto.ca/summit/2014brussels/declaration.html (January 2015).

G7 Energy Ministers (2014). "Rome G7 Energy Initiative for Energy Security." Rome, May 6. http://www.g8.utoronto.ca/energy/140506-rome.html (January 2015).

G8 (2012). "Camp David Declaration." Camp David, May 19. http://www.g8.utoronto.ca/summit/2012campdavid/g8-declaration.html (January 2015).

G8 (2013a). "G8 Lough Erne Communiqué." June 18, Lough Erne, Northern Ireland, United Kingdom. http://www.g8.utoronto.ca/summit/2013lougherne/lough-erne-communique.html (January 2015).

G8 (2013b). "Lough Erne Accountability Report: Keeping Our Promises." June 7. http://www.g8.utoronto.ca/summit/2013lougherne/lough-erne-accountability.html (January 2015).

G8 Research Group (2013). "2012 Camp David G8 Summit Final Compliance Report." 14, June. http://www.g8.utoronto.ca/evaluations/2012compliance/2012compliance.pdf (January 2015).

G20 (2012a). "G20 Leaders Declaration." Los Cabos, June 19. http://www.g20.utoronto.ca/2012/2012-0619-loscabos.html (January 2015).

G20 (2012b). "Los Cabos Growth and Jobs Action Plan." Los Cabos, June 19. http://www.g20.utoronto.ca/2012/2012-0619-loscabos-actionplan.html (January 2015).

G20 (2013a). "G20 Leaders Declaration." St. Petersburg, September 6. http://www.g20.utoronto.ca/2013/2013-0906-declaration.html (January 2015).

G20 (2013b). "St. Petersburg Accountability Report on G20 Development Commitments." August 28. http://www.g20.utoronto.ca/2013/Saint_Petersburg_Accountability_Report_on_G20_Development_Commitments.pdf (January 2015).

G20 (2014). "G20 Leaders' Communiqué." Brisbane, November 16. http://www.g20.utoronto.ca/2014/2014-1116-communique.html (January 2015).

Gillard, Julia (2014). *My Story*. London: Bantam Books.

Haas, Peter M. (2008). "Climate Change Governance after Bali." *Global Environmental Politics* 8(3): 1–7.

Harvey, Fiona (2012). "Doha Climate Gateway: The Reaction." *Guardian*, December 10. http://www.theguardian.com/environment/2012/dec/10/doha-climate-gateway-reaction (January 2015).

Hochstetler, Kathryn (2012). "A Green Economy for the Whole G20." Waterloo ON, May 12. http://www.cigionline.org/publications/green-economy-whole-g20 (January 2015).

Kirton, John J. (2013a). "A Summit of Significant Success: Prospects for the G8 Leaders at Lough Erne." June 12. http://www.g8.utoronto.ca/evaluations/2013lougherne/kirton-prospects-2013.html (January 2015).

Kirton, John J. (2013b). "A Summit of Substantial Success: Prospects for St. Petersburg 2013." In *Russia's G20 Summit: St Petersburg 2013*, John J. Kirton and Madeline Koch, eds. London: Newsdesk, pp. 34–5. http://www.g8.utoronto.ca/newsdesk/stpetersburg (January 2015).

Kirton, John J. (2013c). "A Summit of Very Substantial Success: G20 Governance of the Global Economy and Security at St. Petersburg." September 6. http://www.g20.utoronto.ca/analysis/130906-kirton-performance.html (January 2015).

Kirton, John J. (2014). "The Performance of the G20 Brisbane Summit." December 4, G20 Research Group. http://www.g20.utoronto.ca/analysis/141204-kirton-performance.html (January 2015).

Kirton, John J. and Julia Kulik (2012a). "Delivering a Double Dividend: Prospects for the G20 at Los Cabos Summit." Paper presented at a conference on "Strengthening Sustainable Development by Bringing Business In: From Los Cabos G20 to Rio Plus 20," Global Institute for Sustainability, Tecnológico de Monterrey, June 11, Mexico City. http://www.g20.utoronto.ca/biblio/kirton-kulik-120612.pdf (January 2015).

Kirton, John J. and Julia Kulik (2012b). "A Summit of Significant Success: G8 Governance at Camp David in May 2012." *International Organisations Research Journal* 7(3): 7–24. English-language version of article published in Russian. http://www.g8.utoronto.ca/scholar/kirton-kulik-iori-2012.pdf (January 2015).

Kirton, John J. and Julia Kulik (2012c). "A Summit of Significant Success: G20 Los Cabos Leaders Deliver the Desired Double Dividend." June 19. http://www.g20.utoronto.ca/analysis/120619-kirton-success.html (January 2015).

Low, Melissa (2013). "Losing Canada, Japan and Russia in the Climate Regime: Could the Solution Be in Asia?", April 11. https://unfcccecosingapore.wordpress.com/2013/04/24/losing-canada-japan-and-russia-in-the-climate-regime-could-the-solution-be-in-asia/ (January 2015).

McGrath, Matt (2014). "UN Members Agree Deal at Lima Climate Talks." *BBC News*, December 14. http://m.bbc.com/news/science-environment-30468048 (January 2015).

Merkel, Angela (2012). "Cooperation, Responsibility, Solidarity." In *The G8 Camp David Summit 2012: The Road to Recovery*, John J. Kirton and Madeline Koch, eds. London: Newsdesk, pp. 24–5. http://www.g8.utoronto.ca/newsdesk/campdavid/merkel.html (January 2015).

Mexican Presidency of the G20 (2011). "Priorities of the Mexican Presidency." Mexico City, December 13. http://www.g20.utoronto.ca/2012/2012-111213-priorities-en.html (January 2015).

National Climatic Data Center (2015). "Billion-Dollar Weather and Climate Disasters: Table of Events." http://www.ncdc.noaa.gov/billions/events (January 2015).

Obama, Barack (2012). "A Time of Transformation." In *The G8 Camp David Summit 2012: The Road to Recovery*, John J. Kirton and Madeline Koch, eds. London: Newsdesk, pp. 14–16. http://www.g8.utoronto.ca/newsdesk/campdavid/obama.html (January 2015).

Rafferty, Kevin (2013). "Sunny Spin to an Oily Earth." *Japan Times*, June 4. http://www.japantimes.co.jp/opinion/2013/06/04/commentary/sunny-spin-to-an-oily-earth/#.VCCUWytdVTM (January 2015).

Rooney, Stephen (2013). "The G20 and Energy Sustainability." *Europeinfos*, October. http://www.comece.eu/europeinfos/en/archive/issue164/article/6023.html (January 2015).

Russia's G20 Presidency (2012). "Russian Presidency of the G20: Outline." Moscow, December 28. http://www.g20.utoronto.ca/2013/Brochure_G20_Russia_eng.pdf (January 2015).

Siegel, Matt (2014). "In About Face, Australia PM Joins Global Climate Fund." *Reuters*, December 10. http://in.reuters.com/article/2014/12/10/australia-climatechange-idINL3N0TU1FZ20141210 (January 2015).

United Nations (2012). "Report of the Conference of the Parties on Its Seventeenth Session, Held in Durban from 28 November to 11 December 2011: Addendum. Part Two: Action Taken by the Conference of the Parties at Its Seventeenth Session." FCCC/CP/2011/9/Add.1, March 15. http://unfccc.int/resource/docs/2011/cop17/eng/09a01.pdf (January 2015).

United Nations Conference on Sustainable Development (2012). "The Future We Want." A/CONF.216/L.1, Rio de Janeiro, June 19. http://www.stakeholderforum.org/fileadmin/files/FWWEnglish.pdf (January 2015).

United Nations Framework Convention on Climate Change (2011). "Durban Climate Change Conference – November/December 2011." Bonn. http://unfccc.int/meetings/durban_nov_2011/meeting/6245.php (January 2015).

United States Department of State (2014). "Joint Statement of the U.S.-EU Energy Council." Washington DC, December 3. http://www.state.gov/r/pa/prs/ps/2014/12/234638.htm (January 2015).

White House (2014). "U.S.-China Joint Announcement on Climate Change." Beijing, November 12. https://www.whitehouse.gov/the-press-office/2014/11/11/us-china-joint-announcement-climate-change (January 2015).

Yudaeva, Ksenia (2013). "Russia's Perspective on G20 Summitry." In *Russia's G20 Summit: St Petersburg 2013*, John J. Kirton and Madeline Koch, eds. London: Newsdesk, pp. 36–7. http://www.g20.utoronto.ca/newsdesk/stpetersburg (January 2015).

PART V
Conclusion

Chapter 12
Conclusion

The challenge of controlling climate change has not been met and is still not close to being solved. If the ultimate success of global governance is the actual attainment of the intended, desired, and needed result, then in addressing the climate change challenge, all global regimes, their agents, and leaders since 1975 have failed.

This is not surprising. Controlling climate change is a uniquely difficult global challenge. Its cause and cure involve the daily activities of all humans and other living things on the planet, in the context of the complex physical dynamics in the world and the universe beyond. Unlike other environmental challenges, such as acid rain, ozone depletion, or even the law of the sea, climate change does not encompasses just a single environmental medium or component that can be kept intact by controlling a few central substances or behaviour that causes harm. As a result of its comprehensive, complex, still somewhat uncertain nature, its control requires much greater, often politically painful action across and within the sovereign domains of exclusive, territorial states, above all within those states that possess the greatest relative capability and connectivity. But their ingrained siloed, linear, mechanistic management models and instincts for governance are poorly designed to address the comprehensive, interconnected, complex, uncertain, ever changing dynamics that physically characterize climate change.

It is thus also not surprising that neither the formal multilateral organizations of the United Nations and the Bretton Woods system nor the informal plurilateral summit institutions of the Group of Seven or Eight (G7/8) alone or with the Group of 20 (G20) have thus far failed to conquer or even adequately cope with the challenge, either by working alone or together, or by allowing or supporting either to take the lead. Thus the world relentlessly approaches the critical threshold of its historic, pre-industrial revolution temperature rising more than a dangerous 2°C. And given the range of uncertainty surrounding climate science and the presence of several self-reinforcing, possibly runaway dynamics in the climate system, it is conceivable that it is already too late to control, or even to cope with, the immense damage and death that will result.

Yet this clear failure of global climate governance thus far does not carry with it a conclusion of despair. There are alternative approaches and regimes available. Some have already been employed. The first and most ambitious G7 regime proved promisingly effective when it dominated from 1979 to 1988. And after the long interlude of the great UN-led failure between 1989 and 2004, this G7-pioneered regime has returned with growing force to rival and incrementally replace the UN-generated one.

Thus the analysis in this book brings a message of hope about the prospects for producing an adequate global climate change control regime in time. This analysis begins with the small set of globally predominant democratic major powers acting through their G7 in 1979 to create the most ambitious, effective, inclusive, environment-first climate change control regime to date. It continues with the great detour to the divided, development-first UN regime from 1989 to 2004, and the increasing inadequacy of that regime. It culminates with the return, led first by the G8 plus partners, then by the G20 and the G7, to an expanded, inclusive, environment-first regime from 2005 to 2014. This evolving regime promises to replace the previous one as the globally dominant response and thus to control the problem before the looming climate catastrophe comes with full and irreversible force.

The Cadence and Causes of Global Climate Leadership

Leadership in global climate governance has passed through three phases of creation, retreat, and return, driven largely by changes in the causes highlighted by the models of concert equality and systemic hub governance.

Creation, 1979–1988

In the first phase—creation—from 1979 to 1988, the G7 invented and implemented an effective, ambitious, inclusive, environment-first control regime through direct informal action of its own, initially within the context of energy conservation. It put the environment first by targeting climate change control as a key goal alongside energy security, along with energy conservation, efficiency, and alternatives, as instruments to reach both. It ambitiously set the carbon target and timetable at zero increases immediately in carbon concentrations in the atmosphere. It included all consequential carbon-polluting powers in its institutional actions and impacts, by coming to a consensus on informal, voluntary, but politically obligatory quotas on oil imports and more comprehensive, bottom-up, sectoral measures by G7 members, and by imposing bold economic and security measures to impoverish and defeat the outside carbon powers of the Soviet Union and bloc and the major oil-producing emerging powers of the time. It proved effective in moving toward its goal. The emissions of the members of the Organisation for Economic Co-operation and Development (OECD) declined from 1979 to 1985. The emerging oil powers with their development-first demands for a New International Economic Order (NIEO) were defeated. The Soviet Union and bloc were driven to the brink of collapse by 1988.

During this phase, at their annual summit G7 leaders publicly deliberated on climate change, first in 1979, again in 1985, and increasingly from 1987 to 1988 (see Appendix A-1). Their historic consensus in 1979 was complied with completely in the ensuing years. They made their first decisional commitment in 1985 and another in 1987. They delivered on these commitments, to a substantial degree in the year after their 1985 summit and to a small degree after their 1987 one. They began to develop climate-related global governance, first by creating their own internal energy-conserving official-level institutions—with three in 1979, two in 1980, and one in 1985—and then by giving guidance to two outside international institutions in 1988 (see Appendices F-2 and F-5).

This pattern of performance, carefully charted in the quantitatively measured volume and proportion of G7 behaviour, along with its substantive qualitative content, is consistent with the changes observed in the causes contained in the concert equality and systemic hub models of G8 and G20 governance, and the causal pathway connecting them.

The phase of creation, with its intermittent but increasing performance from 1979 to 1988, was spurred by shocks similar to climate change. These shocks were the oil supply, price, and tanker spill shocks of 1979, following those from 1973, and the famine in Africa and nuclear explosion at Three Mile Island in the United States in 1979. They were aggravated by the visible vulnerability of Europeans and North Americans to atmospherically produced, polluting acid rain. World oil prices spiked from $49.28 a barrel in January 1979 to $59.88 on the eve of the pioneering Tokyo Summit in June, and then peaked at $108.26

in June 1980 (see Appendix G-3). Then they declined to $55.09 by January 1985, as did G7 climate performance (see Appendix A-1). They rose to $61.09 in April 1985 before Bonn Summit where climate change returned, dropped to a new low of $26.82 in April 1986 on the eve of the non-performing Tokyo Summit, but then rose again to $31.75 just before the Toronto Summit in June, as G7 climate governance resumed in 1987 and 1988. In a similar cadence, the annual number of accidents of oil tankers, rigs, and related events rose steadily from 16 in 1976 to peak at 34 in 1979, declined to seven in 1986, and then rose again to 10 each in 1987 and 1988 (see Appendix G-4). During this period natural disasters generally rose, with a notable spike in the number of people affected in 1987 (see Appendix G-5).

The shock of the Soviet invasion of Afghanistan at the end of 1979 and the ensuing new cold war diverted the G7's attention, but ultimately brought carbon-reducing benefits, until the 1985 Bonn Summit linked climate with issues beyond energy. The direct climate shock of the deadly heat wave that afflicted the US in the summer of 1988 gave Americans a glimpse of what global warming would be like. It led the nearby Toronto Summit to develop global climate governance outside the G7 and start the G7/8's continuous and continuing governance of climate change.

The phase of creation was also spurred by the failure of the established multilateral organizations, led by those of the UN. The environment was absent from the UN charter and from the array of dedicated functional organizations created in the 1940s. Neither the old World Meteorological Organization (WMO) from 1950 nor the new United Nations Environmental Programme (UNEP) born in 1972 took up climate change at the political level. No UN summit dealt with climate change. Even at the scientific level, the Intergovernmental Panel on Climate Change (IPCC) delivered its first assessment only in 1990, in part as a result of G7 guidance at the end of this first phase.

The third cause of the phase of creation was the generally growing, globally predominant, and internal equality of the G7 members' capability during this time (see Appendix I). In 1979, G7 predominance was still high, if declining, but after 1980, it soared with the far-reaching decline of the leading countries of the communist East and resource-rich South. More importantly, the G7's substantial internal equality rose through to 1980 led by the strong gains of Japan, Germany, and resource-rich Canada relative to those of a struggling US. It then declined with the "Reagan restoration" and soaring US "superdollar" from 1981 to 1984. But it rose again for the following three years, after the Plaza Accord in September 1985 lowered the value of the US dollar. Accordingly, G7 climate governance arrived, disappeared, then revived in the same temporally tightly matched three-step sequence.

The fourth cause was the continuously complete democratic character of the G7 members from 1975 to 1988 (see Appendix J). While this does not explain the three periods within—invention in 1979, disappearance from 1980 to 1984, and rediscovery and revival in 1985 and 1987–1988—the expanding democratic character of the G7, with the addition of Canada in 1976 and the European Community in 1977, helps explain the invention in 1979, especially with the role played at Tokyo by Canada's Joe Clark and Europe's Crispin Tickell.

The fifth cause was domestic political cohesion. The phase of creation was initiated by the domestic political dominance, personal convictions, public support, and G7 experience of co-founder Helmut Schmidt of Germany, and continued by his successor Helmut Kohl at Bonn in 1985 and beyond. Necessary support came from Canada's Joe Clark, America's Jimmy Carter, and Britain's Margaret Thatcher, with her expertise in chemistry, and through public pressure by European Green parties, in Germany above all. The 1979 invention came at a summit rich in leaders who had been finance ministers and foreign ministers before, while this experience dropped dramatically, as did the G7's climate governance, in the following five years (see Appendix K-2). The domestic electoral security of the summit host also mattered under certain circumstances, by inducing hosts soon facing elections to advance a domestically popular climate and environment agenda. This was done by Japan's Masayoshi Ohira in 1979, Germany's Helmut Kohl in 1985, Italy's Amintore Fanfani in 1987, and Canada's Brian Mulroney in 1988, if not Schmidt in 1978 or Thatcher in 1984 (see Appendix K-1).

As a sixth cause, the G7 created global climate governance from 1979 to 1988 when it was a compact club of six founding members that had slowly added an environmentally committed and competent Canada in 1976 and the European Community in 1977. Its leaders stood at the hub of a network of plurilateral summit institutions (PSIs) for global governance. The network embraced the older institutionalized summits of the Commonwealth (through the UK and Canada) and the North Atlantic Treaty Organization (NATO) (through all G7 members save Japan), with one for la Francophonie (through France and Canada) arriving in 1985. The G7's centrality as a hub was enhanced by the fact that during this era, global plurilateral summits of both ad hoc and institutionalized kinds were still very rare (see Appendix L).

Retreat, 1989–2004

In the second phase of G8 global climate leadership, the retreat from 1989 to 1984, the G7 stepped back from action of its own in favour of shaping and then supporting the emerging UN-led regime. This alternative, divided, development-first regime increasingly failed, as it developed through the United Nations Framework Convention on Climate Change (UNFCCC) created in 1992 and its Kyoto Protocol produced in 1997. Despite the apparent equality between the environment and the economy contained in the name of the UN Conference on Environment and Development (UNCED) in Rio de Janeiro in June 1992, UNCED and the regime it codified put development first, under the defining principle of sustainable development. Environmentalists were given the conditioning adjective while the developing countries and community were given the noun as the central goal.

The UN regime offered a top-down approach based on a distant overall global goal. It initially set few specific targets or timetables, let alone any goals sufficiently ambitious to control the steadily compounding carbon emissions and concentrations and the global warning they brought. Under the central principle of common but differentiated responsibilities, this regime divided the world into one small group of developed countries that were obliged to control their carbon and another, much larger group of emerging and developing countries that were not. The countries in the larger group were thus left free to develop in their prevailing and preferred carbon-intensive way. This they did, creating a serious climate change problem by 2005, the year when a multilaterally unconstrained China surpassed a unilaterally unconstrained US as the biggest carbon polluter in the world. Taking a formal, legal, broadly multilateral convention-protocol approach, the UN's Rio-Kyoto regime began with generalities that were to be progressively but slowly developed into more specific measures at annual meetings of the members that had consented to join—a group that excluded from the precise Kyoto Protocol promises the world's largest carbon polluter and power, namely the United States. The result was an increasing failure to take the needed control action, as the little group of countries that had agreed in 1997 to control their emissions by a little largely did not do so by 2004.

During this phase of retreat, G7/8 climate governance often grew, but to levels usually little higher and at times even lower than those seen during the phase of creation (see Appendix A-1). In domestic political management, all G7 and G8 leaders continued to attend all the summits. But only once, in 2001, did they issue a communiqué compliment regarding climate change to a member, in this case the new Russia (see Appendix B-1). Their public deliberation soared above 400 words in their communiqués in 1989 and 1990, but then, despite spikes above 300 words in 1997, 1998, and 2001, fell to a new low of 53 words in 2002 and below 100 words in 2003 and 2004 (see Appendix A-1). However, in direction setting, at the end of this phase, climate change was given priority placement with three references in 2002 and five in 2003. In its decision making, the G7/8 continuously produced climate commitments but in a cadence very similar to its deliberation, with a normal range of four to seven at each summit, a spike to nine at Denver in 1997 and to 10 at Birmingham in 1998, but a drop to one at Kananaskis in 2002 and three at Sea Island in 2004.

In keeping with the G7/8's shift to shaping and supporting the UN's leadership and regime, these commitments overwhelmingly addressed, in order, the UNFCCC (seven), sustainable development (seven), the Global Environment Facility (GEF) (seven), greenhouse gases (seven), the Committee of the Parties (COP) to the UNFCCC (seven), forests (six), the Rio conference (five), Kyoto (five), research/science (four), and the UN Commission on Sustainable Development (CSD) (three) (see Appendix D-1). Compliance through delivering these decisions varied a great deal, starting at new lows of −0.07 for 1989 and −0.11 for 1990, rising generally steadily to peak at +1.00 for 1998, plummeting to −0.22 for 1999 and 0 for 2001, then rising to about +0.89 for 2002–2004 (see Appendix E-1).

The shift to the UN was also evident in the development of global governance. Inside the G7/8, references to the new environment ministers meeting began to appear in 1988, well after that body had been formed in 1992 (see Appendices A-1 and F-1). Only a few G7/8 official-level climate-related bodies arose: the Officials Group on Forests in 1997, the short-lived Renewable Energy Task Force in 2000, the Energy Officials Follow-up Process in 2002, and the Senior Officials for Science and Technology for Sustainable Development in 2003, followed by a burst of six new bodies at the Sea Island Summit in 2004 (see Appendix F-2). However, references to multilateral organizations outside the G8 came from 1988 to 1993, in 1996, and from 1999 to 2001 (see Appendix F-5). All 16 references were to UN bodies: UNCED (four), the WMO and UNEP (four each), the GEF (three), and COPs and UNFCCC (two each).

This phase of G8 retreat in favour of UN leadership was spurred first by the changing configuration of shock-activated vulnerability, notably the arrival of the diversionary if positive shock of the Cold War victory, the absence of serious oil shocks with sustained effects, and effective action on acid rain, as well as the diversionary Asian-turned-global financial crisis from 1997 to 2002, and the al Qaeda terrorist attack on the US on September 11, 2001. The short-lived oil spike brought by the 1990–1991 Gulf War and the 1996 anniversary of the 1986 Chernobyl nuclear explosion were insufficient to overcome G7/8 confidence that the UN had the formula and time to produce an effective climate change control regime, as designed in 1992.

World oil prices started at a low $33.65 per barrel in January 1989, rose as high as $42.03 during the Gulf War in January 1991, drifted down to a low of $17.20 during the Asian-turned global financial crisis in January 1999, and then climbed to $45.43 by June 2004 (see Appendix G-3). The number of oil-related accidents also declined per year, from a low 13–14 a year in 1989–1990 to five or less from 2000 to 2004 (see Appendix G-4). While the annual number of natural disasters generally rose, their severity was often low, save for short-lived spikes in the number of people affected in 2002 and the damage caused in 1995 and 2004 (see Appendix G-5). Moreover, the chronic, cumulative vulnerability caused by annual greenhouse gas emissions from G7/8 members dropped, from 18,542 metric tons of carbon dioxide equivalent (MtCO2e) in 1990 for G7/8 members to 18,282 MtCO2e in 2004, due primarily to Russia's great decline (see Appendix G-7). Outside the G7, from Russia and the Group of Five (G5) members of Brazil, China, India, Mexico, and South Africa, emissions rose rapidly but to a limited degree: from 9,717 in 1990 to 13,337 in 2004 (see Appendix G-8). The cumulative result did not suggest that G7/8 vulnerability was becoming acute in ways that the new UN-led regime would be unable to control.

Indeed, as a second cause of G8 retreat, the earlier multilateral organizational failure seemed to be replaced by success. In quick succession, the hitherto climate-unconcerned UN system generated IPCC reports in 1990, 1992, 1996, and 2001; it produced the UNCED summit in Rio de Janeiro in 1992 and its follow-ups in 1997 and 2002; and the UNFCCC secretariat began holding annual COPs from 1995 to 2004 (see Appendices H-1, H-2, and H-3).

As a third cause, the G8's global predominance and internal equality slowly declined (see Appendix A-1). Globally the strong, sustained relative capability rise of the G5 and other emerging powers gradually surpassed the G7's addition of Russia in 1998, of East Germany within a now united Germany, and the periodic expansion of the European Union. This trend prevailed despite the short-lived setbacks to Indonesia, Korea, Russia, Brazil, Turkey, and Argentina from the Asian-turned-global financial crisis from 1997 to 2002, from which China and India emerged relatively unscathed. However, the strong economic performance of the US, as the globalized Cold War victor, and the onset of sluggish growth in Japan during its lost decade increased inequality within the G7, and thus increased the influence of a US generally reluctant to be internationally constrained to control its carbon.

Fourth, common democratic principles took hold more broadly and quickly in the multilateral UN than they did in the democratizing G7/8 (see Appendix J). To be sure, the G7 slowly incorporated a reforming, democratizing Russia as a participant and then as a member after 1998, thus becoming the slightly less democratic full G8 club. Yet with post-Cold War globalization and the ensuing global democratic revolution creating a more politically open global community, the G7/8 was content to leave global climate leadership to a more multilateral UN that was now more democratically like-minded than it had been in the former confrontational Cold War–NIEO years. Among the leading middle powers that would soon become systemically significant states, democracy had leapt ahead in Argentina in 1983, Brazil in 1985, Korea and Mexico in 1988, and South Africa in 1994.

As a fifth cause, new leaders with limited convictions about the need to control change arrived in Britain under John Major, Russia under Boris Yeltsin and Vladimir Putin, Japan under Junichiro Koizumi, and the US under George H. Bush from 1989 to 1992 and, especially, George W. Bush from 2001 to 2004. When politically secure leaders who did care hosted the summit, notably Helmut Kohl in 1992 and Bill Clinton in 1997, G8 climate performance temporarily rose (see Appendix K-1). But still no G7/8 leader had ever served as an environment minister before (see Appendix K-2).

As a sixth cause, the retreat in favour of UN leadership coincided with the G7's reduction of constricted participation, due to the increasing inclusion of the Soviet Union, and then Russia, as well as a changing but growing array of invited countries and multilateral organizations. This trend existed even as the network of PSIs centred in the G7/8 increased rapidly in these post-Cold War years (see Appendix L). In 1996 and then, continuously, from 2001 on, the G7/8 started to invite an ever expanding, often changing selection of leaders of countries and international organizations to participate in the summit. Given the particular choice and composition of these groups of guests, the emphasis at the summit moved to development rather than to the environment and climate change control.

Return, 2005–2014

During the third phase—the return to direct G8 climate leadership from 2005 to 2014—the G8 shifted from supporting the increasingly ineffective UN regime toward pioneering an enhanced version of its initial, effective, inclusive, ambitious, environment-first, G8-led regime. The new regime now focused on informally agreed action by the G8 with its new G5, Major Economies Meeting on Energy Security and Climate Change (MEM) and Major Economies Forum (MEF), and G20 partners in direct support and, since 2008, a G20 summit moving to take the lead. It put climate change at the centre, backed by environmental, energy, economic, and other measures in direct support. It secured a consensus from all G8, G5, MEM/MEF, and G20 powers that they would control their carbon, with China in the lead, based on the specific target and timetable of a 50 percent reduction in G8 greenhouse gases by 2050. It did so through informal but politically obligatory sectoral actions by all partners, contentiously but consensually agreed to at G8-centred summits that expanded to include, as equals, all the major carbon-producing powers in the G20. It promised to be effective, as US and G8-wide greenhouse gas emissions declined each year after 2005, and there was new hope that the newly constrained, major emerging carbon powers and polluters would follow this G8 lead.

G8 climate performance soared to sustained high levels by 2009, then declined as the new G20 summit arrived to increasingly assume the lead (see Appendices A-1 and A-2). In domestic political management, the G8's perfect attendance continued until Russian president Vladimir Putin boycotted the 2012 Camp David Summit hosted by Barack Obama and was then not invited to the G7-only summits held in the Hague and then Brussels in 2014. G7 communiqué compliments on climate change were awarded to the US, Japan, Germany, and the UK in 2007, to Japan in 2010, and to France in 2013 (see Appendix B-1). The new G20 summit had perfect attendance in 2008 and 2009, two missing members in June 2010 but only one missing thereafter. It started issuing climate-related communiqué compliments at Pittsburgh in 2009, peaking at two each at Seoul in 2010 and Cannes in 2011 (see Appendix B-2).

In its public deliberation, G8 performance started with 2,667 words on climate change in 2005, significantly more than ever before (see Appendix A-1). It rose to 4,154 in 2007 and peaked at 5,559 in 2009. It then plummeted, declining to below 1,000 words a year since 2012. In an offsetting trend, G20 deliberation started very small at its first summit in 2008, rose rather steadily to peak at 2,018 words at Seoul in November 2010, and remained above 1,000 words, with a rise to 1,697 words at St. Petersburg in 2013, until dropping below 1,000 words at Brisbane in 2014 (see Appendices A-2 and C-2).

In direction setting, the G8 gave priority placement to climate change at every summit from 2005 to 2011, with 10 references each in 2005 and 2007 and a peak of 17 in 2009 (see Appendix A-1). It also began linking climate change to democratic values, starting with six references in 2009 and continuing with two in 2010, one in 2011, and one in 2013. The G20 was also active, giving priority placement to climate eight times and linking it to democratic and human rights values five times.

In its decision making, the G8 made 239 climate commitments in the ten years from 2005 to 2014. It soared to produce 29 commitments in 2005, peaked at 54 in 2008, declined to 5 in 2012 and then rose to 16 as the G7 in 2014. These commitments dealt mainly with technology (22); the UNFCCC (20); climate change in general (17); greenhouse gases (15); reductions (13); the post-Kyoto regime and financing/funding (11 each); mitigation (8); sequestration or the Climate Sequestration Leadership Forum, mid-term goals, and methodological issues (6 each); key sectors such as transport (6) and aviation (4); research/science (2); and emissions profiles (1) (see Appendix D-1). The shift from the UN-centric focus toward a new G8-designed, bottom-up, sectoral approach was clear.

The G20's decision making rose steadily (see Appendix D-2). Its 49 climate commitments overall started with three each at London, Pittsburgh, and Toronto, jumped to eight each at Seoul and Cannes, peaked at 11 at St. Petersburg, but fell to seven at Brisbane in 2014. Their content concentrated on the COP (12), sustainable development (6), and climate change in general, the UNFCCC, and financing/funding (5 each). With a membership broader than the G8, the G20 focused on making COP work and having it adopt a more inclusive regime.

In delivering its climate decisions, G8 performance from 2005 to 2014 was solid. Compliance reached +0.80 in 2005, but then slowly declined to +0.11 in 2012, even with a few rises as in 2009 at +0.64 and 2011 at +0.67 (see Appendix E-1). The G20's compliance was more variable, but on the whole solid to substantial, with an eight-summit average of +0.31 for climate and +0.52 for energy (see Appendix E-2). It peaked on climate for Pittsburgh at +0.86 and on energy for Cannes at +0.61.

In its development of global governance, the G8 started strongly, with three environment ministers' meetings in 2005, but ended both the ministerial meetings and references to them in 2010 until one reference in 2014 (see Appendix F-1). Its official-level governance similarly started strongly, with the Dialogue on Sustainable Energy and the Global Bioenergy Partnership in 2005, and new bodies with an expanding membership through to 2009 (see Appendix F-2). Outside the G8, it soared to make 115 references to 23 different bodies (see Appendix F-4). Its major focus remained the UNFCCC (46), but followed more broadly by the International Energy Agency (IEA) (14), the Global Earth Observation System of Systems (GEOSS) (6), and the Climate Sequestration Leadership Forum and the World Bank (5 each) (see Appendix F-5).

The G20 also performed strongly at its nine summits starting in 2008. It made 23 references to its own climate-relevant institutions (see Appendix F-4). And it made 65 references to outside ones, many beyond the UN system: the UNFCCC (18), World Bank (9), the OECD and COPs (7 each), and the Global Marine Environment Protection initiative (6) (see Appendix F-6).

The first cause of this return to high, inclusive, now expanded G8- and G20-led climate governance was the cadence of increasingly severe, widespread, same and similar, shock-activated vulnerability. These shocks began with deadly heat waves in France, Europe, and India, the Asian tsunami in December 2004, rising oil prices, Hurricane Katrina afflicting the US in late August 2005, and the Fukushima natural and nuclear disaster devastating Japan in March 2011. They were reinforced by rising, diversionary, but soon solidarity-creating terrorist attacks, starting on July 7, 2005, in the capital city of the UK when it was hosting the G8 Gleneagles Summit.

The G8 recognized its own shock-activated vulnerability as never before. It made 21 such references from 2005 to 2014, or two thirds of the 30 it ever made (see Appendix G-1). So did the G20 leaders, who made eight such communiqué-encoded references during this time (see Appendix G-2). During this decade, world oil prices rose sharply, from $55.67 a barrel in January 2005 to a peak of $137.51 by July 2008, followed by a drop to $44.91 in

January 2009, and then remaining around $95.00 a barrel from 2011 until 2014, when they began to drop by the end of the year (see Appendix G-3). Natural disasters also reached new heights in the damage they caused (see Appendix G-5).

Perhaps most importantly, starting in 2005 the greenhouse gas emissions of G7 members began to decline steadily, while those of the G5 members and Russia continued their relentless rise (see Appendices G-7 and G-8). Importantly, in 2005 the emissions of China surpassed those of the US for the first time. It became increasingly clear that the divided UN regime would fail and that the only one that could succeed was a new inclusive, equal, G8- and G20-led regime, with controls applicable to all.

The multilateral organizational failure at the centre of the divided UN regime became shockingly evident with the collapse of the Copenhagen COP in December 2009 and the arrival of the ad hoc plurilateral Copenhagen Accord. IPCC assessments in 2007 and 2014 made the seriousness and urgency of the problem—and thus the UN's failure—increasingly and convincingly clear (see Appendix H-1). Neither the subsequent UN summits nor the annual COPs could repair all the damage done by Copenhagen to the now discredited UN regime (see Appendices H-2 and H-3). But, through PSIs containing all the major carbon-producing powers—beginning with the 2005 Gleneagles Summit and its G5, and then the MEM and MEF summits in 2008–2009, plus the G20 summits beginning in 2008—the G8 returned to climate governance. With the G20, it led in pioneering an expanded, inclusive, equal, environment-first regime that slowly but steadily began to replace the old UN one.

Changes in predominant equalizing capability also contributed to the return of G7/8 and now G20 leadership (see Appendix I). From 2005 to 2014, the relative capability of the G5 and other MEM/MEF and G20 powers rose, together with the G8's growing success in incorporating them as equals in an expanded summit club. This increased capability and expansion allowed and motivated the G8 to return to global climate leadership.

Also contributing was a convergence of members and participants on common democratic principles, both overall and in the form of the greater political openness needed to address the visible, ambient air pollution from hydrocarbons that fuelled climate change (see Appendix J). In general, after 2005 the democratic dominance of the expanded G8 plus G5, MEM/MEF, and G20 allowed the G8 to return to expanded, inclusive, equal global leadership thanks in part to Turkey's opening, even with the constraints of a still communist China, authoritarian Saudi Arabia, and recidivist Russia in 2014.

Political cohesion helped (see Appendix K-1). From 2005 to 2014, the return of G8/ G20 leadership was fuelled by Britain's re-elected Tony Blair as G8 host in 2005. It was then led, as host in 2007, by Germany's Angela Merkel, a physical scientist who had served as Kohl's environment minister before becoming chancellor in 2005 (see Appendix K-2). Her Heiligendamm Summit produced the highest ever performance across all dimensions in G8 climate governance. G8 and G20 leadership was also driven by a climate-committed Obama as US president since 2009 and his supportive G20 colleagues from Korea, Mexico, and, by 2014, China—now led by Xi Jinping.

The return of the G8 was caused, finally, by its controlled, continuing, institutionalized expansion to the G5 and the MEM/MEF, and then to the still-constricted G20 from 2009 to 2014. After 2009, the G8 began restricting the number and increasing the continuity of the guests at its summit, and dispensed with Russia as an attending member in 2014 (see Appendix L). The G20, with its membership fixed since 2008, also institutionalized its invitees, with greater continuity and clubbiness as the result.

Prospects and Possibilities for Future Climate Leadership

The process of replacing the old, divided, development-first, failing, UN-led climate control regime with a new, inclusive, environment-first, effective one led by the G7/8 and the G20 is now well underway. But to fully meet the looming challenge, G7/G20 leadership must convert the UN to the new approach as the latter defines its post-2015 Kyoto regime at the summit-level COP in Paris in December 2015—or replace it outside this process if no conversion comes. The task for the G7 summit that Angela Merkel will host at Schloss Elmau on June 7–8, 2015, and, above all, the G20 summit in Antalya, Turkey, on November 15–16, 2015, is to shape successful solutions. Should Paris become a second Copenhagen, the G7 and G20 must quickly install the replacement regime all by themselves.

It is fully necessary, narrowly probable, and highly desirable for the ecologically as well as economically systemically significant members of the G20 to lead in the politically authoritative, full development of the new, inclusive regime, starting at Antalya and culminating at the G20's Chinese-hosted summit in 2016. Should the UN's Paris COP fail, G7 summitry in 2016 could take the form of Japan—the world's third most powerful country and a climate control pioneer—focusing its summit in the spring on climate change and then China—the second largest power and first-ranked carbon polluter—hosting the G20 summit with climate change as a priority in the autumn of 2016. Both countries have an acute, shock-activated vulnerability–based incentive to lead on climate change. Japan is still reeling from the 2011 Fukushima disaster, and its political capital in Tokyo and all its major economic cities lie on its coastline. They are thus subject to maritime-sourced extreme weather events and gradual sea level rise. China's economic centres are also located on the coast, and its inland areas are increasingly subject to extreme drought.

The G20 is now becoming the leader in effective climate governance. It could even return to its cadence in 2009 and 2010 of having two summits a year to address the compelling global crisis. This return would begin with the 2016 G7 summit taking place in tandem with an ad hoc climate-focused G20 one, as in June 2010. It would do so when US leadership is constrained by the presidential election in November 2016, but when the polls are already showing a shift to a US public more supportive of climate change control. Moreover, it would help Japan and China, as leading Asian rivals, supported by the US and a democratic India, work together on this most compelling global common cause.

Appendices

Appendix A-1 G7/8 Climate Change Performance

Summit	Domestic Political Management — Communiqué compliments #	%	Deliberation — Words #	%	Direction Setting — Priority placement	Democracy	Human rights	Decision Making — # commitments	Delivery — Compliance Score	% assessed	Dev. of Global Governance — Inside	Outside # references	# bodies
1975	0	0	0	0	0	0	0	0	—	—	0	0	0
1976	0	0	0	0	0	0	0	0	—	—	0	0	0
1977	0	0	0	0	0	0	0	0	—	—	0	0	0
1978	0	0	0	0	0	0	0	0	—	—	0	0	0
1979	0	0	28	1.3	0	0	0	0	—	—	0	0	0
1980	0	0	0	0	0	0	0	0	—	—	0	0	0
1981	0	0	0	0	0	0	0	0	—	—	0	0	0
1982	0	0	0	0	0	0	0	0	—	—	0	0	0
1983	0	0	0	0	0	0	0	0	—	—	0	0	0
1984	0	0	0	0	0	0	0	0	—	—	0	0	0
1985	0	0	88	2.9	0	0	0	1	+0.5	100	0	0	0
1986	0	0	0	0	0	0	0	0	—	—	0	0	0
1987	0	0	85	1.5	0	0	0	1	+0.29	100	0	0	0
1988	0	0	140	2.7	0	0	0	0	—	—	0	3	2
1989	0	0	422	6	0	0	0	4	-0.07	100	0	3	2
1990	0	0	491	5.9	0	0	0	7	-0.11	57	0	2	2
1991	0	0	236	2.4	0	0	0	5	+0.38	40	0	1	1
1992	0	0	137	1.8	0	0	0	7	+0.71	43	2	2	1
1993	0	0	154	3.1	0	0	0	4	+0.57	50	0	2	2
1994	0	0	107	2.6	0	0	0	4	+0.71	50	1	0	0
1995	0	0	87	0.7	0	0	0	7	+0.29	14	1	0	0
1996	0	0	167	0.8	0	0	0	3	+0.57	33	1	2	2

1997	0	0	305	1.6	0	0	0	9	+0.29	22	1	0	0
1998	0	0	323	5.3	0	0	0	10	+1.00	30	1	0	0
1999	0	0	198	1.3	0	0	0	4	−0.22	25	1	1	1
2000	0	0	213	1.6	0	0	0	4	+0.44	25	1	1	1
2001	1	11	324	5.2	0	0	0	4	0	100	2	2	2
2002	0	0	53	0.2	3	0	0	1	+0.89	100	1	0	0
2003	0	0	62	0.3	5	0	0	4	+0.88	50	1	0	0
2004	0	0	98	0.3	0	0	0	3	+0.89	67	0	0	0
2005	0	0	2,667	9.3	10	0	0	29	+0.80	17	3	20	6
2006	0	0	1,533	3.1	2	0	0	20	+0.35	45	1	10	5
2007	4	44	4,154	12	10	0	0	44	+0.56	9	1	16	7
2008	0	0	2,568	17.5	8	0	0	54	+0.53	9	2	22	11
2009	0	0	5,559	33.3	17	1	0	42	+0.64	12	1	19	10
2010	1	11	1,282	12	1	1	2	10	+0.26	30	0	5	3
2011	0	0	1,086	5.9	1	1	1	7	+0.67	14	0	7	6
2012	0	0	789	7.1	0	0	0	5	+0.11	20	0	4	3
2013	1	11	525	3.9	0	1	1	12	+0.22	17	0	5	4
2014	0	0	747	14.6	0	0	0	16	N/A		0	7	6
Total	7		24,617		57 (43)	9	1	321	N/A	N/A	21	134	77
Average	0.17	0.02	615.43	4.16	1.4 (1.1)	0.42	0.03	8.0	+0.45	44.0	0.53	3.35	1.95

Notes: All data derived from documents issued in the G7/8 leaders' names at each summit. N/A = not available. Domestic Political Management includes all communiqué compliments related to climate change, i.e., references by name to the G7/8 member(s) that specifically express gratitude in the context of climate change. % indicates how many G7/8 members received compliments in the official documents, depending on the number of full members participating. Deliberation refers to the number of references to climate change. The unit of analysis is the paragraph. % refers to the percentage of the words in each document that relate to climate change. Direction Setting: Priority Placement refers to the percentage of references to climate change in the chapeau or chair's summary; the unit of analysis is the sentence. Democracy refers to the number of references to democracy in relation to climate change. Human Rights refers to the number of references to human rights in relation to climate change. The unit of analysis for democracy and human right references is the paragraph. Decision Making refers to the number of climate change commitments. % assessed refers to percentage of commitments measured for that year. Delivery refers to the overall compliance score for climate change commitments measured. Development of Global Governance: Inside refers to the number of references to G7/8 environment ministers. Outside refers to the number of references to multilateral organizations related to climate change. The unit of analysis is the sentence.

Appendix A-2 G20 Climate Performance

Summit	Domestic Political Management — Communiqué compliments #	%	Deliberation — Words #	%	Direction Setting — Priority placement	Democracy	Human rights	Decision Making — # commitments	Delivery — Compliance Score	% assessed	Development of Global Governance — Inside Ministerials	Official bodies	Outside # references	# bodies
2008 Washington	0	0	64	1.7	0 (0)	0	1	0 (0)	— (−)	— (−)	0	0	0	0
2009 London	0	0	64	1.0	1 (0)	0	0	3 (0)	−0.10 (−)	33 (25)	0	0	1	1
2009 Pittsburgh	1	5	911	9.7	4 (0)	0	0	3 (16)	+0.86 (+0.43)	33 (25)	4	0	10	4
2010 Toronto	1	5	838	7.4	0 (0)	1	0	3 (1)	+0.42 (+0.50)	100 (100)	0	0	3	3
2010 Seoul	2	10	2,018	12.7	2 (0)	1	0	8 (14)	+0.35 (+0.51)	25 (14)	5	3	20	11
2011 Cannes	2	10	1,167	8.2	0 (0)	1	0	8 (18)	— (+0.61)	0 (17)	2	0	11	7
2012 Los Cabos	0	0	1,160	9.1	0 (1)	1	0	6 (10)	+0.38 (+0.58)	40 (10)	1	5	6	5
2013 St. Petersburg	1	5	1,697	5.9	1 (0)	0	0	11 (14)	−0.20 (+0.55)	9 (7)	0	3	10	7
2014 Brisbane	0	0	323	3.5	0 (0)	0	0	7 (16)	N/A	N/A	0	0	4	2
Total	7	—	8,242	—	8 (1)	4	1	49 (89)	—	—	12	11	65	40
Average	0.78	4	916	6.6	0.88 (0.11)	0.4	0.1	5.4 (9.8)	+0.31 (+0.52)	20 (13)	1.3	1.2	7.2	4.4

Notes: All data derived from documents issued in the G20 leaders' names at each summit. N/A = not available. Domestic Political Management includes all communiqué compliments related to climate change, i.e., references by name to the G20 member(s) that specifically express gratitude in the context of climate change. % indicates how many G20 members received compliments in the official documents, depending on the number of full members participating. Deliberation refers to the number of references to climate change. The unit of analysis is the paragraph. % refers to the percentage of the words in each document that relate to climate change. Direction Setting: Priority Placement refers to the number of references to climate change in the chair's summary; the unit of analysis is the sentence. The number in parenthesis refers to the number of references to the environment. Democracy refers to the number of references to democracy in relation to climate change. Human Rights refers to the number of references to human rights in relation to climate change. The unit of analysis is for democracy and human right references is the paragraph. Decision Making refers to the number of climate change commitments. The number in parenthesis refers to the number of energy commitments. Delivery refers the overall compliance score for climate change commitments measured for that year. % assessed refers to percentage of commitments measured. The numbers in parenthesis refer to energy commitments. Development of Global Governance: Inside refers to the number of references to institutions inside the G20 related to climate change. Outside refers to the number of multilateral organizations related to climate change. The unit of analysis is the sentence.

Appendix B-1 G7/8 Communiqué Compliments: Climate Change

	Total	Spread	Canada	France	Germany	Italy	Japan	Russia	United Kingdom	United States	European Union
1975	0	0	0	0	0	0	0	0	0	0	0
1976	0	0	0	0	0	0	0	0	0	0	0
1977	0	0	0	0	0	0	0	0	0	0	0
1978	0	0	0	0	0	0	0	0	0	0	0
1979	0	0	0	0	0	0	0	0	0	0	0
1980	0	0	0	0	0	0	0	0	0	0	0
1981	0	0	0	0	0	0	0	0	0	0	0
1982	0	0	0	0	0	0	0	0	0	0	0
1983	0	0	0	0	0	0	0	0	0	0	0
1984	0	0	0	0	0	0	0	0	0	0	0
1985	0	0	0	0	0	0	0	0	0	0	0
1986	0	0	0	0	0	0	0	0	0	0	0
1987	0	0	0	0	0	0	0	0	0	0	0
1988	0	0	0	0	0	0	0	0	0	0	0
1989	0	0	0	0	0	0	0	0	0	0	0
1990	0	0	0	0	0	0	0	0	0	0	0

Appendix B-1 Continued

	Total	Spread	Canada	France	Germany	Italy	Japan	Russia	United Kingdom	United States	European Union
1991	0	0	0	0	0	0	0	0	0	0	0
1992	0	0	0	0	0	0	0	0	0	0	0
1993	0	0	0	0	0	0	0	0	0	0	0
1994	0	0	0	0	0	0	0	0	0	0	0
1995	0	0	0	0	0	0	0	0	0	0	0
1996	0	0	0	0	0	0	0	0	0	0	0
1997	0	0	0	0	0	0	0	0	0	0	0
1998	0	0	0	0	0	0	0	0	0	0	0
1999	0	0	0	0	0	0	0	0	0	0	0
2000	0	0	0	0	0	0	0	0	0	0	0
2001	1	11%	0	0	0	0	0	1	0	0	0
2002	0	0	0	0	0	0	0	0	0	0	0
2003	0	0	0	0	0	0	0	0	0	0	0
2004	0	0	0	0	0	0	0	0	0	0	0
2005	0	0	0	0	0	0	0	0	0	0	0
2006	0	0	0	0	0	0	0	0	0	0	0
2007	4	44%	0	0	1	0	1	0	1	1	0
2008	0	0	0	0	0	0	0	0	0	0	0
2009	0	0	0	0	0	0	0	0	0	0	0
2010	1	11%	0	0	0	0	1	0	0	0	0
2011	0	0	0	0	0	0	0	0	0	0	0
2012	0	0	0	0	0	0	0	0	0	0	0
2013	1	11%	0	1	0	0	0	0	0	0	0
2014	0	0	0	0	0	0	0	0	0	0	0
Total	7	N/A	0	1	1	0	2	1	1	1	0

Appendix B-2 G20 Communiqué Compliments: Climate Change

Summit	Total	Spread	Argentina	Australia	Brazil	Canada	China	France	Germany	India	Indonesia	Italy	Japan	Korea	Mexico	Russia	Saudi Arabia	South Africa	Turkey	United Kingdom	United States	European Union
2008 Washington	0	0%	0	0	0	0	0	0	0	0	0	0	0	0	0	0	0	0	0	0	0	0
2009 London	0	0%	0	0	0	0	0	0	0	0	0	0	0	0	0	0	0	0	0	0	0	0
2009 Pittsburgh	1	5%	0	0	0	0	0	0	0	0	0	0	0	0	0	0	0	0	0	1	0	0
2010 Toronto	1	5%	0	0	0	0	0	0	0	0	0	0	0	0	1	0	0	0	0	0	0	0
2010 Seoul	2	10%	0	0	0	0	0	0	0	0	0	0	0	0	2	0	0	0	0	0	0	0
2011 Cannes	2	10%	0	0	0	0	0	0	0	0	0	0	0	0	0	0	0	2	0	0	0	0
2012 Los Cabos	0	0%	0	0	0	0	0	0	0	0	0	0	0	0	0	0	0	0	0	0	0	0
2013 St. Petersburg	1	5%	0	0	0	0	0	0	0	0	0	0	0	0	0	0	0	1	0	0	0	0
2014 Brisbane	0	0%	0	0	0	0	0	0	0	0	0	0	0	0	0	0	0	0	0	0	0	0
Total	7	N/A	0	0	0	0	0	0	0	0	0	0	0	0	3	0	0	3	0	1	0	0

Appendix C-1 G7/8 Conclusions on Climate Change, 1975–2014

Year	# Words	% Total Words	# Paragraphs	% Total Paragraphs	# Documents	% Total Documents	Dedicated Documents
1975	0	0	0	0	0	0	0
1976	0	0	0	0	0	0	0
1977	0	0	0	0	0	0	0
1978	0	0	0	0	0	0	0
1979	28	1.3	1	2.6	1	50	0
1980	0	0	0	0	0	0	0
1981	0	0	0	0	0	0	0
1982	0	0	0	0	0	0	0
1983	0	0	0	0	0	0	0
1984	0	0	0	0	0	0	0
1985	88	2.9	1	1.9	1	50	0
1986	0	0	0	0	0	0	0
1987	85	1.5	1	1.1	1	14.3	0
1988	140	2.7	1	1.2	1	33.3	0
1989	422	6	7	5.0	1	9.1	0
1990	491	5.9	5	3.6	1	33.3	0
1991	236	2.4	5	2.8	1	20	0
1992	137	1.8	4	2.5	1	25	0
1993	154	3.1	1	1.2	1	33.3	0
1994	107	2.6	2	2.1	1	50	0
1995	87	0.7	3	1.1	1	25	0
1996	167	0.8	3	1.4	1	14.3	0
1997	305	1.6	5	1.7	1	16.7	0
1998	323	5.3	4	4.1	1	25	0
1999	198	1.3	1	0.3	1	25	0
2000	213	1.6	2	0.5	1	20	0
2001	324	5.2	4	4.4	1	10	0
2002	53	0.2	1	0.7	1	14.3	0
2003	62	0.3	3	2.3	1	5.9	0
2004	98	0.3	2	0.5	1	5	0
2005	2,667	9.3	68	9.9	3	8.1	2
2006	1,533	3.1	26	2.6	3	12	0
2007	4,154	12.0	47	9.0	5	41.7	0
2008	2,568	17.5	21	14.2	3	60	0
2009	5,559	33.3	52	15.8	7	53.8	0
2010	1,282	12	7	7.1	1	33.3	0
2011	1,086	5.9	12	5.6	1	20	0
2012	789	7.1	11	5.9	2	33.3	1
2013	525	3.9	8	3.0	1	25	0
2014	747	14.6	7	10.0	1	100	0
Average	615.7	4.2	7.9	3.1	1.2	21.7	0.075

Notes: Data are drawn from all official documents released by the G7/8 leaders as a group. Charts are excluded. # Words is the number of words in climate change–related passages, excluding document titles and references. The unit of analysis is the paragraph. % Total Words refers to the total number of words in all documents issued by the summit. # Paragraphs is the number of paragraphs containing references to climate change for the summit. Each point is recorded as a separate paragraph. % Total Paragraphs refers to the total number of paragraphs in all documents for the summit. # Documents refers to documents containing references to climate change and excludes dedicated documents. % Total Documents refers to the total number of documents for summit. # Dedicated Documents refers to total number of documents dedicated to climate change issued by the summit.

Appendix C-2 G20 Conclusions on Climate Change, 2008–2014

Summit	# Words	% Total Words	# Paragraphs	% Total Paragraphs	# Documents	% Total Documents	# Dedicated Documents
2008 Washington	64	1.7	2	2.8	1	100	0
2009 London	64	1.0	2	1.2	1	33	0
2009 Pittsburgh	911	9.7	10	11.7	3	100	0
2010 Toronto	838	7.4	11	7.6	1	50	0
2010 Seoul	2,018	12.7	25	11.4	3	60	0
2011 Cannes	1,167	8.2	14	7.2	3	100	0
2012 Los Cabos	1,160	9.1	12	5.9	1	25	0
2013 St. Petersburg	1,697	5.9	17	3.2	6	55	0
2014 Brisbane	323	3.5	5	2.3	2	40	0
Average	916	6.6	10.9	6.0	2.3	63	0

Notes: Data are drawn from all official documents released by the G20 leaders as a group. Charts are excluded. # Words is the number of words in climate change-related passages, excluding document titles and references. The unit of analysis is the paragraph. % Total Words refers to the total number of words in all documents issued by the summit. # Paragraphs is the number of paragraphs containing references to climate change for the summit. Each point is recorded as a separate paragraph. % Total Paragraphs refers to the total number of paragraphs in all documents for the summit. # Documents refers to documents containing references to climate change and excludes dedicated documents. % Total Documents refers to the total number of documents for the summit. # Dedicated Documents refers to total number of documents dedicated to climate change issued by the summit.

Appendix C-3 G20 Finance Conclusions on Climate Change, 1999–2010

Year	# Words	% Total Words	# Paragraphs	% Totals Paragraphs	# Documents	% Total Documents	# Dedicated Documents
1999	50	10.7	1	16.6	1	100	0
2000	0	0	0	0	0	0	0
2001	0	0	0	0	0	0	0
2002	0	0	0	0	0	0	0
2003	0	0	0	0	0	0	0
2004	445	11.3	4	10	3	66.6	0
2005	161	4.7	1	2.8	1	25	0
2006	159	4.5	2	5.5	1	50	0
2007	400	10.3	4	15.3	2	100	0
2008-1[a]	0	0	0	0	0	0	0
2008-2	64	3.6	1	5.8	1	100	0
2009-1[a]	0	0	0	0	0	0	0
2009-2	50	3.6	1	5.8	1	50	0
2009-3	177	13.9	1	8.3	1	100	0
2010-1	99	4.7	3	10	1	100	0
2010-2	0	0	0	0	0	0	0
2010-3	0	0	0	0	0	0	0
Average	94.4	3.9	1.0	4.7	0.7	40.6	0

Notes: [a] Emergency meetings. Data are drawn from all official documents released by the G20 finance ministers and central bank governors as a group. Charts are excluded. # Words is the number of words in climate change–related passages, excluding document titles and references. The unit of analysis is the paragraph. % Total Words refers to the total number of words in all documents issued by the meeting. # Paragraphs is the number of paragraphs containing references to climate change. Each point is recorded as a separate paragraph. % Total Paragraphs refers to the total number of paragraphs in all documents for the meeting. # Documents refers to documents containing references to climate change and excludes dedicated documents. % Total Documents refers to the total number of documents for the meeting. # Dedicated Documents refers to total number of documents dedicated to climate change issued by the meeting.

Appendix D-1 G7/8 Summit Climate Change Commitments by Issue, 1985–2014

Issue	1985	1987	1988	1989	1990	1991	1992	1993	1994	1995	1996	1997	1998	1999	2000	2001	2002	2003	2004	2005	2006	2007	2008	2009	2010	2011	2012	2013	2014	Total
TOTAL	1	1	0	4	7	5	7	4	4	7	3	9	10	4	4[a]	4	1	4	3	29[a]	20	44	54	42	10	7	5	12	16	321
Assist developing countries	0	0	0	0	0	0	0	0	0	0	0	0	0	0	0	0	0	0	0	0	0	2	5	3	0	0	0	0	0	10
Aviation	0	0	0	0	0	0	0	0	0	0	0	0	0	0	0	0	0	0	0	1	0	0	1	1	0	0	0	1	0	4
Avoid consequences	0	0	0	0	0	0	0	0	0	0	0	0	0	0	0	0	0	0	0	0	0	0	1	0	0	0	0	0	0	1
Awareness	0	0	0	0	0	0	0	0	0	0	0	0	0	0	0	0	0	0	0	1[a]	0	0	0	0	0	0	0	0	0	0
Capacity building	0	0	0	0	0	0	0	0	0	0	0	0	0	0	0	0	0	0	0	0	0	0	0	1	1	0	0	0	0	2
Carbon capture and storage	0	0	0	0	0	0	0	0	0	0	0	0	0	0	0	0	0	0	0	0	0	0	2	1	0	0	0	0	0	3
Climate change general	1	0	0	0	0	0	0	0	0	0	1	1	1	0	0	0	0	0	0	0	3	7	2	0	0	1	0	1	1	19
Conference of the Parties	0	0	0	0	0	0	0	0	0	1	0	1	1	0	1[a]	1	0	1	0	0	0	0	0	3	0	1	0	0	1	12
Commission on Sustainable Development	0	0	0	0	0	0	1	1	0	1	0	0	0	0	0	0	0	0	0	0	0	0	0	0	0	0	0	0	0	3
Developing country limits	0	0	0	0	0	0	0	0	0	0	0	1	1	0	0	0	0	0	0	1	0	0	0	0	0	0	0	0	0	3
Developing country technology	0	0	0	0	0	0	0	0	0	0	0	0	1	0	0	0	0	0	0	0	0	1	0	2	0	0	0	0	0	4
Dialogue	0	0	0	0	0	0	0	0	0	0	0	0	0	0	0	0	0	0	0	0	1	0	0	1	0	0	0	0	0	2
Earth Observation System	0	0	0	0	0	0	0	0	0	0	0	0	0	0	0	0	0	0	0	1	0	0	0	0	0	0	0	0	0	1
Emission profiles	0	0	0	0	0	0	0	0	0	0	0	0	0	0	0	0	0	0	0	0	0	1	0	0	0	0	0	0	0	1
Energy alternatives	0	0	0	0	0	0	0	0	0	0	0	0	0	0	0	0	0	0	0	0	0	1	1	0	0	0	0	0	2	5
Energy intensity	0	0	0	0	0	0	0	0	0	0	0	0	0	0	0	0	0	0	0	0	0	0	1	0	0	0	0	0	0	1
Energy use	0	0	0	0	0	0	0	0	0	0	0	0	0	0	0	0	0	0	0	1	1	0	0	0	0	0	0	0	0	2
Environmental problems	0	1	0	0	0	0	0	0	0	0	0	0	0	0	0	0	0	0	0	0	0	0	0	0	0	0	0	0	0	1
Financing/Funding	0	0	0	0	0	0	0	0	0	0	0	0	0	0	0	0	0	0	0	0	0	0	5	1	1	0	0	2	2	11
Forests	0	0	0	3	0	0	0	0	0	0	0	2	1	0	0	0	0	0	0	0	0	4	0	5	0	0	0	0	0	15
Funding least developed countries	0	0	0	0	0	0	0	0	0	0	0	0	0	0	0	0	0	0	0	0	0	0	0	0	1	0	0	0	0	2
Global Climate Observing System	0	0	0	0	0	0	0	0	0	0	0	0	0	0	0	0	0	0	0	1	0	0	0	0	0	0	0	0	0	1
Global Environment Facility	0	0	0	0	0	0	1	1	1	0	0	2	1	0	0	1	0	0	0	0	0	0	0	0	0	0	0	0	0	7

Appendix D-1 Continued

Issue	1985	1987	1988	1989	1990	1991	1992	1993	1994	1995	1996	1997	1998	1999	2000	2001	2002	2003	2004	2005	2006	2007	2008	2009	2010	2011	2012	2013	2014	Total
Global Earth Observation System of Systems	0	0	0	0	0	0	0	0	0	0	0	0	0	0	0	0	0	0	1	1	0	1	0	0	0	0	0	0	0	3
Gleneagles Dialogue	0	0	0	0	0	0	0	0	0	0	0	0	0	0	0	0	0	0	0	2	1	1	0	0	0	0	0	0	0	4
Global warming	0	0	0	0	0	0	0	0	0	0	1	0	0	0	0	0	0	0	0	0	0	0	0	0	0	0	0	0	0	1
Greenhouse gases	0	0	0	0	1	1	0	0	0	0	0	1	2	0	0	0	0	0	0	3[a]	2	5	0	2	0	0	2	0	0	19
Hydrocarbons	0	0	0	0	0	0	0	0	0	0	0	0	0	0	0	0	0	0	0	1	1	0	0	0	0	0	0	0	2	4
Interlinked challenges	0	0	0	0	0	0	0	0	0	0	0	0	0	0	0	0	0	0	0	0	0	1	0	0	0	0	0	0	0	1
Intergovernmental Panel on Climate Change	0	0	0	0	0	0	0	0	0	0	0	0	0	0	0	0	0	0	0	0	0	0	0	0	0	0	0	0	0	0
Kyoto Protocol	0	0	0	0	0	0	0	0	0	0	0	0	3	1	1[a]	0	0	0	0	1	0	0	0	1	0	0	0	0	0	7
Major economies join	0	0	0	0	0	0	0	0	0	0	0	0	0	0	0	0	0	0	0	0	0	1	0	0	1	0	0	0	0	2
Methodological Issues	0	0	0	0	0	0	0	0	0	0	0	0	0	0	0	0	0	0	0	0	0	0	1	0	0	0	0	0	5	6
Mid-term goals	0	0	0	0	0	0	0	0	0	0	0	0	0	0	0	0	0	0	0	0	0	0	3	1	1	1	0	0	0	6
Mitigation	0	0	0	0	0	0	0	0	0	0	0	0	0	0	0	0	0	0	0	0	0	0	6	1	0	0	0	1	0	8
Monitoring	0	0	0	0	0	0	0	0	0	0	0	1	0	1	0	0	0	0	0	0	0	0	0	0	0	0	0	0	0	2
Nairobi work programme	0	0	0	0	0	0	0	0	0	0	0	0	0	0	0	0	0	0	0	0	1	0	0	0	0	0	0	0	0	1
National action plan	0	0	0	0	0	0	0	0	1	0	0	0	0	0	0	0	0	0	0	0	0	1	1	1	0	0	0	0	0	4
Polluter pays	0	0	0	0	0	0	1	0	0	0	0	0	0	0	0	0	0	0	0	0	0	0	0	0	0	0	0	0	0	1
Post-2000 initiatives	0	0	0	0	0	0	1	0	1	0	0	0	0	0	0	0	0	0	0	0	0	0	0	0	0	0	0	0	0	2
Post-Kyoto	0	0	0	0	0	0	0	0	0	0	0	0	0	0	0	0	0	0	0	0	0	0	1	7	0	0	0	0	3	11
Reductions	0	0	0	0	0	0	0	0	0	0	0	0	0	0	0	0	0	0	1	0	1	0	5	2	2	2	0	0	0	13
Renewable energy	0	0	0	0	0	0	0	0	0	0	0	0	0	0	0	0	0	0	0	1	0	3	0	0	0	0	0	0	0	4
Reports/planning	0	0	0	0	0	0	0	0	0	0	0	0	0	0	0	0	0	0	0	0	1	3	0	2	0	0	0	0	0	6
Research/science	0	0	0	0	0	1	0	0	0	0	1	2	0	0	1	0	0	0	0	0	0	0	0	0	0	0	0	1	0	6
Rio conference	0	0	0	0	1	0	0	0	0	1	1	1	0	0	0	0	0	0	0	0	0	0	0	0	0	0	1	0	0	5
Sectoral approaches	0	0	0	0	0	0	0	0	0	0	0	0	0	0	0	0	0	0	0	0	0	0	1	0	0	0	0	0	0	1

Issue area																								Total
Sequestration/Carbon Sequestration Leadership Forum	0	0	0	0	0	0	0	0	0	0	1	0	3[a]	1	2	0	0	0	0	0	0	0	0	4
Sharing practices	0	0	1	0	0	0	0	0	0	0	0	0	0	0	1	1	1	0	1	0	0	0	0	4
Sinks (general)	0	0	0	0	0	0	0	0	0	0	0	0	0	0	0	1	0	0	0	2	1	0	0	3
Sustainable development	0	0	0	0	0	3	1	0	1	1	0	0	0	2	2	1	0	0	2	0	0	0	0	15
Technology	0	0	0	0	0	0	0	0	0	0	1	0	5[a]	1	7	4	5	0	0	0	0	0	0	18
Trade	0	0	0	0	0	0	0	0	0	0	0	0	0	0	2	2	0	0	0	0	0	0	0	2
Transport	0	0	0	0	0	0	0	0	0	0	0	0	1	2	2	0	0	0	0	0	1	0	0	6
United Nations Framework Convention on Climate Change	0	1	1	1	0	0	0	1[a]	0	0	0	1	4	3	3	5	1	1	0	0	0	1	3	26
World Meteorological Organization network	0	0	1	0	0	0	0	0	0	0	0	0	0	0	0	0	0	0	0	0	0	0	0	1

Note: [a] Includes commitments that refer more than one issue area. Totals therefore do not necessarily add up to the total number of commitments calculated for that issue or year.

Appendix D-2 G20 Climate Change Commitments by Issue, 2008–2014

Issue	2008 Washington	2009 London	2009 Pittsburgh	2010 Toronto	2010 Seoul	2011 Cannes	2012 Los Cabos	2013 St. Petersburg	2014 Brisbane	Total
TOTAL	0	3	3	3	8	8	6	11	7	49
Assist developing countries	0	0	0	0	0	0	1	0	0	1
Aviation	0	0	0	0	0	0	0	0	0	0
Avoid consequences	0	0	0	0	0	0	0	0	0	0
Awareness	0	0	0	0	0	0	0	0	0	0
Capacity Building	0	0	0	0	0	0	0	0	0	0
Carbon capture and storage	0	0	0	0	0	0	0	0	0	0
Climate change general	0	0	0	1	3	1	0	1	1	7
Conference of the Parties	0	1	0	1	2	2	2	4	0	12
Commission on Sustainable Development	0	0	0	0	0	0	0	0	0	0
Developing country limits	0	0	0	0	0	0	0	0	0	0
Developing country technology	0	0	0	0	0	0	0	0	0	0
Dialogue	0	0	0	0	0	0	0	0	0	0
Earth Observation System	0	0	0	0	0	0	0	0	0	0
Emission profiles	0	0	0	0	0	0	0	0	0	0
Energy alternatives	0	0	0	0	0	0	0	0	0	0
Energy intensity	0	0	0	0	0	0	0	0	0	0
Energy use	0	0	0	0	0	0	0	0	0	0
Environmental problems	0	0	0	0	0	0	0	0	0	0
Financing/Funding	0	0	1	0	0	2	1	1	2	7
Forests	0	0	0	0	0	0	0	0	0	0
Funding least developed countries	0	0	0	0	0	0	0	0	0	0
Global Climate Observing System	0	0	0	0	0	0	0	0	0	0
Global Environment Fund	0	0	0	0	0	0	0	0	0	0
Global Earth Observation System of Systems	0	0	0	0	0	0	0	0	0	0
Gleneagles Dialogue	0	0	0	0	0	0	0	0	0	0
Global warming	0	0	0	0	0	0	0	0	0	0
Greenhouse gases	0	0	0	0	0	0	0	0	0	0
Hydrocarbons	0	0	0	0	0	0	0	3	0	3
Interlinked challenges	0	0	0	0	0	0	0	0	0	0
Intergovernmental Panel on Climate Change	0	0	0	0	0	0	0	0	0	0
Kyoto Protocol	0	0	0	0	0	0	0	0	0	0
Major economies join	0	0	0	0	0	0	0	0	0	0
Methodological Issues	0	0	0	0	0	0	0	0	0	0

Issue	2008 Washington	2009 London	2009 Pittsburgh	2010 Toronto	2010 Seoul	2011 Cannes	2012 Los Cabos	2013 St. Petersburg	2014 Brisbane	Total
Mid-term goals	0	0	0	0	0	0	0	0	0	0
Mitigation	0	0	0	0	0	0	0	0	0	0
Monitoring	0	0	0	0	0	0	0	0	0	0
Nairobi work programme	0	0	0	0	0	0	0	0	0	0
National action plan	0	0	0	0	0	0	0	0	0	0
Polluter pays	0	0	0	0	0	0	0	0	0	0
Post-2000 initiatives	0	0	0	0	0	0	0	0	0	0
Post-Kyoto	0	0	0	0	0	0	0	0	0	0
Reductions	0	0	0	0	0	0	0	0	0	0
Renewable energy	0	0	0	0	0	0	0	0	0	0
Reports/planning	0	0	0	0	0	0	1	0	0	1
Research/science	0	0	0	0	0	0	0	0	0	0
Rio conference	0	0	0	0	0	1	1	0	0	2
Sectoral approaches	0	0	0	0	0	0	0	0	0	0
Sequestration/Carbon Sequestration Leadership Forum	0	0	0	0	0	0	0	0	0	0
Sharing practices	0	0	0	0	0	0	0	0	0	0
Sinks (general)	0	0	0	0	0	0	0	0	0	0
Sustainable development	0	1	0	0	3	2	0	0	3	9
Technology	0	1	0	0	0	0	0	0	0	1
Trade	0	0	0	0	0	0	0	0	0	0
Transport	0	0	0	0	0	0	0	0	0	0
United Nations Framework Convention on Climate Change	0	0	2	1	0	0	0	2	1	6
World Meteorological Organization network	0	0	0	0	0	0	0	0	0	0

Appendix E-1 G7/8 Climate Change Compliance, 1985–2013

Year[a]	Issue	Average	United States	Japan	Germany	United Kingdom	France	Italy	Canada	Russia	European Union
N=72		+0.45	+0.31	+0.54	+0.62	+0.66	+0.39	+0.10	+0.48	+0.20	+0.79
1985 (1/1) 100%		+0.50	+1.00	+0.00	+1.00	+0.00	+1.00	+0.00	+0.00		+1.00
1985-1	Climate change	+0.5	+1	0	+1	0	+1	0	0		+1
1987 (1/1) 100%		+0.29	+0.00	+0.00	+0.00	+1.00	+0.00	+0.00	+1.00		
1987-32	Environmental protection	+0.29	0	0	0	+1	0	0	+1		
1989 (4/4) 100%		-0.07	-1.00	+0.75	+0.00	+0.50	-0.50	-0.25	+0.00		
1989-1	Greenhouse gases	+0.43	-1	+1	+1	+1	0	0	+1		
1989-2	WMO	-0.43	-1	+1	-1	+1	-1	-1	-1		
1989-3	Forests	-0.29	-1	+1	0	-1	-1	+1	-1		
1989-4	Climate change	0	-1	0	0	+1	0	-1	+1		
1990 (4/7) 57%		-0.11	-1.00	+0.33	+1.00	+0.67	+1.00	+1.00	+1.00		
1990-26	Greenhouse gases	+0.43	-1	0	+1	0	+1	+1	+1		
1990-27-28	Climate change	+0.43	-1	+1	+1	+1	+1	-1	+1		
1990-29	Climate change	-0.29	-1	0	+1	+1	-1	-1	-1		
1990-36	Forests	-1.00	-1	-1	-1	-1	-1	-1	-1		
1991 (2/5) 40%		+0.38	-1.00	+0.00	+1.00	+0.50	+0.50	+0.50	+1.00		
1991-1	UNFCCC	+0.14	-1	0	+1	0	0	0	+1		
1991-4	Greenhouse gases	+0.67	-1		+1	+1	+1	+1	+1		
1992 (3/7) 43%		+0.71	+1.00	+1.00	+1.00	+0.67	+0.67	+0.33	+1.00		
1992-1	UNFCCC	+0.71	+1	+1	+1	+1	0	0	+1		

Year	Topic								
1992-2	GEF	+0.71	-1	+1	+1	+1	+1		
1992-6	Science	+0.71	+1	+1	0	+1	+1		
1993 (2/4) 50%		+0.57	+1.00	+1.00	+0.50	+1.00	+1.00		
1993-1	National plans	+0.14	+1	-1	0	-1	+1		
1993-3	GEF	+1	+1	+1	+1	+1	+1		
1994 (2/4) 50%		+0.71	+1.00	+1.00	+0.50	+1.00	+1.00		
1994-1	National plans	+0.57	+1	+1	0	+1	0		
1994-3	GEF	+0.86	+1	+1	0	+1	0		
1995 (1/7) 14%		+0.29	+1.00	-1.00	+0.00	+1.00	+0.00		
1995-23	COP	+0.29	+1	-1	0	+1	0		
1996 (1/3) 33%		+0.57	+1.00	+1.00	+0.00	+1.00	+0.00		
1996-87	COP	+0.57	+1	+1	0	+1	0		
1997 (2/9) 22%		+0.29	+1.00	+1.00	+0.00	+1.00	+0.00	+1.00	+1.00
1997-8	COP	+0.5	+1	+1	-1	+1	0	+1	+1
1997-9	Greenhouse gases	+0.11	+1	+1	0	+1	-1	-1	+1
1998 (3/10) 30%		+1.00	+1.00	+1.00	+1.00	+1.00	+1.00	+0.00	
1998-32	Kyoto Protocol	+1.00	+1	+1	+1	+1	+1	+1	
1998-34	Kyoto Protocol	+1.00	+1	+1	+1	+1	+1	+1	
1998-35	Trade	+1.00	+1	+1	+1	+1	+1	+1	
1999 (1/4) 25%		-0.22	+0.00	+0.00	-1.00	-1.00	-1.00	-1.00	-1.00
1999-32	Kyoto Protocol	-0.22	+1	0	-1	-1	-1	-1	-1
2000 (1/4) 25%		+0.44	+1.00	+1.00	+1.00	+0.00	+0.00	+0.00	+1.00
2000-86	COP	+0.44	+1	+1	+1	0	0	0	+1

Appendix E-1 Continued

Year[a]	Issue	Average	United States	Japan	Germany	United Kingdom	France	Italy	Canada	Russia	European Union
2001 (4/4) 100%		0	+0.00	+0.00	+0.25	+0.00	+0.00	+0.00	+0.25	+0.00	
2001-xx	GEF	-0.13	0	0	0	0	0	0	0	-1	
2001-xx	Sustainable development	0	0	0	0	0	0	0	0	0	
2001-xx	International agreement	+0.13	0	-1	+1	0	0	0	+1	0	
2001-44	COP	0	0	0	0	0	0	0	0	0	
2002 (1/1) 100%		+0.89	0	+1	+1	+1	+1	+1	+1	+1	+1
2002-8	Sustainable development	+0.89	0	+1	+1	+1	+1	+1	+1	+1	+1
2003 (2/4) 50%		+0.88	+1.00	+1.00	+1.00	+0.50	+1.00	+1.00	+0.50	+1.00	+1.00
2003-75	Renewables	+0.75	+1	+1	+1	0	+1	+1	0	+1	
2003-92	UNFCCC	+1.00	+1	+1	+1	+1	+1	+1	+1	+1	
2004 (2/3) 67%		+0.89	+1.00	+1.00	+1.00	+1.00	+1.00	+0.50	+1.00	+0.50	+1.00
2004-S3	GEOSS	+1.00	+1	+1	+1	+1	+1	+1	+1	+1	+1
2004-S2	Renewables	+0.78	+1	+1	+1	+1	+1	0	+1	0	+1
2005 (5/29) 17%		+0.80	+0.80	+0.60	+1.00	+1.00	+1.00	+0.75	+0.80	+1.00	+1.00
2005-1	UNFCCC	+0.44	0	0	+1	+1	+1	-1	0	+1	+1
2005-A1	Technology	+1.00	+1	+1	+1	+1	+1	+1	+1	+1	+1
2005-A2	Gleneagles	+0.89	+1	+1	+1	+1	+1	0	+1	+1	+1
2005-O9	Gleneagles	+0.67	+1	0	+1	+1	+1	+1	+1	-1	+1
2005-15	Aviation	+1.00	+1	+1	+1	+1	+1	+1	+1	+1	+1
2006 (9/20) 45%		+0.35	+0.71	+0.56	+0.44	+0.75	+0.25	+0.25	+0.43	+0.25	+1.00
2006-62	Sustainable energy	+0.22	0	0	0	+1	0	0	0	0	+1

		+0.33 avg								
2006-99	Energy Intensity	+0.33	0	0	+1	0	0	0	+1	+1
2006-110	Hydrocarbons	-0.11	-1	0	-1	+1	-1	+1	0	-1
2006-116	Transport	+0.44	+1	+1	0	0	0	+1	0	+1
2006-123	Alternative energy	+0.33	+1	+1	0	-1	0	0	0	+1
2006-138	Technology	+0.22	+1	0	+1	0	0	-1	-1	+1
2006-156	Renewables	+0.89	+1	+1	+1	+1	+1	+1	+1	+1
2006-162	Climate change	+0.78	+1	+1	+1	+1	+1	0	0	+1
2006-165	UNFCCC	0	-1	0	+1	0	0	-1	0	+1
2007 (4/44) 9%		+0.56	+1.00	+0.33	+1.00	+0.33	+0.33	+0.50	+0.33	+0.75
2007-35	UNFCCC	+1.00	+1	+1	+1	+1	+1	+1	+1	+1
2007-36	Technology	+0.44	+1	0	+1	0	0	+1	0	0
2007-28	UNFCCC	+0.22	+1	0	+1	0	-1	0	-1	+1
2007-44	Forests	+0.56	+1	+1	+1	0	0	0	0	+1
2008 (5/54) 9%		+0.53	+0.80	+1.00	+1.00	+0.80	-0.20	+0.80	0	+0.80
2008-27	Biofuels	+0.89	+1	+1	+1	+1	+1	+1	0	+1
2008-55	Pollution reduction	+0.78	-1	+1	+1	+1	+1	+1	+1	+1
2008-72	CCS	+0.33	+1	+1	+1	-1	-1	+1	-1	+1
2008-251	General	+0.11	-1	-1	-1	+1	-1	0	0	+1
2008-265	Climate change	+0.56	+1	+1	+1	+1	-1	+1	0	0
2009 (5/42) 12%		+0.64	+0.60	+1.00	+1.00	+1.00	+0.75	+0.80	+0.25	+0.80
2009-49	UNFCCC	+0.67	0	+1	+1	+1	+1	+1	0	0
2009-64	Technology	+1.00	+1	+1	+1	+1	+1	+1	+1	+1
2009-66	Financing	+0.89	+1	+1	+1	+1	+1	+1	0	+1
2009-73	Forests	+0.67	+1	-1	-1	+1	0	0	0	+1
2009-98	CCS	0	0	+1	+1	-1	-1	+1	-1	+1
2010 (3/10) 30%		+0.26	+0.50	+0.33	+0.67	+0.67	+0.50	+0.50	+0.33	+0.67

Appendix E-1 Continued

Year[a]	Issue	Average	United States	Japan	Germany	United Kingdom	France	Italy	Canada	Russia	European Union
2010-26	Pollution reduction	+0.22	−1	−1	+1	+1	+1	0	−1	+1	+1
2010-27	Kyoto Protocol	−0.22	0	−1	0	0	0	−1	0	0	0
2010-55	Disasters	+0.78	+1	+1	0	+1	+1	+1	+1	0	+1
2011 (1/7) 14%		+0.67	+1.00	+1.00	+0.00	+1.00	+1.00	+0.00	+1.00	+1.00	+0.00
2011-51	Pollution reduction	+0.67	+1	+1	0	+1	+1	0	+1	+1	0
2012 (1/5) 20%		+0.11	+0.00	0	+1.00	0	0	−1.00	−1.00	−1.00	+1.00
2012-29	Pollution reduction	+0.11	0	0	+1	0	0	−1	+1	−1	+1
2013 (2/12) 17%		+0.22	+0.50	+0.00	+1.00	+0.50	+0.00	+0.00	+1.00	+1.00	+1.00
2013-145	UNFCCC	+0.56	+1	−1	+1	+1	0	0	+1	+1	+1
2013-150	Financing	−0.11	0	0	+1	0	0	−1	−1	−1	+1

Notes: [a] Numbers in parentheses refer to the number of commitments assessed over the total commitments on climate change identified in summit documents. % refers to the portion of commitments assessed. Each commitment is identified by the year of the summit and numbered as it appears in the list of commitments identified by the G8 Research Group. CCS = carbon capture and storage; COP = Conference of the Parties; GEOSS = Global Earth Observation System of Systems; GEF = Global Environment Facility; UNFCCC = United Nations Framework Convention on Climate Change; WMO = World Meteorological Organization.

Appendix E-2 G20 Summit Compliance on Climate and Energy

Summit (N=22)	Average	Argentina	Australia	Brazil	Canada	China	France	Germany	India	Indonesia	Italy	Japan	Korea	Mexico	Russia	Saudi Arabia	South Africa	Turkey	United Kingdom	United States	European Union
Climate Change (10 commitments)																					
2009 London	-0.10																				
2009L-84	-0.10	-1	0	-1	0	+1	0	0	-1	0	0	0	+1	0	-1	0	0	-1	0	0	+1
2009 Pittsburgh	+0.86																				
2009P-85	+0.86	+1	+1	0	+1	+1	+1	+1	+1	0		+1		+1	+1			+1	+1	+1	
2010 Toronto	+0.42																				
2010T-56	+0.40	+1	+1	+1	+1	+1	0	0	+1	0	+1	0	+1	-1	-1	-1	0	0	+1	+1	+1
2010T-57	-0.06		0	0	0	0	0	0	0	-1	0	0	0	0	0		-1		0	0	+1
2010T-58	+0.89	+1	+1	+1	0	+1	+1	+1	+1	+1	+1	+1	+1	+1	+1		+1		+1	0	+1
2010 Seoul	+0.35																				
2010S-131	+0.25	-1	+1	+1	+1	+1	0	0	0	+1	+1	+1	0	-1	+1	0	-1	-1	-1	+1	+1	+1
2010S-132	+0.47		+1	+1	+1	0	0	+1	+1	-1	-1	+1	0	+1	0		0		+1	+1	+1
2012 Los Cabos	+0.38																				
2012-91	+0.70	+1	+1	+1	+1	+1	+1	+1	+1	-1	0	+1	+1	+1	+1	-1	+1	0	+1	+1	+1
2012-94	+0.05	-1	+1	0	0	+1	0	+1	+1	+1	-1	-1	+1	0	-1	-1	0	0	0	0	0
2013 St. Petersburg	-0.20																				
2013-188	-0.20	-1	-1	-1	-1	-1	+1	+1	+1	-1	+1	+1	+1	+1	-1	-1	-1	0	-1	+1	0	-1
Average	+0.31	-0.14	+0.60	+0.30	+0.40	+0.60	+0.40	+0.60	+0.40	+0.10	+0.22	+0.40	+0.56	+0.30	-0.10	-0.83	0.00	-0.33	+0.70	+0.50	+0.67
Energy (12 commitments)																					
2009 Pittsburgh	+0.43																				
2009P-18	+0.05	0	-1	0	-1	+1	+1	0	0	+1	-1	+1	+1	+1	-1	-1	+1	-1	-1	+1	-1

Appendix E-2 Continued

Summit (N=22)	Average	Argentina	Australia	Brazil	Canada	China	France	Germany	India	Indonesia	Italy	Japan	Korea	Mexico	Russia	Saudi Arabia	South Africa	Turkey	United Kingdom	United States	European Union
2009P–72	+0.45	0	0	0	0	0	+1	+1	0	+1	+1	+1	+1	+1	0	0	0	0	+1	+1	0
2009P–83	+0.44	0	+1	+1	+1	+1	+1		+1	0		+1	0	0	−1	−1	0	0	+1	+1	+1
2009P–84	+0.75	+1	+1	0	+1	+1	+1	+1	+1	0	+1	+1	+1	+1	+1	0	0	+1	+1	+1	0
2010 Toronto	+0.50	0		+1	0	0		+1	+1	−1	0	+1	+1	+1	+1	0	+1	+1	+1	+1	0	−1
2010T–60	+0.50	0	+1	+1	0	0	+1	+1	−1	0	+1	+1	+1	+1	0	+1	+1	+1	+1	0	−1
2010 Seoul	+0.51																				
2010S–127	+0.26	0	+1	+1	+1	−1	0	−1	0	0	+1	0	+1	0	+1	0	+1	−1	+1	0	
2010S–135	+0.75	0	+1	+1	+1	+1	+1	+1	+1	0	+1	+1	+1	+1	+1	−1	+1	0	+1	+1	+1
2011 Cannes	+0.61																				
2011–236	+0.63	0	+1	+1	0	+1	+1	0	+1	0	+1	0	+1	+1	+1	+1	+1	0	+1	0	+1
2011–242	+0.95	+1	+1	+1	+1	+1	+1	+1	+1	+1	+1	−1	+1	+1	+1	−1	+1	0	+1	+1	+1
2011–252	+0.25	0	+1	+1	−1	+1	0	+1	+1	+1	0		+1	+1	0		0	0	+1	−1	0
2012 Los Cabos	+0.58																				
2012–96	+0.58	0	+1	+1	0	+1	+1	0	+1	0	0	0	+1	+1	+1		+1	0	+1	0	+1
2013 St. Petersburg	+0.55																				
2013–12	+0.55	0	−1	+1	0	+1	+1	+1	+1	+1	+1	0	0	+1	+1	0	0	+1	+1	+1	+1
Average	+0.52	+0.17	+0.58	+0.75	+0.25	+0.67	+0.83	+0.55	+0.58	+0.42	+0.55	+0.50	+0.83	+0.83	+0.42	−0.20	+0.58	+0.18	+0.83	+0.50	+0.36
Overall Average	+0.42	+0.05	+0.59	+0.55	+0.32	+0.64	+0.64	+0.57	+0.50	+0.27	+0.40	+0.45	+0.71	+0.59	+0.18	−0.44	+0.33	0.00	+0.77	+0.50	+0.50

Note: Each commitment is identified by the year of the summit, with a letter to indicate the summit in years with more than one, and numbered as it appears in the list of commitments identified by the G20 Research Group. A blank cell indicates no assessment is available.

Appendix F-1 G7/8 Environment Ministers' Meetings

Spring 1992	Germany
June 1992	Meeting of the G7 Environment Ministers, Rio de Janeiro, Brazil (during the Earth Summit)
March 12–13, 1994	Informal Meeting of the G7 Environment Ministers, Florence, Italy
April 29–May 1, 1995	Environment Ministers Meeting, Hamilton, Ontario
May 9–10, 1996	Environment Ministers Meeting, Cabourg, France • Letter to President Chirac from the Chairman of the Meeting of the G7 Environment Ministers • Chairman's Summary
May 5–6, 1997	Environment Leaders' Summit of the Eight, Miami, Florida, 1997 • Declaration of the Environment Leaders of the Eight on Children's Environmental Health • Chair's Summary
April 3–5, 1998	G8 Environment Ministers Meeting, Leeds Castle, Kent, England • Communiqué • Press Release
March 26–28, 1999	G8 Environment Ministers' Meeting, Schwerin, Germany • Communiqué
April 7–9, 2000	G8 Environment Ministers' Meeting, Otsu, Japan • Communiqué
March 2–4, 2001	G8 Environment Ministerial Meeting, Trieste, Italy
April 12–14, 2002	G8 Environment Ministers Meeting, Banff, Canada • Banff Ministerial Statement on the World Summit on Sustainable Development
April 25–27, 2003	G8 Environment Ministers meeting, Paris • Communiqué • Final Communiqué
March 15–16, 2005	International Energy and Environment Ministers Roundtable, London • Press release • Gordon Brown's keynote address • Summary of Proceedings
March 17–18, 2005	G8 Environment and Development Ministers Meeting, Derbyshire, UK • Agenda • Themes • Outreach • Carbon offsetting • Press release
November 1, 2005	Energy and Environment Ministers plus ministers from 12 other countries, London • Chairman's Conclusions • Factsheet: New Developments in UK Activities in Support of Gleneagles Plan of Action • Prime Minister Blair Concludes Climate Change Conference
October 3, 2006	Environment Ministers, Monterrey, Mexico (during a two-day meeting of environment ministers to discuss the G8 Gleneagles Summit's climate action plan)
March 15–17, 2007	G8 Environment Ministers Meeting • Press release from German Ministry for the Environment, Nature Conservation and Nuclear Safety • Potsdam Initiative • Background paper on biodiversity • Transcript of teleconference with John Baird, Minister, Environment Canada

Appendix F-1 Continued

March 14–16, 2008	Gleneagles Dialogue on Climate Change, Clean Energy and Sustainable Development, Chiba • Chairs' Conclusions • Delegation
May 24–26, 2008	G8 Environment Ministers Meeting, Kobe, Japan • Chair's Summary • Kobe 3R Action Plan • Kobe Call for Action for Biodiversity • Japan's New Action Plan towards a Global Zero Waste Society • Japan's Commitments for the implementation of the Kobe Call for Action for Biodiversity
April 22–24, 2009	G8 environment ministers, Siracusa, Italy • Chair's Summary • Siracusa Charter on Biodiversity
2010–2014	No meetings

Appendix F-2   G7/8 Official-Level Bodies

First Summit Cycle (8)

1975	London Nuclear Suppliers Group
1977	International Nuclear Fuel Cycle Evaluation Group
1979	*High Level Group on Energy Conservation and Alternative Energy*
1979	*International Energy Technology Group*
1979	*High Level Group to Review Oil Import Reduction Progress*
1980	*International Team to Promote Collaboration on Specific Projects on Energy Technology*
1980	*High Level Group to Review Result on Energy*

Second Summit Cycle (9)

1985	*Expert Group on Desertification and Dry Zone Grains*
1985	*Expert Group on Environmental Measurement*

Third Summit Cycle (14)

1992	Nuclear Safety Working Group
1993	G8 Non-Proliferation Experts Group

Fourth Summit Cycle (16)

1996	Nuclear Safety Working Group
1997	*Officials Group on Forests*
2000	*Renewable Energy Task Force*
2002	*Energy Officials Follow-up Process*
2002	G8 Global Partnership Review Mechanism
2002	G8 Nuclear Safety and Security Group

Fifth Summit Cycle

2003	*Senior Officials for Science and Technology for Sustainable Development*
2003	G8 Enlarged Dialogue Meeting
2004	Global Partnership Senior Officials Group (GPSOG), January 2004
2004	Global Partnership Working Group (GPWG)
2004	*International Partnership for a Hydrogen Economy (IPHE)*
2004	*IPHE Implementation-Liaison Committee*
2004	*Carbon Sequestration Leadership Forum (CSLF)*
2004	*Renewable Energy and Energy Efficiency Partnership ((REEEP)*
2004	*Generation IV International Forum (GIF)*
2004	*Global Earth Observation System of Systems (GEOSS)*
2005	*Dialogue on Sustainable Energy*

2005	*Global Bioenergy Partnership*
2006	G8 expert group on securing energy infrastructure
2007	Structured High Level Dialogue with major emerging economies (Heiligendamm process) including energy
2008	G8 Experts Group to monitor implementation on food security
2008	*Climate Investment Funds (CIF; CTF; SCF)*
2008	Energy forum
2009	*Major Economies Forum on Energy and Climate*
2009	*Intergovernmental Platform on Biodiversity and Ecosystem Services*
2009	L'Aquila Food Security Initiative
2010	Global Agriculture and Food Security Program

Sixth Summit Cycle
None after 2010

Note: Excludes one-off meeting or conferences, Global Partnership on Weapons and Materials of Mass Destruction, 1990 Brazil Pilot Program on Tropical Forests, and 1992 Global Environment Facility Working Group of Experts. A summit cycle is one rotation of presidencies starting with France and ending with Canada. Most directly relevant bodies are in italics.

Appendix F-3 G7/8 Development of Global Governance Inside: Climate Change

Year	Ministerial	Official	Total
1998 Birmingham	1	0	1
2005 Gleneagles	1	0	1
2006 St. Petersburg	3	3	6
2007 Heiligendamm	0	2	2
2008 Hokkaido	1	2	3
2009 L'Aquila	1	4	5
2010 Muskoka	0	2	2
2011 Deauville	0	0	0
2012 Camp David	0	0	0
2013 Lough Erne	0	0	0
2014 Brussels	1	0	1
Total	8	13	21

Note: No references to environment ministers before 1998 and from 1999 to 2004.

Appendix F-4 G20 Development of Global Governance Inside: Climate Change

Year	Ministerial	Official	Total
2008 Washington	0	0	0
2009 London	0	0	0
2009 Pittsburgh	4	0	4
2010 Toronto	0	0	0
2010 Seoul	5	3	8
2011 Cannes	2	0	2
2012 Los Cabos	1	5	6
2013 St. Petersburg	0	3	3
2014 Brisbane	0	0	0
Total	12	11	23

Appendix F-5 G7/8 Development of Global Governance Outside: Climate Change, 1988–2004

Institution/Year	1988	1989	1990	1991	1992	1993	1994	1995	1996	1997	1998	1999	2000	2001	2002	2003	2004	2005	2006	2007	2008	2009	2010	2011	2012	2013	2014	Total
Total	3	3	2	1	2	2	0	0	2	0	0	1	1	2	0	0	0	20	10	16	22	19	5	7	4	5	7	134
Asian Development Bank																				1								1
Asia Forest Partnership																			1									1
Asia Pacific Economic Cooperation																										1		1
Conference of the Parties									1				1											1			1	4
Carbon Sequestration Leadership Forum																		3	1	1								5
Global Environment Facility						1			1					1				2			1	1						7
Global Earth Observation System of Systems																		3		1	1	1						6
International Atomic Energy Agency																					1	2					1	4
International Civil Aviation Organization																					1	1				1	1	4
International Energy Agency																		4		3	2	1	1	1			2	14
International Maritime Organization																					1	1				1		3
International Partnership for Energy Efficiency Cooperation																					1			1				2
International Renewable Energy Association																											1	1
International Tropical Timber Organization																			1									1
Major Economies Forum																						3			1			4

Organization															Total
Organisation for Economic Co-operation and Development									1			2			3
United Nations Conference on Environment and Development	1	2	1												4
United Nations Environment Programme	2	1	1						1	1	1				7
United Nations Educational, Social, and Cultural Organization						1									1
United Nations Framework Convention on Climate Change	1	1		6	6	8	10	7	3	1	2	2	1		48
United Nations Forum on Forests				1											1
World Meteorological Organization	1	2	1				1			1					5
World Bank	2			1	1	1									5
World Trade Organization				2											2

Note: No references prior to 1988.

Appendix F-6 G20 Summit Development of Global Governance Outside: Climate Change

International Organization	2008 Washington	2009 London	2009 Pittsburgh	2010 Toronto	2010 Seoul	2011 Cannes	2012 Los Cabos	2013 St. Petersburg	2014 Brisbane	Total
Total	0	1	10	3	20	11	6	10	4	65
Conference of the Parties	0	0	0	1	1	0	1	2	2	7
Global Bioenergy Partnership	0	0	0	0	0	0	0	1	0	1
Global Marine Environment Protection Initiative	0	0	0	0	4	1	1	0	0	6
High Level Advisory Group on Climate Change Financing	0	0	0	1	2	0	0	1	0	4
International Association of Drilling Contractors	0	0	0	0	1	0	0	0	0	1
International Atomic Energy Agency	0	0	0	0	0	0	0	1	0	1
International Energy Agency	0	0	1	0	2	1	0	0	0	4
International Monetary Fund	0	0	0	0	0	1	0	1	0	2
International Maritime Organization	0	0	0	0	1	0	0	0	0	1
International Organization of Securities Commissions	0	0	0	0	1	0	0	0	0	1
Organisation for Economic Co-operation and Development	0	0	1	0	2	3	1	0	0	7
Organization of Petroleum Exporting Countries	0	0	0	0	2	1	0	0	0	3
United Nations Framework Convention on Climate Change	0	1	4	1	3	2	2	3	2	18
World Bank	0	0	4	0	1	2	1	1	0	9

Appendix G-1 G7/8 Communiqué References to Climate/Environmental Shocks and Vulnerabilities

Year	# Sentences	# References	Shocks	Vulnerability
1989	3	3	0	3
1990	2	2	0	2
1991	0	0	0	0
1992	0	0	0	0
1993	0	0	0	0
1994	0	0	0	0
1995	0	0	0	0
1996	1	1	0	1
1997	0	0	0	0
1998	1	1	0	1
1999	1	1	0	1
2000	0	0	0	0

Year	# Sentences	# References	Shocks	Vulnerability
2001	1	1	0	1
2002	0	0	0	0
2003	0	0	0	0
2004	0	0	0	0
2005	0	0	0	0
2006	0	0	0	0
2007	3	3	0	3
2008	1	2	1	1
2009	6	9	7	2
2010	4	4	3	1
2011	1	1	0	1
2012	0	0	0	0
2013	0	0	0	0
2014	2	2	1	1
Total	26	30	12	18

Notes: The unit of analysis is the sentence.

\# Sentences refers to the number of sentences that contain at least one reference to a shock or vulnerability in the context of climate change.

\# References refers to the number of individual shocks or vulnerabilities referred to. Inclusion terms for shock are shock, crisis, disaster, emergency/emergencies, and extreme weather. Inclusion terms for vulnerability are vulnerability/vulnerabilities, threat/threaten, exposure, collapse, and extinction in the context of climate, weather, environment, or ecology. Exclusion terms: impact, mitigation, adaptation, variability, risk, resilience, stress, excessive emissions, dangerous, endangered.

Appendix G-2 G20 Communiqué References to Climate/Environmental Shocks and Vulnerabilities

Year	Total	Shock		Vulnerability
		Crisis	Disaster	Threat
2008 Washington	0	0	0	0
2009 London	1	0	0	1
2009 Pittsburgh	3	0	0	3
2010 Toronto	0	0	0	0
2010 Seoul	2	1	0	1
2011 Cannes	1	0	1	0
2012 Los Cabos	1	1	0	0
2013 St. Petersburg	0	0	0	0
2014 Brisbane	0	0	0	0
Total	8	2	1	5

Notes: The unit of analysis is the sentence.

Total refers to the number of sentences that contain at least one reference to a shock or vulnerability in the context of climate change. Inclusion terms for shock are shock, crisis, disaster, emergency/ emergencies, and extreme weather. Inclusion terms for vulnerability are vulnerability/vulnerabilities, threat/threaten, exposure, collapse, and extinction in the context of climate, weather, environment, or ecology. Exclusion terms: impact, mitigation, adaptation, variability, risk, resilience, stress, excessive emissions, dangerous, endangered.

Appendix G-3 World Oil Prices

Date	Crude Oil Price	Summit Month	Crude Oil Price
Jan-75	48.55	Nov-75	45.74
Jan-76	45.50	Jun-76	48.57
Jan-77	53.86	Apr-77	52.51
Jan-78	53.86	Jul-78	51.23
Jan-79	49.28	Jun-79	59.88
Jan-80	94.69	Jun-80	108.26
Jan-81	99.00	Jul-81	89.08
Jan-82	81.36	May-82	85.01
Jan-83	72.29	May-83	68.55
Jan-84	66.04	May-84	66.90
Jan-85	55.09	Apr-85	61.09
Jan-86	47.46	Apr-86	26.82
Jan-87	38.04	Jun-87	40.00
Jan-88	33.62	Jun-88	31.75
Jan-89	33.65	Jul-89	35.79
Jan-90	40.28	Jul-90	32.40
Jan-91	42.03	Jul-91	35.65
Jan-92	30.89	Jun-92	36.18
Jan-93	30.33	Jun-93	29.93
Jan-94	23.26	Jul-94	30.01
Jan-95	27.13	Jun-95	27.38
Jan-96	27.72	Jun-96	29.58
Jan-97	35.86	Jun-97	27.11
Jan-98	23.44	May-98	20.69
Jan-99	17.20	Jun-99	24.40
Jan-00	36.50	Jul-00	39.05
Jan-01	38.29	Jul-01	33.78
Jan-02	25.18	Jun-02	32.15
Jan-03	41.09	May-03	34.76
Jan-04	41.94	Jun-04	45.43
Jan-05	55.67	Jun-05	68.09
Jan-06	74.88	Jul-06	82.88
Jan-07	61.11	May-07	69.17
Jan-08	99.81	Jul-08	137.51
Jan-09	44.81	Jul-09	67.46
Jan-10	81.82	Jun-10	78.36
Jan-11	92.04	May-11	100.90
Jan-12	100.39	May-12	89.52
Jan-13	96.40	Jun-13	93.54
Jan-14	94.53	May-14	98.52

Note: World crude oil prices, US dollars. Source: http://www.macrotrends.net/1369/crude-oil-price-history-chart. For summits that take place in the first week of the month the price of oil from the month prior was used.

Appendix G-4 Oil Tanker, Rigs, and Related Accidents

Year	Spills greater than 700 tonnes
1975	22
1976	16
1977	17
1978	23
1979	34
1980	13
1981	7
1982	4
1983	13
1984	8
1985	8
1986	7
1987	10
1988	10
1989	13
1990	14
1991	7
1992	10
1993	11
1994	9
1995	3
1996	3
1997	10
1998	5
1999	6
2000	4
2001	3
2002	3
2003	4
2004	5
2005	4
2006	5
2007	4
2008	1
2009	2
2010	4
2011	1
2012	0
2013	3

Source: International Tanker Owners Pollution Federation Limited <http://www.itopf.com/fileadmin/data/Documents/Company_Lit/OilSpillstats_2013.pdf>

Appendix G-5 Natural Disasters

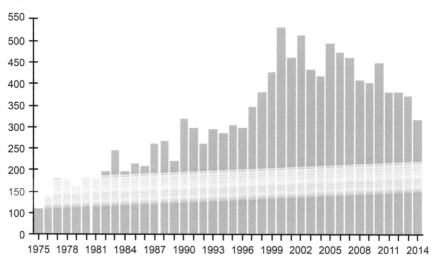

Figure G-5a   Number of Natural Disasters, 1975–2014
Source: EM-DAT: The OFDA/CRED International Disaster Database. www.emdat.be, Université Catholique de Louvain, Brussels, Belgium.

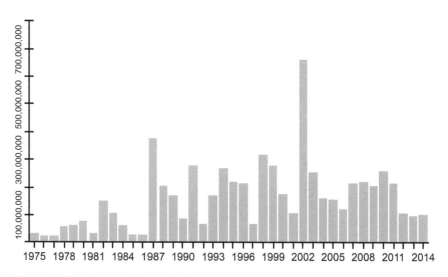

Figure G-5b   People Affected in Disasters, 1975–2014
Source: EM-DAT: The OFDA/CRED International Disaster Database. www.emdat.be, Université Catholique de Louvain, Brussels, Belgium.

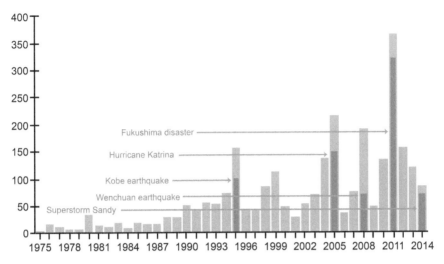

Figure G-5c   Damage Caused by Natural Disasters, 1975–2014
Source: EM-DAT: The OFDA/CRED International Disaster Database. www.emdat.be, Université Catholique de Louvain, Brussels, Belgium.

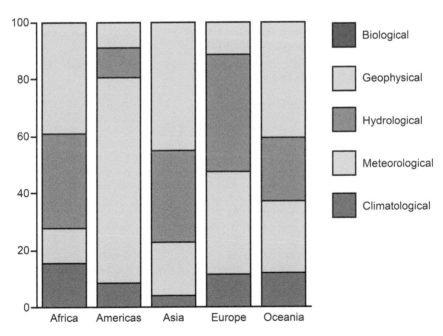

Figure G-5d   Average Annual Damage (US$ billion) Caused by Reported Natural Disasters, by Category 1990–2014
Source: EM-DAT: The OFDA/CRED International Disaster Database. www.emdat.be, Université Catholique de Louvain, Brussels, Belgium.

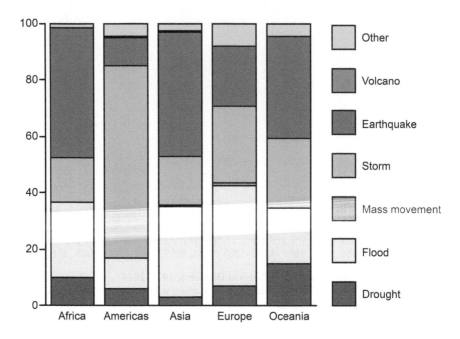

Figure G-5e   Average Annual Damages (US$ billion) Caused by Reported Natural Disasters, by type, 1990–2014

Source: EM-DAT: The OFDA/CRED International Disaster Database. www.emdat.be, Université Catholique de Louvain, Brussels, Belgium.

Appendix G-6 Nuclear Reactor Accidents

Date	Reactor	Deaths	Environmental Impact	Follow Up
1952	NRX, Chalk River, Canada	0	Nil	Closed 1992
1957	Windscale-1, United Kingdom	0	Widespread Contamination	Entombed, being demolished
1961	SL-1, United States	3	Very minor radioactive release	Decommissioned
1966	Fermi-1, United States	0	Nil	Closed 1972
1969	Lucens, Switzerland	0	Very minor radioactive release	Decommissioned
1975	Browns Ferry, United States	0	Nil	Repaired
1979	Three Mile Island–2, United States	0	Short term radiation dose to public	Clean-up program complete, in monitored storage stage of decommissioning
1980	Saint Laurent–A2, France	0	Minor radiation release	Repaired, decommissioned 1992
1986	Chernobyl-4, Ukraine	47	Major radiation release across Eastern Europe and Scandinavia	Entombed
1989	Vandellos-1, Spain	0	Nil	Decommissioned

| 1989 | Greifswald, Spain | 0 | Nil | Decommissioned |
| 2011 | Fukushima 1–3, Japan | 0 | Significant local contamination | Decommissioned |

Source: World Nuclear Association (2015). "Safety of Nuclear Power Reactors. Appendix 2: Serious Nuclear Reactor Accidents." http://www.world-nuclear.org/info/Safety-and-Security/Safety-of-Plants/Appendices/Safety-of-Nuclear-Power-Reactors—-Appendix/ (January 2015).

Appendix G-7 Greenhouse Gas Emissions by G7/8 Member, 1990–2011

Year	Canada	France	Germany	Italy	Japan	United Kingdom	United States	European Union	Subtotal	Russia	Total
1990	596	506	1,107	470	1,123	743	5,696	5,171	15,412	3,130	18,542
1991	600	533	1,078	470	1,135	751	5,668	5,118	15,353	3,080	18,433
1992	621	519	1,037	469	1,146	728	5,730	4,933	15,183	3,035	18,218
1993	631	499	1,026	461	1,143	710	5,859	4,803	15,132	2,796	17,928
1994	659	494	1,011	457	1,200	699	5,947	4,766	15,233	2,522	17,755
1995	680	503	1,004	480	1,216	682	6,007	4,780	15,352	2,418	17,770
1996	684	516	1,025	478	1,233	694	6,168	4,878	15,676	2,384	18,060
1997	688	507	989	483	1,228	666	6,345	4,777	15,683	2,254	17,937
1998	731	532	975	495	1,191	664	6,334	4,767	15,689	2,273	17,962
1999	706	524	937	500	1,231	655	6,363	4,686	15,602	2,287	17,889
2000	708	518	928	503	1,246	656	6,550	4,680	15,789	2,331	18,120
2001	928	470	955	508	1,202	657	6,448	4,698	15,866	2,170	18,036
2002	972	461	940	514	1,234	637	6,376	4,654	15,788	2,173	17,961
2003	996	465	931	530	1,238	645	6,443	4,741	15,989	2,253	18,242
2004	979	463	932	537	1,236	642	6,548	4,750	16,087	2,196	18,283
2005	987	464	901	536	1,243	635	6,538	4,706	16,010	2,215	18,225
2006	838	511	889	533	1,190	637	6,432	4,771	15,801	2,105	17,906
2007	860	505	858	523	1,225	624	6,542	4,721	15,858	2,109	17,967
2008	854	502	862	507	1,136	612	6,374	4,627	15,474	2,220	17,694
2009	801	481	799	456	1,073	562	5,997	4,294	14,463	2,050	16,513
2010	842	488	827	463	1,120	579	6,254	4,386	14,959	2,134	17,093
2011	847	463	806	458	1,170	541	6,135	4,263	14,683	2,217	16,900

Note: All figures refer to million metric tons of carbon dioxide equivalent. Figures include land-use change and forestry. Figures for the European Union include the 28 countries that were members in 2011. Russia joined the Group of Eight in 1998.
Source: CAIT Climate Data Explorer (2015). "Total GHG Emissions Including Land-Use Change and Forestry," January 31. World Resources Institute. http://cait2.wri.org/wri.

Appendix G-8 Greenhouse Gas Emissions by Group of Five Member, 1990–2011

Year	Brazil	China	India	Mexico	South Africa	Subtotal	Russia	Total
1990	1,739	3,047	1,035	434	331	6,586	3,130	9,716
1991	1,749	3,202	1,085	463	327	6,826	3,080	9,906
1992	1,757	3,362	1,120	469	322	7,030	3,035	10,065
1993	1,769	3,614	1,148	473	328	7,332	2,796	10,128
1994	1,783	3,778	1,193	499	334	7,587	2,522	10,109
1995	1,806	4,075	1,261	487	351	7,980	2,418	10,398
1996	1,829	4,151	1,315	511	361	8,167	2,384	10,551
1997	1,850	4,133	1,367	531	376	8,257	2,254	10,511
1998	1,860	4,221	1,387	556	383	8,407	2,273	10,680
1999	1,871	4,139	1,462	555	366	8,393	2,287	10,680
2000	1,884	4,421	1,505	577	373	8,760	2,331	11,091
2001	2,075	4,554	1,414	582	359	8,984	2,170	11,154
2002	2,116	4,846	1,458	598	371	9,389	2,173	11,562
2003	2,149	5,536	1,497	615	398	10,195	2,253	12,448
2004	2,209	6,300	1,586	629	416	11,140	2,196	13,336
2005	2,257	6,966	1,648	657	410	11,938	2,215	14,153
2006	1,844	7,656	1,799	671	413	12,383	2,105	14,488
2007	1,856	8,139	1,910	694	440	13,039	2,109	15,148
2008	1,874	8,350	2,019	704	469	13,416	2,220	15,636
2009	1,400	8,792	2,225	690	451	13,558	2,050	15,608
2010	1,393	9,387	2,304	706	459	14,249	2,134	16,383
2011	1,419	10,260	2,358	723	457	15,217	2,217	17,434

Note: All figures refer to million metric tons of carbon dioxide equivalent. Figures include land-use change and forestry.
Source: CAIT Climate Data Explorer (2015). "Total GHG Emissions Including Land-Use Change and Forestry," January 31. World Resources Institute. http://cait2.wri.org/wri.

Appendix H-1 Reports Published by the Intergovernmental Panel on Climate Change

IPCC Report	Date	Working Group I	Working Group II	Working Group III
First Assessment Report	1990	Scientific Assessment of Climate Change	Impacts Assessment of Climate Change	The IPCC Response Strategies
Supplementary Report	1992			
Second Assessment Report	1996	The Science of Climate Change	Impacts, Adaptations, and Mitigation of Climate Change: Scientific-Technical Analyses	Economic and Social Dimensions of Climate Change

Third Assessment Report	2001	Scientific Aspects of Climate	Vulnerability, Consequence, and Options	Limitation and Mitigation Options
Fourth Assessment Report	2007	The Physical Science Basis	Impacts, Adaptation, and Vulnerability	Mitigation of Climate Change
Fifth Assessment Report	2014	The Physical Science Basis	Impacts, Adaptation, and Vulnerability	Mitigation of Climate Change

Appendix H-2 Conferences of the Parties to the United Nations Framework Convention on Climate Change, 1995–2014

Conference of the Parties Session	Location	Date
1	Berlin, Germany	March 28–April 7, 1995
2	Geneva, Switzerland	July 8–19, 1996
3	Kyoto, Japan	December 1–10, 1997
4	Buenos Aires, Argentina	November 2–13, 1998
5	Bonn, Germany	October 25–November 5, 1999
6	The Hague, Netherlands	November 13–24, 2000
6 (resumed)	Bonn, Germany	July 16–27, 2001
7	Marrakech, Morocco	October 29–November 9, 2001
8	New Delhi, India	October 23–November 1, 2002
9	Milan, Italy	December 1–12, 2003
10	Buenos Aires, Argentina	December 6–17, 2004
11	Montreal, Canada	November 28–December 9, 2005
12	Nairobi, Kenya	November 6–17, 2006
13	Bali, Indonesia	December 3–14, 2007
14	Poznan, Poland	December 1–12, 2008
15	Copenhagen, Denmark	December 7–18, 2009
16	Cancun, Mexico	November 29–December 10, 2010
17	Durban, South Africa	November 28–December 9, 2011
18	Doha, Qatar	November 26–December 7, 2012
19	Warsaw, Poland	November 11–22, 2013
20	Lima, Peru	December 1–12, 2014

Appendix H-3 United Nations Summits Related to Climate Change

Date	Summit	Location
June 3–14, 1992	United Nations Conference on Environment and Development (Earth Summit)	Rio de Janeiro
June 23–27, 1997	Earth Summit Plus Five[a]	New York
August 26–September 4, 2002	World Summit on Sustainable Development	Johannesburg
December 7–18, 2009	United Nations Climate Change Conference	Copenhagen
June 20–22, 2012	United Nations Conference on Sustainable Development	Rio de Janeiro

Note: [a] Attended by the leaders of Italy, Romania, Russia, and Argentina.

Appendix I Overall Capability of G20 Members by Gross Domestic Product

Member	1980	1981	1982	1983	1984	1985	1986	1987	1988	1989	1990	1991
Argentina	209.03	169.77	84.30	104.00	116.77	88.19	106.05	108.73	127.36	81.71	141.35	189.61
Australia	163.73	189.34	187.97	180.35	198.10	175.24	182.37	214.15	272.44	308.27	323.44	324.18
Brazil	148.92	171.14	182.97	146.70	145.99	231.76	268.85	292.63	326.90	448.77	465.01	407.73
Canada	274.37	306.79	314.17	340.61	354.23	362.96	376.39	430.12	508.32	566.84	594.61	610.39
China	303.37	286.98	281.28	301.80	310.69	307.02	297.59	323.97	404.15	451.31	390.28	409.17
France	691.26	608.57	577.68	552.93	523.31	547.83	759.86	918.82	1,003.15	1,007.96	1,247.35	1,249.64

Member	1980	1981	1982	1983	1984	1985	1986	1987	1988	1989	1990	1991
Germany	826.14	695.07	671.16	669.57	630.85	639.70	913.64	1,136.93	1,225.73	1,216.80	1,547.03	1,815.06
India	181.42	195.86	202.86	222.05	215.56	237.62	252.75	283.75	299.65	300.19	326.61	274.84
Indonesia	86.31	96.35	98.92	89.66	91.81	91.53	83.92	79.51	88.28	100.87	112.77	127.44
Italy	470.04	426.26	421.27	437.17	431.92	446.03	631.72	792.88	878.45	913.63	1140.24	1204.45
Japan	1,086.99	1,201.47	1116.84	1,218.11	1,294.61	1,384.53	2,051.06	2,485.24	3,015.39	3,017.05	3,103.70	3,536.80
Korea	64.39	72.40	77.52	85.96	94.95	98.50	113.74	143.38	192.11	246.23	270.41	315.58
Mexico	234.95	301.76	218.99	178.53	210.54	223.42	154.69	169.64	207.53	252.91	298.46	357.80
Russia	N/A	N/A	N/A	N/A	N/A	N/A	N/A	N/A	N/A	N/A	N/A	N/A
Saudi Arabia	163.97	183.56	152.59	128.60	119.05	103.68	86.71	85.41	87.96	95.02	116.69	131.83
South Africa	80.55	82.80	75.94	84.69	74.94	57.27	65.42	85.79	92.24	95.98	112.00	120.24
Turkey	94.26	95.50	86.77	82.91	80.64	90.58	100.48	115.99	121.90	144.09	202.25	203.49
United Kingdom	542.45	520.04	492.33	466.36	441.03	468.96	570.88	704.09	855.78	865.96	1,024.56	1,069.91
United States	2,862.48	3,210.93	3,345.00	3,638.13	4,040.70	4,346.75	4,590.13	4,870.20	5,252.63	5,657.65	5,979.55	6,174.03
European Union	3,654.43	3,270.43	3,157.76	3,072.97	2,948.35	3,053.17	4,119.16	5,052.17	5,652.44	5,748.15	7,047.07	7,498.98

Member	1992	1993	1994	1995	1996	1997	1998	1999	2000	2001	2002	2003
Argentina	228.79	236.52	257.36	258.22	272.08	292.76	298.93	283.69	284.47	269.05	102.74	129.54
Australia	317.65	309.02	353.16	379.72	425.18	426.76	380.56	412.14	399.54	377.36	423.56	539.11
Brazil	390.59	438.30	546.49	769.74	840.05	871.52	844.13	586.92	644.28	554.41	505.71	552.24
Canada	591.33	575.16	575.98	602.00	626.97	650.99	631.45	661.25	724.91	715.44	734.65	865.90
China	488.22	613.22	559.22	727.95	856.08	952.65	1,019.48	1,083.28	1,198.48	1,324.81	1,453.83	1,640.96
France	1,375.83	1,298.40	1,370.63	1,573.08	1,573.13	1,423.13	1,470.90	1,457.07	1,331.59	1,340.27	1,458.20	1,796.68
Germany	2,068.96	2,008.55	2,152.74	2,525.02	2,437.81	2,159.87	2,181.16	2,133.84	1,891.93	1,882.51	2,013.69	2,428.45
India	293.26	284.19	333.01	366.60	399.79	423.19	428.77	453.66	476.35	487.80	510.29	590.97
Indonesia	138.32	158.01	176.89	202.13	227.37	215.75	95.45	140.00	165.02	160.45	195.66	234.86
Italy	1278.10	1027.75	1,060.06	1,132.36	1,266.70	1,199.96	1,226.17	1,209.70	1,107.25	1,124.67	1,229.51	1,517.40
Japan	3,852.79	4,414.96	4,850.35	5,333.93	4,706.19	4,324.28	3,914.58	4,432.60	4,731.20	4,159.86	3,980.82	4,302.94
Korea	338.17	372.21	435.59	531.14	573.00	532.24	357.51	461.81	533.39	504.58	575.93	643.76

Appendix I   Continued

Member												
Mexico	414.93	504.07	527.29	343.78	397.29	480.39	501.96	566.17	671.93	709.94	705.51	700.24
Russia	85.59	183.82	276.90	313.45	391.78	404.95	271.04	195.91	259.70	306.58	345.13	430.29
Saudi Arabia	136.67	137.41	139.65	147.94	163.43	170.88	151.96	161.17	188.69	183.26	188.80	214.86
South Africa	130.53	130.45	135.82	151.12	143.83	148.84	134.22	133.11	132.96	118.56	111.36	168.22
Turkey	213.78	242.47	174.99	227.81	244.39	255.65	269.53	249.82	266.44	195.55	232.28	303.26
United Kingdom	1,112.86	998.35	1,080.84	1,181.01	1,243.17	1,384.54	1,477.97	1,503.12	1,480.15	1,471.10	1,614.41	1,862.27
United States	6,539.28	6,868.70	7,308.70	7,664.05	8,100.15	8,608.48	9,089.13	9,353.50	9,951.48	10,286.18	10,642.30	11,142.23
European Union	8,210.49	7,486.03	7,970.05	9,237.19	9,427.03	8,889.46	9,198.02	9,155.80	8,503.79	8,587.79	9,392.07	11,430.51

Member	2004	2005	2006	2007	2008	2009	2010	2011	2012	2013	2014
Argentina	153.02	183.00	214.03	262.09	328.13	310.35	369.99	447.64	472.82	501.24	523,966.00
Australia	654.98	732.10	777.87	945.60	1,054.59	991.85	1,245.31	1,488.22	1,585.96	1,651.42	1,709.62
Brazil	663.55	881.75	1,089.16	1,366.22	1,650.39	1,622.31	2,142.93	2,492.91	2,449.76	2,520.62	2,690.75
Canada	992.23	1,133.76	1,278.61	1,424.07	1,502.68	1,337.58	1,577.04	1,736.87	1,804.58	1,869.74	1,933.93
China	1,931.65	2,256.92	2,712.92	3,494.24	4,519.95	4,990.53	5,930.39	7,298.15	7,991.74	8,777.20	9,641.85
France	2,055.36	2,137.95	2,259.58	2,586.77	2,842.55	2,631.92	2,562.76	2,776.32	2,712.03	2,786.98	2,884.65
Germany	2,729.92	2,137.95	2,905.45	3,328.59	3,640.73	3,307.20	3,286.45	3,577.03	3,478.77	3,581.13	3,664.33
India	688.74	808.67	908.47	1,152.81	1,251.81	1,253.98	1,597.95	1,676.14	1,779.28	1,961.56	2,163.54
Indonesia	256.84	285.74	364.28	432.19	510.96	538.70	708.35	845.68	928.27	1,055.00	1,214.43
Italy	1,737.80	1,789.38	1,874.72	2,130.24	2,318.16	2,116.63	2,060.89	2,198.73	2,066.93	2,090.25	2,118.57
Japan	4,655.82	4,571.87	4,356.75	4,356.35	4,849.19	5,035.14	5,488.42	5,869.47	5,981.00	6,060.83	6,207.67
Korea	721.98	844.87	951.77	1,049.24	931.41	834.06	1,014.89	1,116.25	1,163.53	1,243.10	1,333.31
Mexico	759.56	848.55	951.79	1,036.32	1,094.03	881.84	1,035.40	1,154.78	1,207.82	1,277.37	1,349.22
Russia	591.18	763.70	989.93	1,299.70	1,660.85	1,222.69	1,487.29	1,850.40	2,021.90	2,310.82	2,473.68
Saudi Arabia	250.67	315.76	356.20	385.20	476.94	377.20	451.39	577.60	651.65	666.81	682.99
South Africa	219.43	246.96	261.18	285.81	274.19	284.24	363.48	408.07	419.93	438.85	460.33
Turkey	392.21	482.69	529.19	649.13	730.32	614.42	734.59	778.09	817.30	878.05	952.62

Member	1984	1985	1986	1987	1988	1989	1990	1991	1992	1993	1994
United Kingdom	2,202.50	2,283.31	2,448.11	2,813.95	2,657.31	2,180.65	2,263.10	2,417.57	2,452.69	2,577.74	2,705.96
United States	11,853.25	12,622.95	13,377.20	14,028.68	14,291.55	13,938.93	14,526.55	15,094.03	15,609.70	16,221.38	16,940.57
European Union	13,185.54	13,773.20	14,689.54	16,994.15	18,341.54	16,360.16	16,259.00	17,577.69	17,070.01	17,590.83	18,162.20

Source: International Monetary Fund (2012). "World Economic Outlook Database." Washington DC, April. http://www.imf.org/external/pubs/ft/weo/2012/01/weodata/index.aspx (January 2015).

Appendix J Common Principles: Polity IV Authority Trends, 1975–2013

Member	1975	1976	1977	1978	1979	1980	1981	1982	1983	1984	1985	1986	1987	1988	1989	1990	1991	1992	1993	1994
Argentina	6	−9	−9	−9	−9	−9	−8	−8	8	8	8	8	8	8	7	7	7	7	7	7
Australia	10	10	10	10	10	10	10	10	10	10	10	10	10	10	10	10	10	10	10	10
Brazil	−4	−4	−4	−4	−4	−4	−4	−3	−3	−3	7	7	7	8	8	8	8	8	8	8
Canada	10	10	10	10	10	10	10	10	10	10	10	10	10	10	10	10	10	10	10	10
China	−8	−7	−7	−7	−7	−7	−7	−7	−7	−7	−7	−7	−7	−7	−7	−7	−7	−7	−7	−7
France	8	8	8	8	8	8	8	8	8	8	8	9	9	9	9	9	9	9	9	9
Germany	10	10	10	10	10	10	10	10	10	10	10	10	10	10	10	10	10	10	10	10
India	7	7	8	8	8	8	8	8	8	8	8	8	8	8	8	8	8	8	8	8
Indonesia	−7	−7	−7	−7	−7	−7	−7	−7	−7	−7	−7	−7	−7	−7	−7	−7	−7	−7	−7	−7
Italy	10	10	10	10	10	10	10	10	10	10	10	10	10	10	10	10	10	10	10	10
Japan	10	10	10	10	10	10	10	10	10	10	10	10	10	10	10	10	10	10	10	10
Korea	−8	−8	−8	−8	−8	−8	−5	−5	−5	−5	−5	−5	−88	6	6	6	6	6	6	6
Mexico	−6	−6	−3	−3	−3	−3	−3	−3	−3	−3	−3	−3	−3	0	0	0	0	0	0	4
Russia	−7	−7	−7	−7	−7	−7	−7	−7	−7	−7	−7	−7	−7	−6	−4	0	0	5	3	3
Saudi Arabia	−10	−10	−10	−10	−10	−10	−10	−10	−10	−10	−10	−10	−10	−10	−10	−10	−10	−10	−10	−10
South Africa	4	4	4	4	4	4	4	4	4	4	4	4	4	4	4	5	5	−88	−88	9
Turkey	9	9	9	9	9	−5	−5	−5	7	7	7	7	7	7	9	9	9	9	8	8
United Kingdom	10	10	10	10	10	10	10	10	10	10	10	10	10	10	10	10	10	10	10	10
United States	10	10	10	10	10	10	10	10	10	10	10	10	10	10	10	10	10	10	10	10

Appendix J Continued

Member	1995	1996	1997	1998	1999	2000	2001	2002	2003	2004	2005	2006	2007	2008	2009	2010	2011	2012	2013
Argentina	7	7	7	7	7	8	8	8	8	8	8	8	8	8	8	8	8	8	8
Australia	10	10	10	10	10	10	10	10	10	10	10	10	10	10	10	10	10	10	10
Brazil	8	8	8	8	8	8	8	8	8	8	8	8	8	8	8	8	8	8	8
Canada	10	10	10	10	10	10	10	10	10	10	10	10	10	10	10	10	10	10	10
China	−7	−7	−7	−7	−7	−7	−7	−7	−7	−7	−7	−7	−7	−7	−7	−7	−7	−7	−7
France	9	9	9	9	9	9	9	9	9	9	9	9	9	9	9	9	9	9	9
Germany	10	10	10	10	10	10	10	10	10	10	10	10	10	10	10	10	10	10	10
India	9	9	9	9	9	9	9	9	9	9	9	9	9	9	9	9	9	9	9
Indonesia	−7	−7	−7	−5	6	6	6	6	6	6	8	8	8	8	8	8	8	8	8
Italy	10	10	10	10	10	10	10	10	10	10	10	10	10	10	10	10	10	10	10
Japan	10	10	10	10	10	10	10	10	10	10	10	10	10	10	10	10	10	10	10
Korea	6	6	6	8	8	8	8	8	8	8	8	8	8	8	8	8	8	8	8
Mexico	4	4	6	6	6	6	8	8	8	8	8	8	8	8	8	8	8	8	8
Russia	3	3	3	3	3	3	6	6	6	6	6	6	4	4	4	4	4	4	4
Saudi Arabia	−10	−10	−10	−10	−10	−10	−10	−10	−10	−10	−10	−10	−10	−10	−10	−10	−10	−10	−10
South Africa	9	9	9	9	9	9	9	9	9	9	9	9	9	9	9	9	9	9	9
Turkey	8	8	7	7	7	7	7	7	7	7	7	7	7	7	7	7	9	9	9
United Kingdom	10	10	10	10	10	10	10	10	10	10	10	10	10	10	10	10	10	10	10
United States	10	10	10	10	10	10	10	10	10	10	10	10	10	10	10	10	10	10	10

Notes: The "Polity Score" refers to the regime authority spectrum, which spans fully institutionalized autocracies through mixed or incoherent authority regimes to fully institutionalized democracies on a 21-point scale from −10 (hereditary monarchy) to +10 (consolidated democracy). Excludes the European Union.

Prior to 1990 figures for West Germany were used.

Prior to 1992 figures for USSR were used.

−88 indicates a transition period.

Source: Polity IV Project: Political Regime Characteristics and Transitions, 1800–2013. Monty G. Marshall, Ted Robert Gurr, and Keith Jaggers, Center for Systemic Peace, Vienna VA. http://www.systemicpeace.org/inscrdata.html (January 2015).

Appendix K-1 G7/8 Hosts' Electoral Shadow of the Future

Summit	Leader	Next Election	Re-election Results	
			Leader, % vote	Party, Seats
1975 Nov 15–17	Valéry d'Estaing, right	Apr 26/May 10, 1981	François Mitterrand, 51.76%	Left, PS=283/491, C=303/491
1976 Jun 27–28	Gerald Ford, right	Nov 2, 1976	Jimmy Carter, 50.1%	Left, Dem=292/435
1977 May 7–8	James Callaghan, left	May 3, 1979	Margaret Thatcher, 43.9%	Right, Cons=339/635
1978 Jul 16–17	Helmut Schmidt, left	Oct 5, 1980	Helmet Schmidt, N/A	Left, SPD=218/497, C=271/497
1979 Jun 28–29	Masayoshi Ohira, right	Oct 7, 1979	Masayoshi Ohira, N/A	Right, LDP=248/511, C=~256/511
1980 Jun 22–23	Francesco Cossiga, left	Jun 26, 1983	Arnaldo Forlani, N/A	Left, N/A
1981 Jul 20–21	Pierre Trudeau, left	Sep 4, 1984	Brian Mulroney, 50.03%	Right, Cons=211/282
1982 Jun 4–6	François Mitterrand, left	Apr 24/May 8, 1988	François Mitterrand, 54.02%	Left, PS=260/575 C=333/575
1983 May 28–30	Ronald Reagan, right	Nov 6, 1984	Ronald Reagan, 58.8%	Left, Dem=253/435
1984 Jun 7–9	Margaret Thatcher, right	Jun 11, 1987	Margaret Thatcher, 42.2%	Right, Cons=376/650
1985 May 2–4	Helmut Kohl, Right	Jan 25, 1987	Helmut Kohl, N/A	Right, CDU=174/497, C=169/497
1986 May 4–6	Yasuhiro Nakasone, right	Jul 6, 1986	Yasuhiro Nakasone, N/A	Right, LDP=304/502
1987 Jun 8–10	Amintore Fanfani, left	Jun 14, 1987	Giovanni Goria, N/A	Left, N/A
1988 Jun 19–21	Brian Mulroney, right	Nov 21, 1988	Brian Mulroney, 43.02%	Right, Cons=169/295
1989 Jul 14–16	François Mitterrand, left	Apr 23/May 7, 1995	Jacques Chirac, 52.64%	Right, RPR=242/577 C=485/577
1990 Jul 9–11	George H. Bush, right	Nov 3, 1992	Bill Clinton, 43.0%	Left, Dem=256/435
1991 Jul 15–17	John Major, right	Apr 9, 1992	John Major, 41.9%	Right, Cons=336/651
1992 Jul 6–8	Helmut Kohl, right	Oct 16, 1994	Helmut Kohl, N/A	Right, CDU=244/672 C=341/672
1993 Jul 7–9	Kiichi Miyazawa, right	Jul 18, 1993	Morihiro Hosokawa, N/A	Left, JNP=35/512 C=~274/512
1994 Jul 8–10	Silvio Berlusconi, right	Apr 21, 1996	Romano Prodi, N/A	Left, Olive Tree=285/630, C=620/630
1995 Jun 15–17	Jean Chrétien, left	Jun 2, 1997	Jean Chrétien, 38.46%	Left, Lib=155/301

Appendix K-1 Continued

Summit	Leader	Next Election	Re-election Results	
			Leader, % vote	Party, Seats
1996 Jun 27–29	Jacques Chirac, right	Apr 21/May 5, 2002	Jacques Chirac, 82.21%	Right, UMP=357/577 C=399/577
1997 Jun 20–22	Bill Clinton, left	Nov 7, 2000	George W. Bush, 47.9%	Right, Rep=223/435
1998 May 15–17	Tony Blair, left	Jun 7, 2001	Tony Blair, 40.7%	Left, Labour=413/659
1999 Jun 18–20	Gerhard Schroeder, left	Sep 22, 2002	Gerhard Schroeder, 38.5%	Left, SPD=251/603 C=306/603
2000 Jul 21–23	Yoshiro Mori, right	Nov 9, 2003	Junichiro Koizumi, 43.85%	Right, LDP=237/480, C=275/480
2001 Jul 20–22	Silvio Berlusconi, right	Apr 9/10, 2006	Romano Prodi, 49.80%	Left, Olive Tree=220/630, C=348/630
2002 Jun 26–27	Jean Chrétien, left	Jun 28, 2004	Paul Martin, 36.73%	Left, Lib=135/308
2003 Jun 1–3	Jacques Chirac, right	Apr 21–22/May 5–6, 2007	Nicolas Sarkozy, 53.06%	Right, UMP=313/577 C=345/577
2004 Jun 8–10	George W. Bush, right	Nov 2, 2004	George W. Bush, 50.7%	Right, Rep=232/435
2005 Jul 6–8	Tony Blair, left	By Jun 3, 2010	N/A	Left, Labour=356/650
2006 Jul 15–17	Vladimir Putin, right	Mar 2, 2008	Dmitry Medvedev, 71%	Right, C=353/450
2007 Jun 6–8	Angela Merkel, right	Sep 27, 2009	Angela Merkel, 33.8%	Right, CDU=194/622
2008 Jul 7–9	Yasuo Fukuda, right	By Sep 2009	Taro Aso, N/A	Right, LDP=296/480, C=317/480
2009 Jul 8–10	Silvio Berlusconi, right	Variable	N/A	Right, PdL=276/630, C=344/630
2010 Jun 25–27	Stephen Harper, right	May 2, 2011	Stephen Harper, 39.62%	Right, Con=166/308
2012 May 18–19	Barack Obama, left	Nov 6, 2012	Barack Obama, 51.1%	Dem = 332/538
2013 Jun 17–18	David Cameron, right	May 7, 2015	N/A	N/A
2014 Jun 4–5	N/A	N/A	N/A	N/A

Notes: N/A = not applicable.
C=coalition; CDU = Christlich-Demokratische Union (Christian Democratic Union); Cons = Conservative Party; Dem = Democratic Party; LDP = Liberal Democratic Party; LIB = Liberal Party; JNP = Japan New Party; PdL = Popolo della Libertà (People of Freedom); PS = Parti socialiste; Rep = Republican Party; RPR = Rassemblement pour la République (Rally for the Republic); SPD = Sozialdemokratische Partei Deutschlands (Social Democratic Party of Germany); UMP = Union pour un mouvement populaire (Union for a Popular Movement).

Appendix K-2 G7/8 Leaders' Competence: Previous Relevant Portfolios

Leader	Total	Finance	Foreign Affairs	Health	Environment	Trade
Canada	4					
Pierre Trudeau (1968–1979)[a]	0					
Joe Clark (1979–1980)	0					
Pierre Trudeau (1980–1984)[a]	0					
Brian Mulroney (1984–1993)	0					
Kim Campbell (1993)	0					
Jean Chrétien (1993–2000)	3	1	1			1
Paul Martin (2003–2006)	1	1				
Stephen Harper (2006–present)	0					
France	2					
Valéry d'Estaing (1974–1981)	1	1				
François Mitterrand (1981–1995)	0					
Jacques Chirac (1995–2007)	0					
Nicolas Sarkozy (2007–2012)	1	1				
François Hollande (2012–present)	0					
Germany	2					
Helmut Schmidt (1974–1982)	1	1				
Helmut Kohl (1982–1998)	0					
Gerhard Schroeder (1998–2005)	0					
Angela Merkel (2005–present)	1				1	
Italy	21					
Aldo Moro (1974–1976)	1		1			
Giulio Andreotti (1976–1979)[a]	2	1	1			
Francesco Cossiga (1979–1980)	0					
Arnaldo Forlani (1980–1981)	1		1			
Giovanni Spadolini (1981–1982)	0					
Amintore Fanfani (1982–1983)[a]	1		1			
Bettino Craxi (1983–1987)	0					
Amintore Fanfani (1987)[a]	0					
Giovanni Goria (1987–1988)	1	1				
Ciriaco De Mita (1988–1989)	1					1
Giulio Andreotti (1989–1992)[a]	2	1	1			
Giuliano Amato (1992–1993)[a]	0					
Carlo Azeglio Ciampi (1993–1994)	0					
Silvio Berlusconi (1994–1995)[a]	2		1	1		
Lamberto Dini (1995–1996)	1		1			
Romano Prodi (1996–1998)[a]	0					

Appendix K-2 Continued

Leader	Total	Finance	Foreign Affairs	Health	Environment	Trade
Massimo D'Alema (1998–2000)	1	1				
Giuliano Amato (2000–2001)[a]	2	1	1			
Silvio Berlusconi 2001–2006)[a]	2		1	1		
Romano Prodi (2006–2008)[a]	0					
Silvio Berlusconi (2008–2011)[a]	2		1	1		
Mario Monti (2011–2013)	2	1	1			
Enrico Letta (2013–2014)	0					
Matteo Renzi (2014–present)	0					
Japan	26					
Takeo Miki (1974–1976)	0					
Takeo Fukuda (1976–1978)	2	1	1			
Masayoshi Ohira (1979–1980)	2	1	1			
Zenko Suzuki (1980–1982)	1			1		
Yasuhiro Nakasone (1982–1987)	1					1
Noboru Takeshita (1987–1989)	1	1				
Sosuke Uno (1989)	2		1			1
Toshiki Kaifu (1989–1991)	0					
Kiichi Miyazawa (1991–1993)	2	1				1
Morihiro Hosokawa (1993–1994)	0					
Tsutomi Hata (1994)	3	1	1			1
Tomiichi Murayama (1994–1996)	0					
Ryutaro Hashimoto (1996–1998)	3	1		1		1
Keizo Obuchi (1998–2000)	1		1			
Yoshiro Mori (2000–2001)	1					1
Junichiro Koizumi (2001–2006)	2		1	1		
Shinzo Abe (2006–2007)[a]	0					
Yasuo Fukuda (2007–2008)	0					
Taro Aso (2008–2009)	1		1			
Yukio Hatoyama (2009–2010)	2	1		1		
Naoto Kan (2010–2011)	1	1				
Yoshihiko Noda (2011–2012)	1	1				
Shinzo Abe (2012–present)[a]	0					
Russia	0					
Mikhail Gorbachev (1985–1991)	0					
Boris Yeltsin (1991–1999)	0					
Vladimir Putin (2000–2008)[a]	0					
Dmitry Medvedev (2008–2012)	0					
Putin (2012–present)[a]	0					

Leader	Total	Finance	Foreign Affairs	Health	Environment	Trade
United Kingdom	5					
Harold Wilson (1974–1976)	0					
James Callaghan (1976–1979)	2	1	1			
Margaret Thatcher (1979–1990)	0					
John Major (1990–1997)	2	1	1			
Tony Blair (1997–2007)	0					
Gordon Brown (2007–2010)	1	1				
David Cameron (2010–present)	0					
United States	0					
Gerald Ford (1974–77)	0					
Jimmy Carter (1977–1981)	0					
Ronald Reagan (1981–1989)	0					
George H. Bush (1989–1993)	0					
Bill Clinton (1993–2001)	0					
George W. Bush (2001–2008)	0					
Barack Obama (2009–present)	0					
Total	60	23	21	7	1	8

Notes: [a] Indicates that the individual served as leader more than once.
Includes individuals who attended G8 summits as leaders of their country. Numbers indicate whether a leader served in that ministry but does not account for time frame or number of times served. Leaders may have served in other portfolios not included. No leaders have served as development minister.

Appendix K-3 G20 Hosts' Electoral Shadow of the Future

Date	Leader	Next Election	Re-election Results	
			Leader, % vote	Party, Seats
2008 Nov 14–15	Barack Obama	Nov 6, 2012	Barack Obama, 51.1%	Democratic, left, 332
2009 Apr 1–2	Gordon Brown	May 6, 2010	David Cameron, 36.1%	Conservative, right, 307
2009 Sep 24–25	Barack Obama	Nov 6, 2012	Barack Obama, 51.1%	Democratic, left, 332
2010 Jun 26–27	Stephen Harper	May 2, 2011	Stephen Harper, 39.62%	Conservative, right, 166
2010 Nov 11–12	Kim Hwang-sik	Feb 26, 2013	Jung Hong-won	Conservative, right
2011 Nov 3–4	Nicolas Sarkozy	Apr 22, 2012	François Hollande, 51.6%	Left, Parti socialiste, 280
2012 Jun 18–19	Felipe Calderon	Jul 1, 2012	Enrique Peña Nieto, 38.2%	PRI, Centrist
2013 Sep 5–6	Dmitry Medvedev	Mar 4, 12	Vladimir Putin, 63.6%	United Russia, N/A
2014 Nov 15–16	Tony Abbott	By Jan 14, 2017	N/A	N/A

Note: PRI = Partido Revolucionario Institucional (Institutional Revolutionary Party). N/A = not applicable.

Appendix K-4 G20 Leaders' Competence: Previous Portfolios

Leader	In Office	Total	Finance	Foreign Affairs	Health	Environment	Trade
Argentina							
Cristina Kirschner	2007–present	0	0	0	0	0	0
Australia							
Kevin Rudd	2007–2010, 2013	1	0	1	0	0	0
Julia Gillard	2010–2013	0	0	0	0	0	0
Tony Abbott	2013–present	1	0	0	1	0	0
Brazil							
Luiz Inácio Lula da Silva	2003–2010	0	0	0	0	0	0
Dilma Rousseff	2010–present	0	0	0	0	0	0
China							
Hu Jintao	2002–2012	0	0	0	0	0	0
Xi Jinping	2012–present	0	0	0	0	0	0
Canada							
Stephen Harper	2006–present	0	0	0	0	0	0
France							
Nicolas Sarkozy	2007–2012	1	1	0	0	0	0
François Hollande	2012–present	0	0	0	0	0	0
Germany							
Angela Merkel	2005–present	1	0	0	1	0	0
India							
Manmohan Singh	2004–2014	1	1	0	0	0	0
Narendra Modi	2014–present	0	0	0	0	0	0
Indonesia							
Susilo Yudhayono	2004–2014	0	0	0	0	0	0
Joko Widodo	2014–present	0	0	0	0	0	0
Italy							
Silvio Berlusconi	2008–2011	2	0	1	1	0	0
Mario Monti	2011–2013	2	1	1	0	0	0
Enrico Letta	2013–2014	0	0	0	0	0	0
Matteo Renzi	2014–present	0	0	0	0	0	0
Japan							
Taro Aso	2008–2009	2	1	1	0	0	0
Yukio Hatoyama	2009–2010	0	0	0	0	0	0
Naoto Kan	2010–2011	1	1	0	0	0	0

Leader	In Office	Total	Finance	Foreign Affairs	Health	Environment	Trade
Yoshihiko Noda	2011–2012	1	1	0	0	0	0
Shinzo Abe	2012–present	0	0	0	0	0	0
Korea							
Lee Myung-bak	2008–2013	0	0	0	0	0	0
Park Geun-hye	2013–present	0	0	0	0	0	0
Mexico							
Felipe Calderon	2006–2012	0	0	0	0	0	0
Enrique Pena Nieto	2012–present	0	0	0	0	0	0
Russia							
Dmitry Medvedev	2008–2012	0	0	0	0	0	0
Vladimir Putin	2012–present	0	0	0	0	0	0
Saudi Arabia							
King Abdullah bin Abdulaziz	2008–2015	0	0	0	0	0	0
King Salman	2015–present	0	0	0	0	0	0
South Africa							
Thabo Mbeki	1999–2008	0	0	0	0	0	0
Kgalema Mottanthe	2008–2009	0	0	0	0	0	0
Jacob Zuma	2009–2014	0	0	0	0	0	0
Turkey							
Recep Erdogan	2003–2014	0	0	0	0	0	0
Ahmet Davutoglu	2014–Present	1	0	1	0	0	0
United Kingdom							
Gordon Brown	2007–2010	1	1	0	0	0	0
David Cameron	2010–Present	0	0	0	0	0	0
United States							
George W. Bush	2001–2009	0	0	0	0	0	0
Barack Obama	2009–Present	0	0	0	0	0	0
Total		15	7	5	3	0	0

Appendix L-1 G7/8 Summit Attendees

Year	Total	G8	G5	MEM/MEF	G20	Country/IO
1975	6	6	0	0	0	0
1976	7	7	0	0	0	0
1977	8	8	0	0	0	0
1978	8	8	0	0	0	0
1979	8	8	0	0	0	0
1980	8	8	0	0	0	0
1981	8	8	0	0	0	0
1982	9	9	0	0	0	0
1983	8	8	0	0	0	0
1984	8	8	0	0	0	0
1985	8	8	0	0	0	0
1986	9	9	0	0	0	0
1987	9	9	0	0	0	0
1988	9	9	0	0	0	0
1989	8	8	0	0	0	0
1990	9	9	0	0	0	0
1991	9	9	0	0	0	0
1992	9	9	0	0	0	0
1993	9	9	0	0	0	0
1994	9	9	0	0	0	0
1995	9	9	0	0	0	0
1996	13	9	0	0	0	0/4[a]
1997	10	10	0	0	0	0
1998	9	9	0	0	0	0
1999	9	9	0	0	0	0
2000	9	9	0	0	0	0
2001	21	10	1[b]	0	0	6/4[c]
2002	15	10	1[b]	0	0	3/1[d]
2003	27	10	5	0	1[e]	8/4[f]
2004	22	10	1[b]	0	1[g]	5/5[h]
2005	26	9	5	0	0	6/6[i]
2006	24	10	5	0	0	0/9[j]
2007	27	9	5	0	0	6/7[k]
2008	25	9	5	3[l]	0	6/2[m]
2009	41	10	5	4[l]	2[n]	9/11[o]
2010	20	10	1[b]	0	0	9[p]
2011	22	10	1[b]	0	0	7/4[q]
2012	15	9	0	0	0	4/2[r]
2013	16	10	0	0	0	5/1[s]
2014	9	9	0	0	0	0

Notes: G5 = Group of Five (Brazil, China, India, Mexico, South Africa). MEM/MEF = Major Economies Meeting/Forum on Energy Security and Climate Change. IO = international organization.
[a] International Bank for Reconstruction and Development, International Monetary Fund, World Trade Organization, United Nations.

[b] South Africa.
[c] Algeria, Bangladesh, El Salvador, Mali, Nigeria, Senegal / United Nations, World Bank, World Health Organization, World Trade Organization.
[d] Algeria, Nigeria, Senegal / United Nations.
[e] Saudi Arabia.
[f] Algeria, Egypt, Morocco, Malaysia, Nigeria, Saudi Arabia, Senegal, Switzerland / International Monetary Fund, United Nations, World Bank, World Trade Organization.
[g] Turkey.
[h] Algeria, Ghana, Nigeria, Senegal, Uganda / Afghanistan, Bahrain, Iraq, Jordan, Yemen.
[i] Algeria, Ethiopia, Ghana, Nigeria, Senegal, Tanzania / African Union, International Energy Agency, International Monetary Fund, United Nations, World Bank, World Trade Organization.
[j] African Union; Commonwealth of Independent States; International Atomic Energy Agency; International Energy Agency; United Nations Educational, Scientific, and Cultural Organization; United Nations; World Bank; World Health Organization; World Trade Organization.
[k] Algeria, Egypt, Ethiopia, Ghana, Nigeria, Senegal / African Union, International Energy Agency, International Monetary Fund, Organisation for Economic Co-operation and Development, United Nations, World Bank, World Trade Organization.
[l] Australia, Indonesia, Korea.
[m] Algeria, Ethiopia, Ghana, Nigeria, Senegal, Tanzania / United Nations, World Bank.
[n] Spain, Turkey.
o Algeria, Angola, Denmark, Egypt, Ethiopia, Liberia, Netherlands, Nigeria, Senegal / African Union, Food and Agriculture Organization, International Energy Agency, International Fund for Agricultural Development, International Labour Organization, International Monetary Fund, Organisation for Economic Co-operation and Development, United Nations, World Bank, World Food Programme, World Trade Organization.
[p] Algeria, Colombia, Egypt, Ethiopia, Haiti, Jamaica, Malawi, Nigeria, Senegal.
[q] Algeria, Egypt, Equatorial Guinea, Ethiopia, Senegal, Tunisia / African Union, Arab League, New Partnership for Africa's Development, United Nations.
[r] Benin, Ethiopia, Ghana, Tanzania / African Development Bank, African Union.
[s] Ethiopia, Liberia, Libya, Senegal, Somalia / African Union.

Appendix L-2 G20 Summit Attendees

Summit	Total	Countries	International Orgs
2008 Washington	2	Netherlands, Spain	0
2009 London	8	Netherlands, Spain, Thailand (ASEAN)	Financial Stability Forum, International Monetary Fund, New Partnership for Africa's Development, United Nations, World Trade Organization
2009 Pittsburgh	9	Ethiopia (NEPAD), Netherlands, Spain, Thailand (ASEAN)	Financial Stability Board, International Monetary Fund, Organisation for Economic Co-operation and Development, United Nations, World Trade Organization
2010 Toronto	13	Ethiopia (NEPAD), Malawi, Netherlands, Nigeria, Spain, Vietnam	African Union, Association for Southeast Asian Nations, Financial Stability Board, International Labour Organization, Organisation for Economic Co-operation and Development, United Nations, World Trade Organization

Appendix L-2 Continued

Summit	Total	Countries	International Orgs
2010 Seoul	11	Singapore, Spain, Vietnam	African Union, Association of Southeast Asian Nations, Financial Stability Board, International Labour Organization, New Partnership for Africa's Development, Organisation for Economic Co-operation and Development, United Nations, World Trade Organization
2011 Cannes	13	Ethiopia, Singapore, Spain	African Union, Basel Committee on Banking Supervision, Cooperation Council for the Arab States of the Gulf, European Central Bank, Financial Stability Board, International Labour Organization, New Partnership for Africa's Development, Organisation for Economic Co-operation and Development, United Nations, World Trade Organization
2012 Los Cabos	12	Benin (African Union), Cambodia (ASEAN), Chile, Colombia, Ethiopia, Spain	Financial Stability Board, Food and Agriculture Organization, International Labour Organization, Organisation for Economic Co-operation and Development, United Nations, World Trade Organization
2013 St. Petersburg	11	Brunei (ASEAN), Ethiopia (AU), Singapore, Kazakhstan, Senegal, Spain, Switzerland	Financial Stability Board, Organisation for Economic Co-operation and Development, United Nations, World Trade Organization
2014 Brisbane	11	Mauritania (AU), Myanmar (ASEAN), New Zealand, Senegal, Singapore, Spain	Financial Stability Board, International Labour Organization, Organisation for Economic Co-operation and Development, United Nations, World Trade Organization

Bibliography

Abbott, Kenneth W., Robert Keohane, Andrew Moravcsik, et al. (2000). "The Concept of Legalization." *International Organization* 54(3): 401–20.

Afionis, Stavros (2009). "The Role of the G8/G20 in International Climate Change Negotiations." *In-Spire Journal of Law, Politics, and Societies* 4(2): 1–12. http://www.academia.edu/721847/The_Role_of_the_G-8_G-20_in_International_Climate_Change_Negotiations (January 2015).

Agence Europe (2007). "European Commission Expectations for Heiligendamm Summit." Brussels, June 1.

Agence France Presse (2007). "Japan Aims to Lead Post-Kyoto Climate Change Fight." Tokyo, March 20.

Agence France Presse (2007). "Merkel Says UN Report Shows Need for Action of Climate Change." April 6.

Agenzia ANSA (1994). "From Tokyo to Naples, the Stages: The Florence Meeting on the Environment." ANSA Dossier (folder), Rome.

Agidee, Yinka (2012). "Climate Change: Has the Momentum Slowed." In *The G8 Camp David Summit 2012: The Road to Recovery*, John J. Kirton and Madeline Koch, eds. London: Newsdesk, p. 185. http://www.g8.utoronto.ca/newsdesk/g8campdavid2012.pdf (January 2015).

Aguilar, Soledad (2007). "Elements for a Robust Climate Regime Post-2012: Options for Mitigation." *Review of European Community and International Environmental Law* 16(3): 356–67.

Aldy, Joseph E. and Robert N. Stavins (2008). "Climate Policy Architectures for the Post-Kyoto World." *Environment* 50(3): 7–17.

Anderson, Stephen O. and K. Madhava Sarma (2002). *Protecting the Ozone Layer: The United Nations History*. London: Earthscan.

Asahi (2006). July 18, p. 8.

Associated Press (2001). "US Won't Follow Climate Treaty Provisions, Whitman Says." *New York Times*, March 28. http://www.nytimes.com/2001/03/28/politics/28WHIT.html (January 2015).

Associated Press (2006). "Merkel Pledges to Make Climate Protection Key Part of Germany's G8, EU Presidencies." December 1.

Barnes, James (1994). *Promises, Promises! A Review: G7 Economic Summit Declarations on Environment and Development*. Washington DC: Friends of the Earth.

Barroso, José Manuel (2014). "Remarks by President Barroso on the Results of the G7 Summit in Brussels." Closing press conference, Brussels, June 5. http://www.g8.utoronto.ca/summit/2014brussels/barroso.html (January 2015).

Bayne, Nicholas (1997). "Impressions of the Denver Summit." G8 Research Group, Denver. http://www.g8.utoronto.ca/evaluations/1997denver/impression/index.html (January 2015).

Bayne, Nicholas (1998). "Impressions of the Birmingham Summit." May 19, G8 Research Group, Birmingham. http://www.g8.utoronto.ca/evaluations/1998birmingham/impression/index.html (January 2015).

Bayne, Nicholas (1999). "Continuity and Leadership in an Age of Globalisation." In *The G8's Role in the New Millennium*, Michael R. Hodges, John J. Kirton, and Joseph P. Daniels, eds. Aldershot: Ashgate, pp. 21–44.

Bayne, Nicholas (2000). "First Thoughts on the Okinawa Summit, 21–23 July 2000." July 27. http://www.g8.utoronto.ca/evaluations/2000okinawa/bayne.html (January 2015).

Bayne, Nicholas (2000). *Hanging In There: The G7 and G8 Summit in Maturity and Renewal*. Aldershot: Ashgate.

Bayne, Nicholas (2001). "Impressions of the Genoa Summit, 20–22 July 2001." July 28, G8 Research Group, Genoa. http://www.g8.utoronto.ca/evaluations/2001genoa/assess_summit_bayne.html (January 2015).

Bayne, Nicholas (2002). "Impressions of the Kananaskis Summit, 26–27 June 2002." Kananaskis, July 23. http://www.g8.utoronto.ca/evaluations/2002kananaskis/assess_baynea.html (January 2015).

Bayne, Nicholas (2003). "Impressions of the Evian Summit, 1–3 June 2003." June 3. http://www.g8.utoronto.ca/evaluations/2003evian/assess_bayne030603.html (January 2015).

Bayne, Nicholas (2004). "Impressions of the 2004 Sea Island Summit." June 29. http://www.g8.utoronto.ca/evaluations/2004seaisland/bayne2004.html (January 2015).

Bayne, Nicholas (2005). "Overcoming Evil with Good: Impressions of the Gleneagles Summit, 6–8 July 2005." G8 Research Group, July 18. http://www.g8.utoronto.ca/evaluations/2005gleneagles/bayne2005-0718.html (January 2015).

Bayne, Nicholas (2005). *Staying Together: The G8 Summit Confronts the 21st Century*. Aldershot: Ashgate.

Bayne, Nicholas (2010). *Economic Diplomat*. Durham UK: Memoir Club.

Bayne, Nicholas (2013). "The UK's Role in the G7 and G8." In *The UK Summit: The G8 at Lough Erne 2013*, John J. Kirton and Madeline Koch, eds. London: Newsdesk, pp. 40–41. http://www.g8.utoronto.ca/newsdesk/g8lougherne2013.pdf (January 2015).

BBC News (2006). "PM Praises Business Pledge on CO2." June 6. http://news.bbc.co.uk/2/hi/uk/5050774.stm (January 2015).

Bell, Ruth Greenspan (2006). "What to Do about Climate Change?" *Foreign Affairs* 85: 105–13. http://www.foreignaffairs.com/articles/61710/ruth-greenspan-bell/what-to-do-about-climate-change (January 2015).

Benedick, Richard E. (1991). *Ozone Diplomacy: New Directions in Safeguarding the Planet*. Cambridge MA: Harvard University Press.

Benoit, Bertrand (2007). "Bush Faces Isolation on Climate at G8 Talks." *Financial Times*, June 4, p. 1. http://www.ft.com/intl/cms/s/0/96e5bc2c-1237-11dc-b963-000b5df10621.html (January 2015).

Benoit, Bertrand, Jean Eaglesham, Fiona Harvey, et al. (2007). "Merkel Accepts Defeat in Bid to Win US Emissions Pledge." *Financial Times*, June 7, p. 6.

Benoit, Bertrand, Andrew Ward, and Hugh Williamson (2007). "Cheers All Round for 'Winner' Merkel." *Financial Times*, June 9, p. 6. http://www.ft.com/intl/cms/s/0/7274c3b6-15e7-11dc-a7ce-000b5df10621.html (January 2015).

Benoit, Bertrand and John Williamson (2007). "Merkel to Push Bush on Climate Change." *Financial Times*, June 4, p. 2. http://www.ft.com/intl/cms/s/0/e8f229d8-1237-11dc-b963-000b5df10621.html#axzz3Sm9i1Ukk (January 2015).

Bergsten, C. Fred and C. Randall Henning (1996). *Global Economic Leadership and the Group of Seven*. Washington DC: Institute for International Economics.

Bernstein, Steven (2000). "Ideas, Social Structure, and the Compromise of Liberal Environmentalism." *European Journal of International Relations* 6(4): 464–512. doi: 10.1177/1354066100006004002.

Bernstein, Steven (2001). *The Compromise of Liberal Environmentalism*. New York: Columbia University Press.

Biermann, Frank and Steffen Bauer, eds. (2005). *A World Environmental Organization: Solution or Threat for Effective International Environmental Governance*. Aldershot: Ashgate.

Black, Richard (2006). "UK Appoints 'Climate Ambassador'." June 8. http://news.bbc.co.uk/2/hi/science/nature/5057678.stm (January 2015).

Blair, Tony (2010). *A Journey*. London: Random House.

Blix, Hans (1986). "The Post-Chernobyl Outlook for Nuclear Power: A View on Responses to the Accident from an International Perspective." *IAEA Bulletin*, Autumn, pp. 9–12. http://www.iaea.org/Publications/Magazines/Bulletin/Bull283/28304780912.pdf (January 2015).

Bodansky, Daniel (2001). "The History of the Global Climate Change Regime." In *International Relations and Global Climate Change*, Urs Luterbacher and Detlef F. Sprinz, eds. Cambridge MA: MIT Press, pp. 23–39.

Brainard, Lael (2004). "G7/8 Oral History: Interview." G8 Research Group, February 6. http://www.g8.utoronto.ca/oralhistory/ (January 2015).

Briffa, Keith R., Thomas S. Bartholin, Dieter Eckstein, et al. (1990). "A 1,400-Year Tree-Ring Record of Summer Temperatures in Fennoscandia." *Nature* 346(6283): 434-39. doi: 10.1038/346434a0.

Brown, Paul, Patrick Wintour, and Michael White (2005). "Leaked G8 Draft Angers Green Groups." *Guardian*, May 28. http://www.theguardian.com/society/2005/may/28/environment.greenpolitics (January 2015).

Bueckert, Dennis (2005). "Martin Blamed for Cool U.S. Response to Climate Talks: Washington Angered by PM's Critical Comments." *Calgary Herald*, December 9, p. A11.

Burney, Derek (2005). *Getting It Done: A Memoir*. Montreal: McGill-Queen's University Press.

Busby, Joshua W. (2010). "After Copenhagen: Climate Governance and the Road Ahead." Working paper, August, Council on Foreign Relations, New York. http://www.cfr.org/climate-change/after-copenhagen/p22726 (January 2015).

BusinessGreen (2011). "Durban Climate Deal Impossible, say US and EU Envoys." *Guardian*, April 28. http://www.guardian.co.uk/environment/2011/apr/28/durban-climate-deal-impossible (January 2015).

CAFOD (2010). "CAFOD Responses to Seoul G20 Communiqué." Notice distributed at the 2010 Seoul Summit, Seoul, November 12. http://www.bond.org.uk/data/files/CAFOD_response_to_Seoul_G20_communique.doc (January 2015).

Calderón, Felipe (2011). "The Current Challenges for Global Economic Growth." December 13. http://www.g20.utoronto.ca/2012/2012-111213-calderon-en.html (January 2015).

Calderón, Felipe (2012). "Mexico's Vision of Innovation and Cooperation at the Los Cabos Summit." In *The G20 Mexico Summit 2012: The Quest for Growth and Stability*,

John J. Kirton and Madeline Koch, eds. London: Newsdesk, pp. 14–15. http://www.g8.utoronto.ca/newsdesk/loscabos/calderon.html (January 2015).

Cameron, David (2012). "In Fight for Open World, G8 Still Matters." *Globe and Mail*, November 20. http://www.theglobeandmail.com/report-on-business/economy/david-cameron-in-fight-for-open-world-g8-still-matters/article5508595/ (January 2015).

Campbell, Jennifer (2006). "Tamil Tigers Blacklisting Lauded." April 12.

Canada (1996, February). 1995 Canada's Year as G7 Chair: The Halifax Summit Legacy. G7/G8 Research Collection in the John W. Graham Library, Trinity College, University of Toronto. Locator ID: 1.14.

Canada (2000). "Government of Canada Action Plan 2000 on Climate Change." Ottawa.   http://env.chass.utoronto.ca/env200y/ESSAY2001/gofcdaplan_eng2.pdf (January 2015).

Canada. Department of Foreign Affairs and International Trade (1995). "Background Information: The   Environment."   Ottawa.   http://www.g8.utoronto.ca/environment/1995hamilton/hamenvi7.html (January 2015).

Canada. Department of Foreign Affairs and International Trade (1995). "The Halifax Summit, June 15–17, 1995: Background Information." Ottawa.

Canada. Department of Foreign Affairs and International Trade (1996). "The Lyon Summit, June 27–29, 1996: Background Information." Ottawa.

Canada. Environment Canada (1995). "Canada's National Action Plan on Climate Change." Ottawa.   http://web.archive.org/web/20020420084029/http://www.ec.gc.ca/climate/resource/cnapcc/indexe.html (January 2015).

Canada. Prime Minister's Office (2007). "The 2007 G8 Summit." Heiligendamm, June 8. http://www.pm.gc.ca/eng/news/2007/06/08/2007-g8-summit (January 2015).

Canada. Prime Minister's Office (2007). "Prime Minister Stephen Harper Calls for International Consensus on Climate Change." June 4, Berlin. http://pm.gc.ca/eng/news/2007/06/04/prime-minister-stephen-harper-calls-international-consensus-climate-change (January 2015).

Canada. Prime Minister's Office (2008). "PM Hails Breakthrough on Climate Change at 2008 G8 Summit." Toyako, July 9. http://www.pm.gc.ca/eng/news/2008/07/09/pm-hails-breakthrough-climate-change-2008-g8-summit (January 2015).

Canada. Standing Committee on Foreign Affairs and International Trade (1995). "From Bretton Woods to Halifax and Beyond: Towards a 21st Summit for the 21st Century." Report of the House of Commons Standing Committee on Foreign Affairs and International Trade on the Issues of International Financial Institutions for the Agenda of the June 1995 G7 Halifax Summit, House of Commons, Ottawa. http://www.g8.utoronto.ca/governmental/hc25/index.html (January 2015).

Cappe, Mel (1995). "From Hamilton to Halifax." In *The Halifax Summit, Sustainable Development, and International Institutional Reform*, John J. Kirton and Sarah Richardson, eds. Ottawa: National Round Table on the Environment and the Economy. http://www.g8.utoronto.ca/scholar/kirton199503/cappe/document.html (January 2015).

Carin, Barry and Alan Mehlenbacher (2010). "Constituting Global Leadership: Which Countries Need to Be Around the Summit Table for Climate Change and Energy Security?" *Global Governance* 16(1): 21–38.

Carr, Bob (2014). *Diary of a Foreign Minister*. Sydney: NewSouth Publishing.

Carter, Jimmy (1979). "'Crisis of Confidence' Speech." July 15. http://millercenter.org/president/speeches/speech-3402 (January 2015).

Carter, Jimmy (1982). *Keeping Faith: Memoirs of a President*. New York: Bantam Books.

CBC News (2007). "Critics Target PM's G8 Climate Change Message." June 5. http://www.cbc.ca/news/canada/critics-target-pm-s-g8-climate-change-message-1.640238 (January 2015).

Center for Climate and Energy Solutions (2006). "Twelfth Session of the Conference of the Parties to the UN Framework Convention on Climate Change and Second Meeting of the Parties to the Kyoto Protocol." http://www.c2es.org/international/negotiations/cop-12/summary (January 2015).

Chang, Jae-soon (2012). "Lee, Mexican President Cite 'Green Growth' as Solution to Global Crisis." *Yonhap English News*, May 22. http://english.yonhapnews.co.kr/national/2012/05/22/91/0301000000AEN20120522000700315F.HTML (January 2015).

Chengappa, Raj (2009). "The Earth in ICU." *India Today*, December 24. http://indiatoday.intoday.in/story/The+earth+in+ICU/1/76465.html (January 2015).

Christopher, Warren (1996). "American Diplomacy and the Global Environment Challenges of the 21st Century." Speech to alumni and faculty of Stanford University, April 9, Palo Alto CA. http://1997-2001.state.gov/wviww/global/oes/speech.html (January 2015).

Climate and Clean Air Coalition to Reduce Short-Lived Climate Pollutants (2014). "Country Partners." Paris. http://www.unep.org/ccac/Partners/CountryPartners/tabid/130289/Default.aspx (January 2015).

Climate and Clean Air Coalition to Reduce Short-Lived Climate Pollutants (2014). "Definitions: Short-Lived Climate Pollutants." Paris. http://www.unep.org/ccac/Short-LivedClimatePollutants/Definitions/tabid/130285/Default.aspx (January 2015).

Climatic Impact Assessment Program (1975). "Impacts of Climatic Change on the Biosphere." Monograph 5, DOT-TST-75-55, September, United States Department of Transportation, Washington DC.

Clinton, Bill (2004). *My Life*. New York: Knopf.

Clinton, Hillary Rodham (2007). "Security and Opportunity for the Twenty-First Century." *Foreign Affairs*, November/December. http://www.foreignaffairs.com/articles/63005/hillary-rodham-clinton/security-and-opportunity-for-the-twenty-first-century (January 2015).

Coon, Charli E. (2001). "Why President Bush Is Right to Abandon the Kyoto Protocol." May 11, Heritage Foundation, Washington DC. http://www.heritage.org/research/reports/2001/05/president-bush-right-to-abandon-kyoto-protocol (January 2015).

Cooper, Andrew F. and Agata Antkiewicz, eds. (2008). *Emerging Powers in Global Governance: Lessons from the Heiligendamm Process*. Waterloo ON: Wilfrid Laurier University Press.

Daniels, Joseph P. (1993). *The Meaning and Reliability of Economic Summit Undertakings* New York: Garland Publishing.

Davenport, Coral (2010). "Cancun Climate Talks Reach Crucial Stage." *National Journal*, December 9.

Davenport, Coral (2014). "A Climate Accord Based on Global Peer Pressure." *New York Times*, December 14. http://www.nytimes.com/2014/12/15/world/americas/lima-climate-deal.html (January 2015).

Davenport, Deborah (2008). "The International Dimension of Climate Policy." In *Turning Down the Heat: The Politics of Climate Policy in Affluent Democracies*, Hugh Compston and Ian Bailey, eds. Basingstoke: Palgrave Macmillan.

Der Bund – die Tageszeitung (1993). "Auf dem Gipfel der Tatenlosigkeit." July 7.

Dessai, Suraje (2001). "The Climate Regime from The Hague to Marrakech: Saving or Sinking the Kyoto Protocol?" Working Paper 12, December, Tyndall Centre for Climate

Change Research. http://www.tyndall.ac.uk/content/climate-regime-hague-marrakech-saving-or-sinking-kyoto-protocol (January 2015).

Deutscher Bundestag (1979, 4 July). Stenographischer Bericht, Plenarprotokoll Nr. 08/167. Deutscher Bundestag. http://dipbt.bundestag.de/doc/btp/08/08167.pdf (January 2015).

Dimitrov, Radoslav S. (2010). "Inside Copenhagen: The State of Climate Governance." *Global Environmental Politics* 10(2): 18–24. doi: 10.1162/glep.2010.10.2.18.

Dimitrov, Radoslav S. (2010). "Inside UN Climate Change Negotiations: The Copenhagen Conference." *Review of Policy Research* 27(6): 795–821.

Dobson, Hugo (2004). *Japan and the G7/8, 1975–2002*. London: RoutledgeCurzon.

Downing Street (2005). "Prime Minister Blair Concludes Climate Change Conference." London, November 1. http://www.g8.utoronto.ca/environment/env_energy051101-blair.htm (January 2015).

Dubash, Navroz K. (2009). "Copenhagen: Climate of Mistrust." *Economic and Political Weekly* 44(52): 8–11.

DW Staff (2006). "Merkel to Target Climate Change as G8, EU Leader." September 28. http://dw.de/p/9BHk (January 2015).

Earth Negotiations Bulletin (1992, 12 June). "Plenary." 2(11). http://www.iisd.ca/vol02/0211000e.html (January 2015).

Earth Negotiations Bulletin (1998, 16 November). "Report of the Fourth Conference of the Parties to the UN Framework Convention on Climate Change: 2–13 November 1998." 12(97). http://www.iisd.ca/download/pdf/enb1297e.pdf (January 2015).

Earth Negotiations Bulletin (1999, 8 November). "Report of the Fifth Conference of the Parties to the UN Framework Convention on Climate Change: 25 October–5 November 1999." 12(123). http://www.iisd.ca/download/pdf/enb12123e.pdf (January 2015).

Earth Negotiations Bulletin (2000, 27 November). "Report of the Sixth Conference of the Parties to the Framework Convention on Climate Change: 13–25 November 2000." 12(163). http://www.iisd.ca/download/pdf/enb12163e.pdf (January 2015).

Earth Negotiations Bulletin (2001, 30 July). "Summary of the Resumed Sixth Session of the Conference of the Parties to the UN Framework Convention on Climate Change: 16–27 July 2001." 12(176). http://www.iisd.ca/download/pdf/enb12176e.pdf (January 2015).

Earth Negotiations Bulletin (2001, 12 November). "Summary of the Seventh Session of the Conference of the Parties to the UN Framework Convention on Climate Change: 30 October–10 November 2001." 12(189). http://www.iisd.ca/download/pdf/enb12189e.pdf (January 2015).

Earth Negotiations Bulletin (2003, 15 December). "Summary of the Ninth Conference of the Parties to the UN Framework Convention on Climate Change: 1–12 December 2003." 12(231). http://www.iisd.ca/download/pdf/enb12231e.pdf (January 2015).

Earth Negotiations Bulletin (2004, 20 December). "Summary of the Tenth Conference of the Parties to the UN Framework Convention on Climate Change: 6–18 December." 12(260). http://www.iisd.ca/download/pdf/enb12260e.pdf (January 2015).

Earth Negotiations Bulletin (2005, 12 December). "Summary of the Eleventh Conference of the Parties to the UN Framework Convention on Climate Change and First Conference of the Parties Serving as the Meeting of the Parties to the Kyoto Protocol: 28 November–10 December 2005." 12(291). http://www.iisd.ca/download/pdf/enb12291e.pdf (January 2015).

Earth Negotiations Bulletin (2006, 20 November). "Summary of the Twelfth Conference of the Parties to the UN Framework Convention on Climate Change and Second Meeting of the Parties to the Kyoto Protocol: 6–17 November 2006." 12(318). http://www.iisd. ca/download/pdf/enb12318e.pdf (January 2015).

Earth Negotiations Bulletin (2007, 18 December). "Summary of the Thirteenth Conference of Parties to the UN Framework Convention on Climate Change and Third Meeting of the Parties to the Kyoto Protocol: 3–15 December." 12(354). http://www.iisd.ca/ download/pdf/enb12354e.pdf (January 2015).

Earth Negotiations Bulletin (2012, 11 December). "Summary of the Doha Climate Change Conference." 12(567). http://www.iisd.ca/vol12/enb12567e.html (January 2015).

Earth Negotiations Bulletin (2013, 26 November). "Summary of the Warsaw Climate Change Conference." 12(594). http://www.iisd.ca/vol12/enb12594e.html (January 2015).

Earth Summit 2012 Stakeholder Forum (2012). "Earth Summit 2012: Background." http:// www.earthsummit2012.org/about-us (January 2015).

Economist (2013). "Central European Floods: A Hard Lesson Learned." Eastern Approaches: Ex-communist Europe (Blog). June 6. http://www.economist.com/blogs/ easternapproaches/2013/06/central-european-floods (January 2015).

Elliott, Larry and Patrick Wintour (2006). "Blair Wants Developing Nations in New G13 to Help Secure Key Deals." *Guardian*, July 13. http://www.theguardian.com/ politics/2006/jul/13/uk.topstories3 (January 2015).

Environment Canada (2013). "Canada Invests in Global Climate and Clean Air Solutions." Ottawa, April 10. http://www.ec.gc.ca/default.asp?lang=En&n=714D9AAE-1&news=47727622-982C-4AE8-91CF-F9F50F81FF91 (January 2015).

Erlanger, Steven (2008). "France Sticks With Nuclear Power Oil Prices and Climate Change, It Says, Vindicate '50s Choice." *International Herald Tribune*, August 18.

European Council (2014). "Background Note and Facts about the EU's Role and Action." Brussels, June 3. http://www.g8.utoronto.ca/summit/2014brussels/background_140603. html (January 2015).

European Union (2001). "EU Reaction to the Speech by US President Bush on Climate Change." IP/01/821, Brussels, June 12. http://europa.eu/rapid/press-release_IP-01-821_en.htm (January 2015).

Fauver, Robert (2003). "G7/8 Oral History: Interview." G8 Research Group, March 13. http://www.g8.utoronto.ca/oralhistory/ (January 2015).

Fox News (2004). "Bush, Kerry Still Closely Matched." June 10. http://www.foxnews.com/ story/2004/06/10/06102004-bush-kerry-still-closely-matched/ (January 2015).

Franklin, Claire A., Richard T. Burnett, Richard J.P. Paolini, et al. (1985). "Health Risks from Acid Rain: A Canadian Perspective." *Environmental Health Perspectives* 63: 155–68. http://www.ncbi.nlm.nih.gov/pmc/articles/PMC1568495/pdf/envhper00446-0153.pdf (January 2015).

French Presidency of the G20 and G8 (2011). "Dossier de Presse." Paris, January 24. http://www.g8.utoronto.ca/summit/2011deauville/2011-g20-g8_dossier_ presse.pdf (January 2015).

Fujioka, Chisa (2008). "Britain Dismisses Japan Climate Change Plan." March 14, Reuters. http://www.reuters.com/article/2008/03/14/us-climate-g-idUSSP6817320080314 (January 2015).

Fukuda, Yasuo (2008). "Special Address on the Occasion of the Annual Meeting of the World Economic Forum." Davos, January 26. http://japan.kantei.go.jp/ hukudaspeech/2008/01/26speech_e.html (January 2015).

G7 (1975). "Declaration of Rambouillet." Rambouillet, November 17. http://www.g8.utoronto.ca/summit/1975rambouillet/communique.html (January 2015).

G7 (1976). "Joint Declaration of the International Conference." San Juan, June 28. http://www.g8.utoronto.ca/summit/1976sanjuan/communique.html (January 2015).

G7 (1977). "Appendix to Downing Street Summit Conference." London, May 8. http://www.g8.utoronto.ca/summit/1977london/appendix.html (January 2015).

G7 (1978). "Declaration." Bonn, July 17. http://www.g8.utoronto.ca/summit/1978bonn/communique (January 2015).

G7 (1979). "Declaration." Tokyo, June 29. http://www.g8.utoronto.ca/summit/1979tokyo/communique.html (January 2015).

G7 (1980). "Declaration." Venice, June 23. http://www.g8.utoronto.ca/summit/1980venice/communique (January 2015).

G7 (1983). "Williamsburg Declaration on Economic Recovery." Williamsburg, May 30. http://www.g8.utoronto.ca/summit/1983williamsburg/communique.html (January 2015).

G7 (1984). "London Economic Declaration." London, June 9. http://www.g8.utoronto.ca/summit/1984london/communique.html (January 2015).

G7 (1985). "The Bonn Economic Declaration: Towards Sustained Growth and Higher Employment." Bonn, May 4. http://www.g8.utoronto.ca/summit/1985bonn/communique.html (January 2015).

G7 (1986). "Statement on the Implications of the Chernobyl Nuclear Accident." Tokyo, May 5. http://www.g8.utoronto.ca/summit/1986tokyo/chernobyl.html (January 2015).

G7 (1986). "Tokyo Economic Declaration." Tokyo, May 6. http://www.g8.utoronto.ca/summit/1986tokyo/communique.html (January 2015).

G7 (1987). "Venezia Economic Declaration." Venice, June 10. http://www.g8.utoronto.ca/summit/1987venice (January 2015).

G7 (1988). "Toronto Economic Summit Economic Declaration." Toronto, June 21. http://www.g8.utoronto.ca/summit/1988toronto/communique.html (January 2015).

G7 (1989). "Economic Declaration." Paris, July 16. http://www.g8.utoronto.ca/summit/1989paris/communique (January 2015).

G7 (1990). "Houston Economic Declaration." Houston, July 11. http://www.g8.utoronto.ca/summit/1990houston/declaration.html (January 2015).

G7 (1991). "Economic Declaration: Building a World Partnership." London, July 17. http://www.g8.utoronto.ca/summit/1991london/communique (January 2015).

G7 (1991). "Political Declaration: Strengthening the International Order." London, July 16. http://www.g8.utoronto.ca/summit/1991london/political.html (January 2015).

G7 (1992). "Economic Declaration: Working Together for Growth and a Safer World." July 8. http://www.g8.utoronto.ca/summit/1992munich/communique (January 2015).

G7 (1993). "Tokyo Summit Political Declaration: Striving for a More Secure and Humane World." Tokyo, July 8. http://www.g8.utoronto.ca/summit/1993tokyo/political.html (January 2015).

G7 (1994). "G7 Communiqué." Naples, July 9. http://www.g9.utoronto.ca/summit/1994naples/communique (January 2015).

G7 (1995). "Halifax Summit Communiqué." Halifax, June 16. http://www.g8.utoronto.ca/summit/1995halifax/communique/index.html (January 2015).

G7 (1996). "Chairman's Statement: Toward Greater Security and Stability in a More Cooperative World." June 29, Lyon. http://www.g8.utoronto.ca/summit/1996lyon/chair.html (January 2015).

G7 (1996). "Economic Communiqué: Making a Success of Globalization for the Benefit of All." Lyon, June 29. http://www.g8.utoronto.ca/summit/1996lyon/chair.html (January 2015).

G7 (1996). "A New Partnership for Development." Released after a Work Session with the Secretary-General of United Nations, the International Monetary Fund Managing Director, the President of the World Bank and the World Trade Organization Director-General, Lyon, June 29. http://www.g8.utoronto.ca/summit/1996lyon/partner.html (January 2015).

G7 (1997). "Communiqué." Denver, June 22. http://www.g8.utoronto.ca/summit/1997denver/g8final.htm (January 2015).

G7 (2014). "G7 Brussels Summit Declaration." Brussels, June 5. http://www.g8.utoronto.ca/summit/2014brussels/declaration.html (January 2015).

G7 Energy Ministers (2014). "Rome G7 Energy Initiative for Energy Security." Rome, May 6. http://www.g8.utoronto.ca/energy/140506-rome.html (January 2015).

G7 Environment Ministers (1995). "Chairperson's Highlights." Hamilton, Canada, May 1.

G7 Environment Ministers (1996). "Chairman's Summary." Cabourg, France, May 10. http://www.g8.utoronto.ca/environment/1996cabourg/summary_index.html (January 2015).

G7 Environment Ministers (1997). "Chair's Summary." Miami, May 6. http://www.g8.utoronto.ca/environment/1997miami/summary.html (January 2015).

G7 Finance Ministers (1999). "Report of G7 Finance Ministers on the Köln Debt Initiative to the Köln Economic Summit." Cologne, June 18. http://www.g8.utoronto.ca/finance/fm061899.htm (January 2015).

G8 (1997). "Communiqué." June 22, Denver. http://www.g8.utoronto.ca/summit/1997denver/g8final.htm (January 2015).

G8 (1998). "Communiqué." Birmingham, May 15. http://www.g8.utoronto.ca/summit/1998birmingham/finalcom.htm (January 2015).

G8 (1999). "Communiqué." Cologne, June 20. http://www.g8.utoronto.ca/summit/1999koln/finalcom.htm (January 2015).

G8 (2000). "G8 Communiqué Okinawa 2000." Okinawa, July 23. http://www.g8.utoronto.ca/summit/2000okinawa/finalcom.htm (January 2015).

G8 (2001). "Communiqué." Genoa, July 22. http://www.g8.utoronto.ca/summit/2001genoa/finalcommunique.html (January 2015).

G8 (2001). "Statement by the G8 Leaders (Death in Genoa)." Genoa, July 21. http://www.g8.utoronto.ca/summit/2001genoa/g8statement1.html (January 2015).

G8 (2002). "The Kananaskis Summit Chair's Summary." Kananaskis, June 27. http://www.g8.utoronto.ca/summit/2002kananaskis/summary.html (January 2015).

G8 (2003). "Chair's Summary." Evian, June 3. http://www.g8.utoronto.ca/summit/2003evian/communique_en.html (January 2015).

G8 (2003). "Marine Environment and Tanker Safety: A G8 Action Plan." Evian, June 3. http://www.g8.utoronto.ca/summit/2003evian/marine_en.html (January 2015).

G8 (2003). "Science and Technology for Sustainable Development: A G8 Action Plan." Evian, June 2. http://www.g8.utoronto.ca/summit/2003evian/sustainable_development_en.html (January 2015).

G8 (2003). "Water: A G8 Action Plan." Evian, June 2. http://www.g8.utoronto.ca/summit/2003evian/water_en.html (January 2015).

G8 (2004). "Chair's Summary." Sea Island, June 10. http://www.g8.utoronto.ca/summit/2004seaisland/summary.html (January 2015).

G8 (2004). "Science and Technology for Sustainable Development: "3R" Action Plan and Progress on Implementation." Sea Island. http://www.g8.utoronto.ca/summit/2004seaisland/sd.html (January 2015).

G8 (2005). "Chair's Summary." Gleneagles, July 8. http://www.g8.utoronto.ca/summit/2005gleneagles/summary.html (January 2015).

G8 (2005). "Climate Change, Clean Energy, and Sustainable Development." Gleneagles, July 8. http://www.g8.utoronto.ca/summit/2005gleneagles/climatechange.html (January 2015).

G8 (2005). "Gleneagles Plan of Action: Climate Change, Clean Energy, and Sustainable Development." Gleneagles, July 8. http://www.g8.utoronto.ca/summit/2005gleneagles/climatechangeplan.html (January 2015).

G8 (2006). "Global Energy Security." St. Petersburg, July 16. http://www.g8.utoronto.ca/summit/2006stpetersburg/energy.html (January 2015).

G8 (2007). "Chair's Summary." Heiligendamm, June 8. http://www.g8.utoronto.ca/summit/2007heiligendamm/g8-2007-summary.html (January 2015).

G8 (2007). "Growth and Responsibility in the World Economy." Heiligendamm, June 7. http://www.g8.utoronto.ca/summit/2007heiligendamm/g8-2007-economy.html (January 2015).

G8 (2008). "Chair's Summary." Hokkaido, July 8. http://www.g8.utoronto.ca/summit/2008hokkaido/2008-summary.html (January 2015).

G8 (2008). "Environment and Climate Change." Hokkaido, July 8. http://www.g8.utoronto.ca/summit/2008hokkaido/2008-climate.html (January 2015).

G8 (2008). "G8 Hokkaido Toyako Summit Leaders' Declaration." Hokkaido, July 8. http://www.g8.utoronto.ca/summit/2008hokkaido/2008-declaration.html (January 2015).

G8 (2009). "Responsible Leadership for a Sustainable Future." L'Aquila, Italy, July 8. http://www.g8.utoronto.ca/summit/2009laquila/2009-declaration.html (January 2015).

G8 (2010). "G8 Muskoka Declaration: Recovery and New Beginnings." Huntsville, Canada, June 26. http://www.g8.utoronto.ca/summit/2010muskoka/communique.html (January 2015).

G8 (2010). "Muskoka Accountability Report." June 20. http://www.g8.utoronto.ca/summit/2010muskoka/accountability (January 2015).

G8 (2011). "G8 Declaration: Renewed Commitment for Freedom and Democracy." Deauville, May 27. http://www.g8.utoronto.ca/summit/2011deauville/2011-declaration-en.html (January 2015).

G8 (2012). "Camp David Declaration." Camp David, May 19. http://www.g8.utoronto.ca/summit/2012campdavid/g8-declaration.html (January 2015).

G8 (2013). "G8 Lough Erne Communiqué." June 18, Lough Erne, Northern Ireland, United Kingdom. http://www.g8.utoronto.ca/summit/2013lougherne/lough-erne-communique.html (January 2015).

G8 (2013). "Lough Erne Accountability Report: Keeping Our Promises." June 7. http://www.g8.utoronto.ca/summit/2013lougherne/lough-erne-accountability.html (January 2015).

G8 Energy Ministers (2002). "G8 Energy Ministers Meeting: Co-chairs' Statement." Detroit, May 3. http://www.g8.utoronto.ca/energy/energy0702.html (January 2015).

G8 Energy Ministers (2006). "Chair's Statement of G8 Energy Ministerial Meeting." Moscow, March 16. http://www.g8.utoronto.ca/energy/energy060316.html (January 2015).

G8 Environment Ministers (1999). "G8 Environment Ministers Communiqué." Schwerin, Germany, March 28. http://www.g8.utoronto.ca/environment/1999schwerin/communique.html (January 2015).

G8 Environment Ministers (2000). "G8 Environment Ministers Communiqué." Otsu, Japan, April 9. http://www.g8.utoronto.ca/environment/2000otsu/communique.html (January 2015).

G8 Environment Ministers (2001). "G8 Environment Ministers Communiqué." Trieste, Italy, March 4. http://www.g8.utoronto.ca/environment/2001trieste/communique.html (January 2015).

G8 Environment Ministers (2002). "Banff Ministerial Statement on the World Summit on Sustainable Development." Banff, April 14. http://www.g8.utoronto.ca/environment/020415.html (January 2015).

G8 Environment Ministers (2003). "G8 Environment Ministers Communiqué." Paris, April 27. http://www.g8.utoronto.ca/environment/2003paris/env_communique_april_2003_eng.html (January 2015).

G8 Foreign Ministers (1999). "Conclusions of the Meeting of the G8 Foreign Ministers." Cologne, June 10. http://www.g8.utoronto.ca/foreign/fm9906010.htm (January 2015).

G8 Research Group (2001). "From Okinawa 2000 to Genoa 2001: Issue Performance Assessment – Environment." http://www.g8.utoronto.ca/evaluations/2001genoa/assessment_environment.html (January 2015).

G8 Research Group (2006). "2005 Gleneagles Final Compliance Report." June 12. http://www.g8.utoronto.ca/evaluations/2005compliance_final/index.html (January 2015).

G8 Research Group (2006). "St. Petersburg Performance." July 17. http://www.g8.utoronto.ca/evaluations/2006stpetersburg/performance.html (January 2015).

G8 Research Group (2008). "2007 Heiligendamm G8 Summit Final Compliance Report." Toronto. http://www.g8.utoronto.ca/evaluations/2007compliance_final/ (January 2015).

G8 Research Group (2009). "2008 Hokkaido-Toyako G8 Summit Final Compliance Report." June 30. http://www.g8.utoronto.ca/evaluations/2008compliance-final/index.html (January 2015).

G8 Research Group (2010). "2010 Canada Summit Expanded Dialogue Country Assessment Report." June 27. http://www.g8.utoronto.ca/evaluations/2010muskoka/2010-muskoka-ed.pdf (January 2015).

G8 Research Group (2010). "Climate Change and the Environment at the G8 and G20." June 11. http://www.g8.utoronto.ca/briefs/ccenv-100611.pdf (January 2015).

G8 Research Group (2011). "2010 Muskoka G8 Summit Final Compliance Report." May 24. http://www.g8.utoronto.ca/evaluations/2010compliance-final (January 2015).

G8 Research Group (2013). "2012 Camp David G8 Summit Final Compliance Report." 14, June. http://www.g8.utoronto.ca/evaluations/2012compliance/2012compliance.pdf (January 2015).

G20 (2008). "Declaration of the Summit on Financial Markets and the World Economy." Washington DC, October 15. http://www.g20.utoronto.ca/2008/2008declaration1115.html (January 2015).

G20 (2009). "G20 Leaders Statement: The Pittsburgh Summit." Pittsburgh, September 25. http://www.g20.utoronto.ca/2009/2009communique0925.html (January 2015).

G20 (2009). "Global Plan for Recovery and Reform." London, April 2. http://www.g20.utoronto.ca/2009/2009communique0402.html (January 2015).

G20 (2010). "The G20 Seoul Summit Document." Seoul, November 12. http://www.g20.utoronto.ca/2010/g20seoul-doc.html (January 2015).

G20 (2011). "Cannes Summit Final Declaration – Building Our Common Future: Renewed Collective Action for the Benefit of All." Cannes, November 4. http://www.g20.utoronto.ca/2011/2011-cannes-declaration-111104-en.html (January 2015).

G20 (2012). "G20 Leaders Declaration." Los Cabos, June 19. http://www.g20.utoronto.ca/2012/2012-0619-loscabos.html (January 2015).

G20 (2012). "Los Cabos Growth and Jobs Action Plan." Los Cabos, June 19. http://www.g20.utoronto.ca/2012/2012-0619-loscabos-actionplan.html (January 2015).

G20 (2013). "G20 Leaders Declaration." St. Petersburg, September 6. http://www.g20.utoronto.ca/2013/2013-0906-declaration.html (January 2015).

G20 (2013). "St. Petersburg Accountability Report on G20 Development Commitments." August 28. http://www.g20.utoronto.ca/2013/Saint_Petersburg_Accountability_Report_on_G20_Development_Commitments.pdf (January 2015).

G20 (2014). "G20 Leaders' Communiqué." Brisbane, November 16. http://www.g20.utoronto.ca/2014/2014-1116-communique.html (January 2015).

Gates, Bill (2011). "Innovation with Impact: Financing 21st Century Development." Report to G20 leaders at the Cannes Summit, November. http://www.gatesfoundation.org/~/media/GFO/Documents/2011%20G20%20Report%20PDFs/Full%20Report/g20reportenglish.pdf (January 2015).

Gehring, Thomas (1992). *Dynamic International Regimes: Institutions for International Environmental Governance*. Frankfurt: Peter Lang.

German Presidency of the G8 (2007). "Pressekonferenz zum G8-Gipfel." Audio available in English, Heiligendamm, June 8. http://www.g8.utoronto.ca/summit/2007heiligendamm/g8-2007-merkel.html (January 2015).

Gill, Stephen (2000). "The Constitution of Global Capitalism." Paper presented at the annual convention of the International Studies Association, March 15, Los Angeles.

Gillard, Julia (2014). *My Story*. London: Bantam Books.

Goldemberg, José (2007). "Energy Choices Toward a Sustainable Future." *Environment* 49(10): 7–11.

Goldenberg, Suzanne (2009). "Obama's Arrival Expected to Inject Fresh Momentum into Copenhagen Talks." *Guardian*, December 17. http://www.theguardian.com/environment/2009/dec/17/barack-obama-copenhagen-hillary-clinton (January 2015).

Goldfarb Consultants (1996). "The Goldfarb Report 1996." Toronto.

González-Aller Jurado, Cristóbal and Karl Falkenberg (2010). "Letter: Expression of Willingness to Be Associated with the Copenhagen Accord and Submission of the Quantified Economy-Wide Emissions Reduction Targets for 2020." January 28, European Union and European Commission, Brussels. http://unfccc.int/files/meetings/cop_15/copenhagen_accord/application/pdf/europeanunioncphaccord_app1.pdf (January 2015).

Green, Andrew (2008). "Bringing Institutions and Individuals into a Climate Policy for Canada." In *A Globally Integrated Climate Policy for Canada*, Stephen Bernstein, Jutta Brunnée, David Duff, et al., eds. Toronto: University of Toronto Press, pp. 246–57.

Grubb, Michael (1990). *Energy Policies and the Greenhouse Effect*. Vol. 1: Policy Appraisal. Aldershot: Dartmouth.

Guebert, Jenilee, Zaria Shaw, and Sarah Jane Vassallo (2011). "G8 Conclusions on Climate Change, 1975–2011." Toronto, June 20. http://www.g8.utoronto.ca/conclusions/climatechange.pdf (January 2015).

Haas, Peter M. (1992). "Introduction: Epistemic Communities and International Policy Coordination." *International Organization* 46(1): 35.

Haas, Peter M. (2002). "UN Conferences and Constructivist Governance of the Environment." *Global Governance* 8(1): 73–91.

Haas, Peter M. (2008). "Climate Change Governance after Bali." *Global Environmental Politics* 8(3): 1–7.

Haas, Peter M., ed. (2008). *International Environmental Governance*. Aldershot: Ashgate.

Hainsworth, Susan (1990). "Coming of Age: The European Community and the Economic Summit." Country Study No. 7, G7 Research Group. http://www.g8.utoronto.ca/scholar/hainsworth1990/index.html (January 2015).

Hajnal, Peter I. (1989). *The Seven-Power Summit: Documents from the Summits of Industrialized Countries, 1975–1989*. Millwood NY: Kraus International Publishers.

Hajnal, Peter I., ed. (1997). *International Information: Documents, Publications, and Electronic Information of International Governmental Organizations*. Englewood CO: Libraries Unlimited.

Hajnal, Peter I. (1999). *The G7/G8 System: Evolution, Role, and Documentation*. Aldershot: Ashgate.

Hajnal, Peter I. (2008). "Meaningful Relations: The G8 and Civil Society." In *G8: Hokkaido Toyako Summit 2008*, Maurice Fraser, ed. London: Newsdesk, pp. 206–08.

Hajost, Scott, Bruce Rich, and Todd Goldman (1993). "Struggles of Developing Nations a Job for G7: G7, World Bank Bear Burden of Environmental Reform." *Christian Science Monitor*, July 6, p. 19.

Harper, Stephen (2007). "Prime Minister Harper outlines Agenda for a Stronger, Safer, Better Canada." Ottawa, February 6. http://pm.gc.ca/eng/news/2007/02/06/prime-minister-harper-outlines-agenda-stronger-safer-better-canada (January 2015).

Harper, Stephen (2009). "The 2010 Muskoka Summit." In *The 2009 G8 Summit: From La Maddalena to L'Aquila*, John J. Kirton and Madeline Koch, eds. London: Newsdesk, pp. 18–19. http://www.g8.utoronto.ca/newsdesk/harper-2009.html (January 2015).

Harper, Stephen (2010). "Canada's G8 Priorities." Ottawa, January 26. http://www.g8.utoronto.ca/summit/2010muskoka/harper-priorities.html (January 2015).

Harper, Stephen (2010). "Statement by the Prime Minister of Canada at the 2010 World Economic Forum." Davos, January 28. http://www.g8.utoronto.ca/summit/2010muskoka/harper-davos.html (January 2015).

Harris, Kathleen (2010). "G8 Leaders Take Heat for Failing to Act on Global Warming." *Toronto Sun*, June 26. http://www.torontosun.com/news/g20/2010/06/26/14527901.html (January 2015).

Harris, Paul (2001). *International Equity and Global Environmental Politics: Power and Principles in US Foreign Policy*. Aldershot: Ashgate.

Harvey, Fiona (2012). "Doha Climate Gateway: The Reaction." *Guardian*, December 10. http://www.theguardian.com/environment/2012/dec/10/doha-climate-gateway-reaction (January 2015).

Harvey, Fiona and John Vidal (2011). "Durban COP17: Connie Hedegaard Puts Pressure on China, US, and India." *Guardian*, December 11. http://www.guardian.co.uk/environment/2011/dec/09/durban-climate-change-connie-hedegaard (January 2015).

Hattori, Takashi (2007). "The Rise of Japanese Climate Change Policy: Balancing the Norms of Economic Growth, Energy Efficiency, International Contribution, and Environmental Protection." In *The Social Construction of Climate Change: Power, Knowledge, Norms, Discourses*, Mary Pettenger, ed. Aldershot: Ashgate, pp. 75–97.

Hecht, Alan and Dennis Tirpak (1995). "Framework Agreement on Climate Change: A Scientific and Policy History." *Climatic Change* 29(4): 371–402.

Hedegaard, Connie (2010). "Cancún Must Take Us Towards a Global Climate Deal." *European View* 9(2): 175–79.

Heeney, Timothy (1988). "Canadian Foreign Policy and the Seven Power Summits." Country Study No. 1, Centre for International Studies, University of Toronto. http://www.g8.utoronto.ca/scholar/heeney1988/heenl.htm (January 2015).

Hill, Dilys M. (1994). "Domestic Policy." In *The Bush Presidency: Triumphs and Adversities*, Dilys M. Hill and Phil Williams, eds. London: Macmillan.

Hochstetler, Kathryn (2012). "A Green Economy for the Whole G20." Waterloo ON, May 12. http://www.cigionline.org/publications/green-economy-whole-g20 (January 2015).

Hodges, Michael R. (1999). "The G8 and the New Political Economy." In *The G8's Role in the New Millennium*, Michael R. Hodges, John J. Kirton, and Joseph P. Daniels, eds. Vol. 69–74. Aldershot: Ashgate, pp. 69–74.

Hodges, Michael R., John J. Kirton, and Joseph P. Daniels, eds. (1999). *The G8's Role in the New Millennium*. Aldershot: Ashgate.

Hoffmann, Matthew (2005). *Ozone Depletion and Climate Change: Constructing a Global Response*. Albany: SUNY Press.

Ifill, Gwen (1992). "Clinton Cites Bush 'Errors'." *New York Times*, June 13.

Ikenberry, G. John (1988). "Market Solutions for State Problems: The International and Domestic Politics of American Oil Decontrol." *International Organization* 42(1): 151–77.

Ikenberry, G. John (2003). *Strategic Reactions to American Preeminence: Great Power Politics in the Age of Unipolarity*. Washington DC: National Intelligence Council.

Information Unit on Climate Change (1993). "The Bergen Conference and Its Proposals for Addressing Climate Change." Châtelaine, Switzerland. http://unfccc.int/resource/ccsites/senegal/fact/fs220.htm (January 2015).

Intergovernmental Panel on Climate Change (1995). "IPCC Second Assessment: Climate Change 1995." Geneva. http://www.ipcc.ch/pdf/climate-changes-1995/ipcc-2nd-assessment/2nd-assessment-en.pdf (January 2015).

Japan. Ministry of Foreign Affairs (1994). "Japan's ODA Annual Report (Summary) 1994: The 40th Anniversary of Japan's ODA: Accomplishments and Challenges–3. Perspectives of Japan's ODA Towards the 21st Century." Tokyo. http://www.mofa.go.jp/policy/oda/summary/1994/3.html (January 2015).

Japan. Ministry of Foreign Affairs (2008). "Financial Mechanism for 'Cool Earth Partnership'." Tokyo, November. http://www.mofa.go.jp/policy/economy/wef/2008/mechanism.html (January 2015).

Jeffery, Simon (2001). "Protester Shot Dead in Genoa Riot." *Guardian*, July 20. http://www.theguardian.com/world/2001/jul/20/globalisation.usa (January 2015).

Johnson, Pierre Marc and John J. Kirton (1995). "Sustainable Development and Canada at the G7 Summit." In *The Halifax Summit, Sustainable Development, and International Institutional Reform*, John J. Kirton and Sarah Richardson, eds. Ottawa: National Round Table on the Environment and the Economy. http://www.g8.utoronto.ca/scholar/kirton199503/johnson/index.html (January 2015).

Johnson, Rachel (1991). "G7 Summit in London: Kohl and Bush Focus on Environment." *Financial Times*, July 16.

Johnson, Rachel (1991). "G7 Summit in London: Summiteers Backpedal on the Environment." *Financial Times*, July 17.

Johnson, Rachel (1991). "Summiteers Get Ready to Head for the Forest." *Financial Times*, July 12.

Kaiser, Karl, John J. Kirton, and Joseph P. Daniels, eds. (2000). *Shaping a New International Financial System: Challenges of Governance in a Globalizing World.* Aldershot: Ashgate.

Karlsson, Christer, Mattias Hjerpe, Charles Parker, et al. (2012). "The Legitimacy of Leadership in International Climate Change Negotiations." *Ambio* 41(1): 46–55. doi: 10.1007/s13280-011-0240-7.

Kennan, George (1970). "To Prevent a World Wasteland: A Proposal." *Foreign Affairs* 48 (April): 401–13. http://www.foreignaffairs.com/articles/24149/george-f-kennan/to-prevent-a-world-wasteland (January 2015).

Keohane, Robert O. (1978). "The International Energy Agency: State Influence and Transgovernmental Politics." *International Organization* 32(4): 929–51.

Keohane, Robert O. and Joseph S. Nye (1977). *Power and Interdependence: World Politics in Transition.* Boston: Little, Brown.

Kilbourn, Peter T. (1989). "Environment Is Becoming Priority Issue." *New York Times*, May 14. http://www.nytimes.com/1989/05/15/business/environment-is-becoming-priority-issue.html (January 2015).

Kim Jae-kyoung (2010). "Seoul to Bring G20 Leaders' Attention to Green Growth." *Korea times*, August 22. http://www.koreatimes.co.kr/www/news/biz/2011/03/301_71799.html (January 2015).

Kirschbaum, Erik (2006). "Germany to Put Global Warming Back on G8 Agenda." *Reuters*, September 26.

Kirton, John J. (1989). "Contemporary Concert Diplomacy: The Seven-Power Summit and the Management of International Order." Paper presented at the annual convention of the International Studies Association, 29 March–1 April, London. http://www.g8.utoronto.ca/scholar/kirton198901 (January 2015).

Kirton, John J. (1990). "Sustainable Development at the Houston Seven Power Summit." Paper prepared for the Foreign Policy Committee, National Round Table on the Environment and the Economy, September 6. http://www.g8.utoronto.ca/scholar/kirton199001/index.html (January 2015).

Kirton, John J. (1993). "The Seven Power Summits as a New Security Institution." In *Building a New Global Order: Emerging Trends in International Security*, David Dewitt, David Haglund, and John J. Kirton, eds. Toronto: Oxford University Press, pp. 335–57.

Kirton, John J. (1996). "The G7 Has Finally Reached Adulthood." *Financial Post*, June 22.

Kirton, John J. (1999). "An Assessment of the 1999 Cologne G7/G8 Summit by Issue Area." June 20, G8 Research Group, Cologne. http://www.g8.utoronto.ca/evaluations/1999koln/issues/kolnperf.htm (January 2015).

Kirton, John J. (1999). "Explaining G8 Effectiveness." In *The G8's Role in the New Millennium*, Michael R. Hodges, John J. Kirton, and Joseph P. Daniels, eds. Aldershot: Ashgate, pp. 45–68.

Kirton, John J. (2000). "Creating Peace and Human Security: The G8 and Okinawa Summit Contribution." May 26, G8 Research Group. http://www.g8.utoronto.ca/scholar/kirton200002/ (January 2015).

Kirton, John J. (2001). "Generating Genuine Global Governance: Prospects for the Genoa G8 Summit." July 15, G8 Research Group, Genoa. http://www.g8.utoronto.ca/evaluations/2001genoa/prospects_kirton.html (January 2015).

Kirton, John J. (2001). "Guiding Global Economic Governance: The G20, the G7, and the International Monetary Fund at Century's Dawn." In *New Directions in Global*

Economic Governance: Managing Globalisation in the Twenty-First Century, John J. Kirton and George M. von Furstenberg, eds. Aldershot: Ashgate, pp. 143–67.

Kirton, John J. (2002). "Embedded Ecologism and Institutional Inequality: Linking Trade, Environment, and Social Cohesion in the G8." In *Linking Trade, Environment, and Social Cohesion: NAFTA Experiences, Global Challenges*, John J. Kirton and Virginia W. Maclaren, eds. Aldershot: Ashgate, pp. 45–72.

Kirton, John J. (2002). "The Promise of the Kananaskis Summit." *Calgary Herald*, June 26. http://www.g8.utoronto.ca/evaluations/2002kananaskis/assess_promise.html (January 2015).

Kirton, John J. (2002). "A Summit of Historic Significance: A Gold Medal for the Kananaskis G8." *Calgary Herald*, June 27. http://www.g8.utoronto.ca/evaluations/2002kananaskis/assess_goldmedal.html (January 2015).

Kirton, John J. (2002). "Winning Together: The NAFTA Trade-Environment Record." In *Linking Trade, Environment, and Social Cohesion: NAFTA Experiences, Global Challenges*, John J. Kirton and Virginia W. Maclaren, eds. Aldershot: Ashgate, pp. 79–99.

Kirton, John J. (2004). "Generating Effective Global Environmental Governance: The North's Need for a World Environmental Organization." In *A World Environmental Organization: Solution or Threat for Effective International Environmental Governance?*, Frank Biermann and Steffen Bauer, eds. Aldershot: Ashgate, pp. 145–74.

Kirton, John J. (2004). "Prospects for the G8 Sea Island Summit Seven Weeks Hence." Paper prepared for a seminar at Armstrong Atlantic State University, April 22, Savannah. http://www.g8.utoronto.ca/scholar/kirton2004/kirton_seaisland_040426.html (January 2015).

Kirton, John J. (2005). "America at the G8: From Vulnerability to Victory at the Sea Island Summit." In *New Perspectives on Global Governance: Why America Needs the G8*, Michele Fratianni, John J. Kirton, Alan M. Rugman, et al., eds. Aldershot: Ashgate, pp. 31–50.

Kirton, John J. (2005). "Gleneagles G8 Boosts Blair at Home." August 1. http://www.g8.utoronto.ca/evaluations/2005gleneagles/coverage.html (January 2015).

Kirton, John J. (2005). "Gleneagles Performance: Insittutions Created." July 9. http://www.g8.utoronto.ca/evaluations/2005gleneagles/2005institutions.html (January 2015).

Kirton, John J. (2005). "What to Watch For at Gleneagles." G8 Research Group, July 6. http://www.g8.utoronto.ca/evaluations/2005gleneagles/kirton2005-0706.html (January 2015).

Kirton, John J. (2006). "Explaining Compliance with G8 Finance Commitments: Agency, Institutionalization, and Structure." *Open Economies Review* 17(4): 459–75.

Kirton, John J. (2006). "The G8 and Global Energy Governance: Past Performance, St. Petersburg Opportunities." Paper presented at a conference on "The World Dimension of Russia's Energy Security," Moscow State Institute of International Relations, April 21, Moscow. http://www.g8.utoronto.ca/scholar/kirton2006/kirton_energy_060623.pdf (January 2015).

Kirton, John J. (2006). "Shocked into Success: Prospects for the St. Petersburg Summit." Paper prepared for the G8 Pre-Summit Conference on "G8 Performance, St. Petersburg Possibilities," Moscow State Institute on International Relations, Moscow, June 29. http://www.g8.utoronto.ca/conferences/2006/mgimo/kirton_prospects_060717.pdf (January 2015).

Kirton, John J. (2007). *Canadian Foreign Policy in a Changing World*. Toronto: Thomson Nelson.

Kirton, John J. (2007). "Mulroney's 1988 Toronto Summit." Unpublished manuscript.

Kirton, John J. (2007). "Prospects for the 2007 G8 Summit." In *G8 Summit 2007: Growth and Responsibility*, Maurice Fraser, ed. London: Newsdesk, pp. 18–21. http://www.g8.utoronto.ca/newsdesk/G8-2007.pdf (January 2015).

Kirton, John J. (2008–09). "Consequences of the 2008 US Elections for America's Climate Change Policy, Canada and the World." *International Journal* 64(1): 153–62.

Kirton, John J. (2009). "Governing Global Trucks and Buses: The G20 and G8 Roles." Keynote address, Truck and Bus Forum, May 12, Lyon. http://www.g8.utoronto.ca/scholar/kirton-lyon-090511.pdf (January 2015).

Kirton, John J. (2009). "Prospects for the 2010 Muskoka G8 Summit." Paper prepared for a conference on "The 2009 G8's Sustainable Development Challenge: Initiative and Implementation," Aspen Institute Italia, July 1, Rome. http://www.g8.utoronto.ca/evaluations/2010muskoka/2010prospects090702.html (January 2015).

Kirton, John J. (2010). "Working Together for G8-G20 Partnership: The Muskoka-Toronto Twin Summits, June 2010." *International Organisations Research Journal* 5(5): 14–20. http://iorj.hse.ru/data/2011/03/15/1211462316/4.pdf (January 2015).

Kirton, John J. (2012). "Energy Security amidst Disaster: The Global Governance Response to March 11, 2011." Paper prepared for the Japan Futures Initiative Spring Symposium 2012, Balsillie School of International Affairs, University of Waterloo, March 14, Waterloo ON.

Kirton, John J. (2013). *G20 Governance for a Globalized World*. Farnham: Ashgate.

Kirton, John J. (2013). "A Summit of Significant Success: Prospects for the G8 Leaders at Lough Erne." June 12. http://www.g8.utoronto.ca/evaluations/2013lougherne/kirton-prospects-2013.html (January 2015).

Kirton, John J. (2013). "A Summit of Substantial Success: Prospects for St. Petersburg 2013." In *Russia's G20 Summit: St Petersburg 2013*, John J. Kirton and Madeline Koch, eds. London: Newsdesk, pp. 34–35. http://www.g8.utoronto.ca/newsdesk/stpetersburg (January 2015).

Kirton, John J. (2013). "A Summit of Very Substantial Success: G20 Governance of the Global Economy and Security at St. Petersburg." September 6. http://www.g20.utoronto.ca/analysis/130906-kirton-performance.html (January 2015).

Kirton, John J. (2014). "The Performance of the G20 Brisbane Summit." December 4, G20 Research Group. http://www.g20.utoronto.ca/analysis/141204-kirton-performance.html (January 2015).

Kirton, John J. and Madeline Boyce (2009). "The G8's Climate Change Performance." August 12. http://www.g8.utoronto.ca/evaluations/g8climateperformance.pdf (January 2015).

Kirton, John J., Courtney Brady, and Janel Smith (2005). "Gleneagles Performance: Money Mobilized." July 8. http://www.g8.utoronto.ca/evaluations/2005gleneagles/2005money.html (January 2015).

Kirton, John J., Joseph P. Daniels, and Andreas Freytag, eds. (2001). *Guiding Global Order: G8 Governance in the Twenty-First Century*. Aldershot: Ashgate.

Kirton, John J. and Ella Kokotsis (2003). "Impressions of the G8 Evian Summit." Evian, June 3. http://www.g8.utoronto.ca/evaluations/2003evian/assess_kirton_kokotsis.html (January 2015).

Kirton, John J. and Julia Kulik (2012). "Delivering a Double Dividend: Prospects for the G20 at Los Cabos Summit." Paper presented at a conference on "Strengthening Sustainable Development by Bringing Business In: From Los Cabos G20 to Rio Plus 20," Global Institute for Sustainability, Tecnológico de Monterrey, June 11, Mexico City. http://www.g20.utoronto.ca/biblio/kirton-kulik-120612.pdf (January 2015).

Kirton, John J. and Julia Kulik (2012). "A Summit of Significant Success: G8 Governance at Camp David in May 2012." *International Organisations Research Journal* 7(3): 7-24. English-language version of article published in Russian. http://www.g8.utoronto.ca/scholar/kirton-kulik-iori-2012.pdf (January 2015).

Kirton, John J. and Julia Kulik (2012). "A Summit of Significant Success: G20 Los Cabos Leaders Deliver the Desired Double Dividend." June 19. http://www.g20.utoronto.ca/analysis/120619-kirton-success.html (January 2015).

Kirton, John J., Julia Kulik, and Caroline Bracht (2014). "The Political Process in Global Health and Nutrition Governance: The G8's 2010 Muskoka Initiative on Maternal, Child, and Newborn Health." *Annals of the New York Academy of Sciences* 40: 1–15. doi: 10.1111/nyas.12494.

Kirton, John J., Julia Kulik, and Caroline Bracht (2014, 21 July). "The Political Process in Global Health and Nutrition Governance: The G8's 2010 Muskoka Initiative on Maternal, Child, and Newborn Health." *Annals of the New York Academy of Sciences*. doi: 10.1111/nyas.12494.

Kirton, John J., Julia Kulik, Caroline Bracht, et al. (2014). "Connecting Climate Change and Health Through Global Summitry." *World Medical and Health Policy* 6(1): 73-100. doi: 10.1002/wmh3.83.

Kirton, John J., Marina V. Larionova, and Paolo Savona, eds. (2010). *Making Global Economic Governance Effective: Hard and Soft Law Institutions in a Crowded World.* Farnham: Ashgate.

Kirton, John J. and Victoria Panova (2003). "Coming Together: Prospects for the G8 Evian Summit." Conference on "Governing Globalization: Corporate, Public, and G8 Governance", May 27, Fontainebleau. http://www.g8.utoronto.ca/scholar/kirton2003/kirton_prospects_030520.html (January 2015).

Kirton, John J. and Sarah Richardson (1995). "The Halifax Summit, Sustainable Development and International Institutional Reform: Preliminary Discussion Paper and Background Material." February 27, National Round Table on the Environment and the Economy, Ottawa.

Kirton, John J. and Sarah Richardson (1995). "Introduction: The Halifax Summit, Sustainable Development, and International Institutional Reform." In *The Halifax Summit, Sustainable Development, and International Institutional Reform*, John J. Kirton and Sarah Richardson, eds. Ottawa: National Round Table on the Environment and the Economy. http://www.g8.utoronto.ca/scholar/kirton199503/rt95ind.htm (January 2015).

Kirton, John J., Nikolai Roudev, Michael Lehan, et al. (2005). "Shocks from Terrorism to G8 Countries and Citizens: Major Incidents." G8 Research Group, July 20. http://www.g8.utoronto.ca/evaluations/factsheet/factsheet_terrorism.html (January 2015).

Kluger, Jeffrey (2001). "A Climate of Despair." *Time*, April 1. http://content.time.com/time/magazine/article/0,9171,104596,00.html (January 2015).

Kokotsis, Eleanore (1995). "Keeping Sustainable Development Commitments: The Recent G7 Record." http://www.g8.utoronto.ca/scholar/kirton199503/kokotsis/index.html (January 2015).

Kokotsis, Eleanore (1999). *Keeping International Commitments: Compliance, Credibility, and the G7, 1988–1995*. New York: Garland.

Kopvillem, Peeter (1989). "A Keener Earth Watch." *Macleans*, January 2, p. 18.

Krasner, Stephen (1983). *International Regimes*. Ithaca, NY: Cornell University Press.

Kremlin (2007). "President Vladimir Putin Held a Press Conference Following the End of the G8 Summit." Heiligendamm, June 8. http://eng.kremlin.ru/news/15023 (January 2015).

Kuipers, Dean (2011). "Progress at End of Durban COP17 Climate talks." *Los Angeles Times*, December 12. http://articles.latimes.com/2011/dec/12/local/la-me-gs-progress-at-end-of-durban-cop17-climate-talks-20111212 (January 2015).

Lanchberry, John (1996). "The Rio Earth Summit." In *Diplomacy at the Highest Level: The Evolution of International Summitry*, David Dunn, ed. London: Palgrave Macmillan, pp. 220–43.

Larionova, Marina V. and John J. Kirton, eds. (2015). *The G8-G20 Relationship in Global Governance*. Farnham: Ashgate.

Lascelles, David (1992). "Getting Down to Earth in Rio." *Financial Times*, June 3.

Lee, Jesse (2009). "'A Meaningful and Unprecedented Breakthrough Here in Copenhagen'." December 18. http://www.whitehouse.gov/blog/2009/12/18/a-meaningful-and-unprecedented-breakthrough-here-copenhagen (January 2015).

Leiserowitz, Anthony (2007). "American Opinions on Global Warming." Yale University/Gallup/ClearVision Institute Poll, Yale School of Forestry and Environmental Studies, New Haven CT.

Lesage, Dries, Thijs Van de Graaf, and Kirsten Westphal (2010). *Global Energy Governance in a Multipolar World*. Farnham: Ashgate.

Lewington, Jenniffer and Ross Howard (1990). "Ottawa Promises Action This Fall on Carbon-Dioxide Emissions." *Globe and Mail*, April 19.

Lewis, Paul (1992). "U.S. at the Earth Summit: Isolated and Challenged." *New York Times*, June 10. http://www.nytimes.com/1992/06/10/world/us-at-the-earth-summit-isolated-and-challenged.html (January 2015).

Louet, Sophie (2006). "France's Chirac Warns Mankind Faces Climate Volcano." *Reuters*, July 16.

Lovell, Jeremy (2007). "G8 Set for Transatlantic Clash on Climate." *Reuters*, May 11. http://in.reuters.com/article/2007/05/11/idINIndia-29779420070511 (January 2015).

Low, Melissa (2013). "Losing Canada, Japan and Russia in the Climate Regime: Could the Solution Be in Asia?", April 11. https://unfcccecosingapore.wordpress.com/2013/04/24/losing-canada-japan-and-russia-in-the-climate-regime-could-the-solution-be-in-asia/ (January 2015).

Luckhurst, Jonathan (2013). "Building Cooperation between the BRICS and Leading Industrialized States." *Latin American Policy* 4(2): 251-68. doi: 10.1111/lamp.12018.

MacKinnon, Mark (2001). "Canada Threatens to Reject Kyoto Pact." *Globe and Mail*, July 19. http://www.theglobeandmail.com/news/national/canada-threatens-to-reject-kyoto-pact/article4150728/ (January 2015).

Maclean's (1989). "Looking Ahead." *Macleans*, January 2.

MacNeill, Jim (1995). "UN Agencies and the OECD." In *The Halifax Summit, Sustainable Development, and International Institutional Reform*, John J. Kirton and Sarah Richardson, eds. Ottawa: National Round Table on the Environment and the Economy. http://www.g8.utoronto.ca/scholar/kirton199503/rt95mac.htm (January 2015).

Maddox, Bronwen (1993). "The World on His Shoulders." *Financial Times*, June 30.

Major Economies Meeting on Energy Security and Climate Change (2008). "Declaration of Leaders Meeting of Major Economies on Energy Security and Climate Change." Hokkaido, Japan, July 9. http://www.g8.utoronto.ca/summit/2008hokkaido/2008-mem.html (January 2015).

Massai, Leonardo (2010). "The Long Way to the Copenhagen Accord: Climate Change Negotiations in 2009." *Review of European Community and International Environmental Law* 19(1): 104–21.

Max, Arthur (2007). "Delegates Debate Urgency of Climate Change in Key Policy Report." *Associated Press*, April 4.

Max, Arthur (2011). "Ban Ki-Moon, UN Chief, Doubts Climate Deal." *Associated Press*, December 6.

McGrath, Matt (2014). "UN Members Agree Deal at Lima Climate Talks." *BBC News*, December 14. http://m.bbc.com/news/science-environment-30468048 (January 2015).

Meadows, Donella, Dennis Meadows, and Jorgan Randers (1992). *Beyond the Limits: Confronting Global Collapse Envisioning a Sustainable Future*. White River Junction VT: Chelsea Green.

Merkel, Angela (2012). "Cooperation, Responsibility, Solidarity." In *The G8 Camp David Summit 2012: The Road to Recovery*, John J. Kirton and Madeline Koch, eds. London: Newsdesk, pp. 24–25. http://www.g8.utoronto.ca/newsdesk/campdavid/merkel.html (January 2015).

Mexican Presidency of the G20 (2011). "Priorities of the Mexican Presidency." Mexico City, December 13. http://www.g20.utoronto.ca/2012/2012-111213-priorities-en.html (January 2015).

Milinski, Manfred, Ralf D. Sommerfeld, Hans-Jürgen Krambeck, et al. (2008). "The Collective-Risk Social Dilemma and the Prevention of Simulated Dangerous Climate Change." *Proceedings of the National Academy of Sciences* 105(7): 2291–94. doi: 10.1073/pnas.0709546105.

Miller, Pamela A. (1999). "Exxon Valdez Oil Spill: Ten Years Later." Technical background paper for Alaska Will League. *Arctic Connections* 3/99. http://arcticcircle.uconn.edu/SEEJ/Alaska/miller2.htm (January 2015).

Minutes (1979, 28 June). G7 Tokyo Summit (Session 1). Margaret Thatcher Foundation. http://www.margaretthatcher.org/document/112029 (January 2015).

Mitterrand, François (1982). *The Wheat and the Chaff*. New York: Seaver Books.

Mori, Yoshiro (2000). "Address by Prime Minister Yoshiro Mori at the Discussion Group on the Kyushu-Okinawa Summit, Okinawa Summit." June 5. http://www.ioc.u-tokyo.ac.jp/~worldjpn/documents/texts/summit/20000605.S1E.html (January 2015).

Mulroney, Brian (1989, 16 July). Opening Statement by Prime Minister Brian Mulroney at his Final Press Conference, Summit of the Arch. G7/G8 Research Collection in the John W. Graham Library, Trinity College, University of Toronto. Locator ID: 22.665.

Mulroney, Brian (2007). *Memoirs*. Toronto: McClelland and Stewart.

Nankivell, Neville (1995). "Summit Produced Real Substance." *Financial Post*, June 30.

National Climatic Data Center (2015). "Billion-Dollar Weather and Climate Disasters: Table of Events." http://www.ncdc.noaa.gov/billions/events (January 2015).

Nature (1979, 3 May). "Costs and Benefits of Carbon Dioxide." 279(5708): 1. Editorial. doi: 10.1038/279001a0.

Nau, Henry (2004). "G7/8 Oral History: Interview." G8 Research Group, May 7. http://www.g8.utoronto.ca/oralhistory/nau040507.html (January 2015).

Nau, Henry and David Shambaugh, eds. (2004). *Divided Diplomacy and the Next Administration: Conservative and Liberal Alternatives*. Washington DC: Elliott School of International Affairs, George Washington University.

Norman, Peter (1991). "A Table Piled High with Problems: The Group of Seven Summit in London Will Expose Both Harmony and Division Between Industrial Nations." *Financial Times*, July 12.

Oaks, John (1992). "An Environmentalist? Bush? Forget It." *New York Times*, May 8. http://www.nytimes.com/1992/05/08/opinion/an-environmentalist-bush-forget-it.html (January 2015).

Obama, Barack (2012). "A Time of Transformation." In *The G8 Camp David Summit 2012: The Road to Recovery*, John J. Kirton and Madeline Koch, eds. London: Newsdesk, pp. 14–16. http://www.g8.utoronto.ca/newsdesk/campdavid/obama.html (January 2015).

Oberthür, Sebastian and Hermann Ott (1999). *The Kyoto Protocol: International Climate Policy for the 21st Century*. Berlin: Springer.

Obuchi, Keizo (2000). "Statement by Prime Minister Keizo Obuchi Discussion Group on the Kyushu-Okinawa Summit." February 28. http://www.mofa.go.jp/announce/announce/2000/2/228-2.html (January 2015).

Oosthoek, Jan (2005). "The Gleneagles G8 Summit and Climate Change: A Lack of Leadership."*Globalizations*2(3):443–46.doi:dx.doi.org/10.1080/14747730500409447.

Owen, Henry (1979, 22 June). G7 Tokyo Summit: Henry Owen Brief for President Carter. Margaret Thatcher Foundation (original source: Carter Library). http://www.margaretthatcher.org/document/111681 (January 2015).

Oxfam (2011). "Sarkozy Renews Calls for FTT, Signals Ambitious G20 Agenda on Food Price Volatility." Ottawa, January 24. http://www.oxfam.ca/news-and-publications/news/sarkozy-renews-calls-ftt-signals-ambitious-g20-agenda-food-price-volatili (January 2015).

Paarlberg, Robert L. (1992). "Ecodiplomacy: U.S. Environmental Policy Goes Abroad." In *Eagle in a New World: American Grand Strategy in the Post-Cold War Era*, Kenneth A. Oye, Robert J. Lieber, and Donald S. Rothschild, eds. New York: HarperBusiness.

Park, Chris C. (1989). *Chernobyl: The Long Shadow*. Abdingdon: Routledge.

Pew Center on Global Climate Change (2007). "Thirteenth Session of the Conference of the Parties to the UN Framework Convention on Climate Change and Third Session of the Meeting of the Parties to the Kyoto Protocol." Bali. http://www.c2es.org/docUploads/Pew%20Center_COP%2013%20Summary.pdf (January 2015).

Pomerance, Rafe (1989). "The Dangers from Climate Warming: A Public Awakening." In *The Challenge of Global Warming*, Dean Abrahamson, ed. Washington DC: Island Press, pp. 259–69.

Prodi, Romano (2001). "EU-US Summit, Göteborg 14 June 2001 Statement of Romano Prodi President of the European Commission." IP/01/849, Göteborg, June 15. http://europa.eu/rapid/press-release_IP-01-849_en.htm (January 2015).

Putin, Vladimir (2006). "Speech at Meeting with the G8 Energy Ministers." Moscow, March 16. http://www.g8.utoronto.ca/energy/energy_putin060316.html (January 2015).

Putnam, Robert (1988). "Diplomacy and Domestic Politics: The Logic of Two-Level Games." *International Organization* 42(3): 427–60.

Putnam, Robert and Nicholas Bayne (1984). *Hanging Together: Co-operation and Conflict in the Seven-Power Summit*. 1st ed. Cambridge MA: Harvard University Press.

Putnam, Robert and Nicholas Bayne (1987). *Hanging Together: Co-operation and Conflict in the Seven-Power Summit*. 2nd ed. London: Sage Publications.

Rafferty, Kevin (2013). "Sunny Spin to an Oily Earth." *Japan Times*, June 4. http://www.japantimes.co.jp/opinion/2013/06/04/commentary/sunny-spin-to-an-oily-earth/#. VCCUWytdVTM (January 2015).

Rajamani, Lavanya (2009). "'Cloud' over Climate Negotiations: From Bangkok to Copenhagen and Beyond." *Economic and Political Weekly* 44(43): 11–15.

Raustiala, Kal and David G. Victor (2004, April). "The Regime Complex for Plant Genetic Resources." *International Organization* 58(2): 277–309.

Reagan, Ronald (1990). *An American Life*. New York: Pocket Books.

Rennie, Steve (2010). "Tories Put Climate Change on G8 Agenda after Pressure from World Leaders." *Globe and Mail*, June 14. http://www.theglobeandmail.com/news/politics/tories-put-climate-change-on-g8-agenda-after-pressure-from-world-leaders/article1211785/ (January 2015).

Reuters (2001). "G8 Leaders to Tackle African Crisis at Summit." March 7.

Richardson, Sarah and John J. Kirton (1995). "Conclusion." In *The Halifax Summit, Sustainable Development, and International Institutional Reform*, John J. Kirton and Sarah Richardson, eds. Ottawa: National Round Table on the Environment and the Economy. http://www.g8.utoronto.ca/scholar/kirton199503/conclusion (January 2015).

Riddell, Peter (1989). "Thatcher Warns of Insidious Threat to Planet: The UK Prime Minister's Wide-Ranging Speech to the UN." *Financial Times*, November 9.

Risbud, Sheila (2006). "Civil Society Engagement: A Case Study of the 2002 G8 Environment Ministers Meeting." In *Sustainability, Civil Society, and International Governance: Local, North American, and Global Contributions*, John J. Kirton and Peter I. Hajnal, eds. Aldershot: Ashgate, pp. 337–42.

Rooney, Stephen (2013). "The G20 and Energy Sustainability." *Europeinfos*, October. http://www.comece.eu/europeinfos/en/archive/issue164/article/6023.html (January 2015).

Rosenberg, Norman, Joel Darmstadter, and Pierre Crosson (1989). "Overview: Climate Change." *Environment* 31(3): 2.

Rosewicz, Barbara and Michel McQueen (1989). "Clearing the Air: Bush, Resolving Clash In Campaign Promises, Tilts to Environment." *Wall Street Journal*, June 13.

Rowlands, Ian H. (1995). "The Climate Change Negotiations: Berlin and Beyond." *Journal of Environment & Development* 4(2): 145–63. doi: 10.1177/107049659500400207.

Russia's G8 Presidency (2006). "Final Press Briefing with President Putin." St. Petersburg, July 17. http://www.g8.utoronto.ca/summit/2006stpetersburg/putin060717.html (January 2015).

Russia's G20 Presidency (2012). "Russian Presidency of the G20: Outline." Moscow, December 28. http://www.g20.utoronto.ca/2013/Brochure_G20_Russia_eng.pdf (January 2015).

Schelling, Thomas C. (2002). "What Makes Greenhouse Sense?" *Foreign Affairs*, May/June. http://www.foreignaffairs.com/articles/58002/thomas-c-schelling/what-makes-greenhouse-sense (January 2015).

Schmidt, Helmut (1979). "An Interview with Helmut Schmidt." *Time*, June 11.

Schmidt, Helmut (1979, 4 July). Policy Statement Made by the Chancellor of the Federal Republic of Germany Herr Helmut Schmidt before the German Bundestag. G7/G8 Research Collection in the John W. Graham Library, Trinity College, University of Toronto. Locator ID: 28.1062.

Schmidt, Helmut (1989). *Men and Powers: A Political Retrospective.* Translated by Ruth Hein. New York: Random House.

Schneider, Keith (1992). "Gore Bringing Environmental Policy to Fore." December 15. http://www.nytimes.com/1992/12/15/us/the-transition-gore-bringing-environmental-policy-to-fore.html (January 2015).

Schoon, Nicholas and Andrew Marshall (1991). "Environmental Issues Given 15 Minutes at London Summit." *Independent*, July 18.

Sebenius, James K. (1991). "Designing Negotiations Toward a New Regime: The Case of Global Warming." *International Security* 15(4): 110–48.

Shabecoff, Philip (1989). "U.S. to Urge Joint Environmental Effort at Summit." *New York Times*, July 6. http://www.nytimes.com/1989/07/06/world/us-to-urge-joint-environmental-effort-at-summit.html (January 2015).

Shi Jiangtao (2011). "Durban Delivers Road Map on Climate Change Treaty." *South China Morning Post*, December 12.

Sich, Alexander R. (1996). "The Denial Syndrome." *Bulletin of the Atomic Scientists* 52(3): 38–39.

Siegel, Matt (2014). "In About Face, Australia PM Joins Global Climate Fund." *Reuters*, December 10. http://in.reuters.com/article/2014/12/10/australia-climatechange-idINL3N0TU1FZ20141210 (January 2015).

Simpson, Jeffrey (1989). "The Greening of the G7." *Globe and Mail*, July 19.

Smith, Gordon (1995). "Canada and the Halifax Summit." In *The Halifax Summit, Sustainable Development, and International Institutional Reform*, John J. Kirton and Sarah Richardson, eds. Ottawa: National Round Table on the Environment and the Economy. http://www.g8.utoronto.ca/scholar/kirton199503/smith/document.html (January 2015).

Smith, Gordon (2008). "Canada's Foreign Policy Priorities." *Globe and Mail*, April 23.

Smith, Heather (2008). "Canada and Kyoto: Independence or Indifference?" In *An Independent Foreign Policy for Canada: Challenges and Choices for the Future*, Brian Bow and Patrick Lennox, eds. Toronto: University of Toronto, pp. 207–21.

Speth, Gustav (2004). "Red Sky at Morning: America and the Crisis of the Global Environment." Interview with Joanne Myers, April 22. http://www.carnegiecouncil.org/studio/multimedia/20040422/index.html (January 2015).

Spini, Valdo (1994). "The Floreince Meeting on the Environment." G7 Napoli Summit '94 (dossier), Agenzia ANSA, Rome.

Stavins, Robert N. and Robert C. Stowe (2010). "What Hath Copenhagen Wrought? A Preliminary Assessment." *Environment* 52(3): 8–13. http://www.environmentmagazine.org/Archives/Back%20Issues/May-June%202010/what-wrath-full.html (January 2015).

Stevens, William (1993). "Gore Promises U.S. Leadership On Sustainable Development Path." *New York Times*, June 15. http://www.nytimes.com/1993/06/15/news/gore-promises-us-leadership-on-sustainable-development-path.html (January 2015).

Stewart, Edison (1989). "PM, Bush Set Sights on Acid Rain Accord." *Toronto Star*, February 11.

Strong, Maurice (1973). "One Year After Stockholm: An Ecological Approach to Management." *Foreign Affairs*, July, pp. 690–707. http://www.mauricestrong.net/images/docs/one%20year%20after%20stockholm%281973%29.pdf (January 2015).

Sustainable Energy Development Center (2006). *Global Energy Security: Summary of Russia's G8 Presidency.* Moscow: Sustainable Energy Development Center.

Tanaka, Miya (2008). "Japanese PM to Seek Bush's Cooperation on North Korea, Climate Change." *Kyodo*, July 6. BBC Monitoring Asia Pacific.

Thatcher, Margaret (1989, 8 November). Speech to United Nations General Assembly. Margaret Thatcher Foundation. http://www.margaretthatcher.org/document/107817 (January 2015).

Thatcher, Margaret (1993). *The Downing Street Years*. New York: HarperCollins.

Tickell, Crispin (1977). *Climatic Change and World Affairs*. Cambridge MA: Center for International Affairs, Harvard University.

Timberlake, Cotten (1991). "On Environment, Communique Is Long on Words, Short on Specifics." *Associated Press*, July 17.

Tollefson, Jeff (2010, 3 December). "Cancún Week One: A Climate of Confusion." *Nature*. doi:10.1038/news.2010.653

Toronto Star (1989). "Green Revolution Arrives in Paris." July 14.

Underdal, Arild (1994). "Leadership Theory: Rediscovering the Arts of Management." In *International Multilateral Negotiation: Approaches to the Management of Complexity*, I. William Zartman, ed. San Francisco: Jossey-Bass, pp. 178–97.

United Nations (1985). "Protocol to the 1979 Convention on Long-Range Transboundary Air Pollution on the Reduction of Sulphur Emissions or Their Transboundary Fluxes by at Least 30 Per Cent." Entered into force 2 September 1987. http://www.unece.org/fileadmin/DAM/env/documents/2012/EB/1985.Sulphur.e.pdf (January 2015).

United Nations (1999). "Report of the Conference of the Parties on Its Fourth Session, Held at Buenos Aires from 2 to 14 November 1998." Buenos Aires. http://unfccc.int/resource/docs/cop4/16.pdf (January 2015).

United Nations (2006). "Citing 'Frightening Lack of Leadership' on Climate Change, Secretary-General Calls Phenomenon an All-Encompassing Threat in Address to Nairobi Talks." Press release, New York NY, November 15. http://www.un.org/press/en/2006/sgsm10739.doc.htm (January 2015).

United Nations (2008). "G8 Summit Good Start But Further Action Needed to Tackle Global Crisis, Says Ban." July 9. http://www.un.org/apps/news/story.asp?NewsID=27312#.VPC7obPF-hk (January 2015).

United Nations (2008). "Report of the Conference of the Parties on Its Thirteenth Session, Held in Bali from 3 to 15 December 2007: Addendum. Part Two: Action Taken by the Conference of the Parties at Its Thirteenth Session." FCCC/CP/2007/6/Add.1, March 14. http://unfccc.int/resource/docs/2007/cop13/eng/06a01.pdf (January 2015).

United Nations (2010). "Report of the Conference of the Parties on Its Fifteenth Session, Held in Copenhagen from 7 to 19 December 2009. Addendum. Part Two: Action Taken by the Conference of the Parties at Its Fifteenth Session " FCCC/CP/2009/11/Add.1, March 30. http://unfccc.int/resource/docs/2009/cop15/eng/11a01.pdf (January 2015).

United Nations (2012). "Report of the Conference of the Parties on Its Seventeenth Session, Held in Durban from 28 November to 11 December 2011: Addendum. Part Two: Action Taken by the Conference of the Parties at Its Seventeenth Session." FCCC/CP/2011/9/Add.1, March 15. http://unfccc.int/resource/docs/2011/cop17/eng/09a01.pdf (January 2015).

United Nations Climate Change Secretariat (2007). "UNFCCC Executive Secretary: G8 Document Reenergises Multilateral Climate Change Process under the United Nations." Bonn, June 7. http://unfccc.int/files/press/news_room/press_releases_and_advisories/application/pdf/20070607_g8_press_release_english.pdf (January 2015).

United Nations Conference on Sustainable Development (2012). "The Future We Want." A/ CONF.216/L.1, Rio de Janeiro, June 19. http://www.stakeholderforum.org/fileadmin/ files/FWWEnglish.pdf (January 2015).

United Nations Framework Convention on Climate Change (1995). "Decisions Adopted by the Conference of the Parties." Report of the Conference of the Parties on Its First Session, Held at Berlin from 28 March to 7 April 1995. FCCC/CP/1995/7/Add.1, June 6, Berlin. http://unfccc.int/resource/docs/cop1/07a01.pdf (January 2015).

United Nations Framework Convention on Climate Change (1998). "Report of the Conference of the Parties on Its Third Session, Held at Kyoto from 1 to 11 December 1997." FCCC/CP/1997/7, March 24, Kyoto. http://unfccc.int/resource/docs/ cop3/07.pdf (January 2015).

United Nations Framework Convention on Climate Change (2009). "Decisions Adopted by the Conference of the Parties." Report of the Conference of the Parties on its 15th session, Copenhagen. http://unfccc.int/resource/docs/2009/cop15/eng/11a01. pdf (January 2015).

United Nations Framework Convention on Climate Change (2009). "Fact Sheet: 10 Frequently Asked Questions about the Copenhagen Deal." November. http://unfccc.int/ files/press/fact_sheets/application/pdf/10_faqs_copenhagen_deal.pdf (January 2015).

United Nations Framework Convention on Climate Change (2011). "Durban Climate Change Conference – November/December 2011." Bonn. http://unfccc.int/meetings/ durban_nov_2011/meeting/6245.php (January 2015).

United States Central Intelligence Agency (1976). "USSR: The Impact of Recent Climate Change on Grain Production." Report ER 76-10577 U, October, Washington DC. http:// www.foia.cia.gov/sites/default/files/document_conversions/89801/DOC_0000499885. pdf (January 2015).

United States Department of Energy (2009). "President Obama Sets a Target for Cutting U.S. Greenhouse Gas Emissions." Washington DC, December 2. http://apps1.eere. energy.gov/news/news_detail.cfm/news_id=15650 (January 2015).

United States Department of State (1992). "US Environment Initiatives and the UN Conference on Environment and Development." *US Department of State Dispatch Supplement* 3(4). http://dosfan.lib.uic.edu/ERC/briefing/dispatch/1992/html/ Dispatchv3Sup4.html (January 2015).

United States Department of State (1993). "Group of Seven (G7) 1993 Economic Summit. Fact Sheet: Global Environmental Issues." *US Department of State Dispatch Supplement* 4(3). http://dosfan.lib.uic.edu/ERC/briefing/dispatch/1993/html/ Dispatchv4Sup3.html (January 2015).

United States Department of State (2003). "Press Availability in St. Petersburg." Konstantin Palace, St. Petersburg, June 1. http://2001-2009.state.gov/p/eur/rls/rm/2003/21113.htm (January 2015).

United States Department of State (2007). "A New International Climate Change Framework." Washington DC, May 31. http://2001-2009.state.gov/g/oes/rls/ fs/2007/92156.htm (January 2015).

United States Department of State (2014). "Joint Statement of the U.S.-EU Energy Council." Washington DC, December 3. http://www.state.gov/r/pa/prs/ps/2014/12/234638.htm (January 2015).

United States National Academy of Sciences (1979). "Carbon Dioxide and Climate: A Scientific Assessment." Report of an Ad Hoc Study Group on Carbon Dioxide and Climate, Woods Hole, Massachusetts, July 23-27, to the Climate Research Board,

Assembly of Mathematical and Physical Sciences, National Research Council, Washington DC. http://web.atmos.ucla.edu/~brianpm/download/charney_report.pdf (January 2015).

Vabulas, Felicity and Duncan Snidal (2013). "Organization Without Delegation: Informal Intergovernmental Organizations (IIGOs) and the Spectrum of Intergovernmental Arrangements." *Review of International Organizations* 8(2): 193–220.

Victor, David G. (2011). "Why the UN Can Never Stop Climate Change." *Guardian*, April 4. http://www.guardian.co.uk/environment/2011/apr/04/un-climate-change (January 2015).

Victor, David G., Amy M. Jaffe, and Mark H. Hayes, eds. (2006). *Natural Gas and Geopolitics: From 1970 to 2040*. Cambridge: Cambridge University Press.

Vidal, John (2010). "Cameron Refuses to Attend UN Climate Change Talks." *Guardian*, November 29. http://www.theguardian.com/environment/2010/nov/29/cameron-cancun-climate-change-summit?INTCMP=ILCNETTXT3487 (January 2015).

Vidal, John (2010). "Kyoto Protocol May Suffer Fate of Julius Caesar at Durban Climate Talks." *Guardian*, November 29. http://www.guardian.co.uk/environment/blog/2011/nov/29/kyoto-protocol-julius-caesar-durban (January 2015).

Vidal, John (2011). "Climate Change Conference in Trouble as China Rejects Proposals for New Treaty." *Guardian*, December 9. http://www.theguardian.com/environment/2011/dec/09/climate-change-conference-durban-treaty (January 2015).

Vig, Norman J. (1994). "Presidential Leadership and the Environment: From Reagan to Bush to Clinton." In *Environmental Policy in the 1990s: Toward a New Agenda*, Norman J. Vig and Michael E. Kraft, eds. Washington DC: CQ Press, pp. 71–95.

Wall Street Journal (1990). "Review and Outlook: Environmental Balance." July 13, p. A8.

Weiss, Edith Brown and Harold K. Jacobson, eds. (2000). *Engaging Countries: Strengthening Compliance with International Environmental Accords*. Cambridge MA: MIT Press.

Welch, David (2005). *Painful Choices: A Theory of Foreign Policy Change*. Princeton: Princeton University Press.

White House (1989, 14 July). Press Briefing by White House Chief of Staff, Governor John Sununu. G7/G8 Research Collection in the John W. Graham Library, Trinity College, University of Toronto. Locator ID: 22.677.

White House (1993). "Press Briefing by Counselor to the President David Gergen, Under-Secretary of State Joan Spero, Under-Secretary of Treasury Larry Summers, and Special Assistant to the President Bob Fauver." Tokyo, July 9. http://www.presidency.ucsb.edu/ws/?pid=60165 (January 2015).

White House (1997). "Remarks by President Clinton in Presentation of the Final Communiqué of the Denver Summit of the Eight." Denver, June 22. http://www.g8.utoronto.ca/summit/1997denver/denclintp.htm (January 2015).

White House (1997). "Transcript: President Clinton's Final Denver Summit Press Conference." Denver, June 22. http://www.g8.utoronto.ca/summit/1997denver/clint22.htm (January 2015).

White House (2001). "Press Briefing by Ari Fleischer." Washington DC, March 28. http://www.presidency.ucsb.edu/ws/?pid=47500 (January 2015).

White House (2014). "U.S.-China Joint Announcement on Climate Change." Beijing, November 12. https://www.whitehouse.gov/the-press-office/2014/11/11/us-china-joint-announcement-climate-change (January 2015).

Whitney, Craig R. (1989). "80 Nations Favor Ban to Help Ozone." *New York Times*, May 3. http://www.nytimes.com/1989/05/03/world/80-nations-favor-ban-to-help-ozone.html (January 2015).

Williams, Jody (2010). "Why Is Canada Not Putting Climate Change on the G20 Agenda?" *Globe and Mail*, June 11. http://www.theglobeandmail.com/globe-debate/why-is-canada-not-putting-climate-change-on-the-g20-agenda/article1372784/ (January 2015).

Williamson, Hugh (2007). "G8 Summit: Merkel Keeps Focus Broad." *Financial Times*, June 4, p. 5.

Wintour, Patrick and Larry Elliott (2008). "Bush Signs G8 Deal to Halve Greenhouse Gas Emissions by 2050." *Guardian*, July 8. http://www.theguardian.com/world/2008/jul/08/g8 (January 2015).

Woods, Allan (2010). "Climate Change Gets Short Shrift from G8 Leaders." *Toronto Star*, June 26. http://www.thestar.com/news/world/g8/article/829116--climate-change-gets-short-shrift-from-g8-leaders (January 2015).

World Commission on Environment and Development (1987). *Our Common Future* (Brundtland Report). Oxford: Oxford University Press.

World Meteorological Organization, United Nations Environment Programme, and Government of Canada (1988). "The Changing Atmosphere: Implications for Global Society." June 27–30, Toronto. http://cmosarchives.ca/History/ChangingAtmosphere1988e.pdf (January 2015).

World Wildlife Fund (2011). "Small Steps on Climate Finance in Cannes Will Require Giant Leaps in Durban." Cannes, November 4. http://wwf.ca/newsroom/?uNewsID=10082 (January 2015).

Young, Oran R. (1991). "Political Leadership and Regime Formation: On the Development of Institutions in International Society." *International Organization* 45(3): 281–308. doi: 10.1017/S0020818300033117.

Yudaeva, Ksenia (2013). "Russia's Perspective on G20 Summitry." In *Russia's G20 Summit: St Petersburg 2013*, John J. Kirton and Madeline Koch, eds. London: Newsdesk, pp. 36–37. http://www.g20.utoronto.ca/newsdesk/stpetersburg (January 2015).

Index

GLOBAL ENVIRONMENTAL GOVERNANCE SERIES

Full series list

Moving Health Sovereignty in Africa
Disease, Governance, Climate Change
Edited by John J. Kirton, Andrew F. Cooper,
Franklyn Lisk and Hany Besada

Africa's Health Challenges
Sovereignty, Mobility of People
and Healthcare Governance
Edited by Andrew F. Cooper, John J. Kirton,
Franklyn Lisk and Hany Besada

Corporate Responses to
EU Emissions Trading
Resistance, Innovation or Responsibility?
Edited by Jon Birger Skjærseth and
Per Ove Eikeland

Renewable Energy Policy
Convergence in the EU
The Evolution of Feed-in Tariffs in
Germany, Spain and France
David Jacobs

The EU as International
Environmental Negotiator
Tom Delreux

Global Energy Governance in a
Multipolar World
Dries Lesage, Thijs Van de Graaf
and Kirsten Westphal

Innovation in Global Health Governance
Critical Cases
Edited by Andrew F. Cooper
and John J. Kirton

Environmental Skepticism
Ecology, Power and Public Life
Peter J. Jacques

Transatlantic Environment and Energy Politics
Comparative and International Perspectives
Edited by Miranda A. Schreurs, Henrik Selin,
and Stacy D. VanDeveer

The Legitimacy of International Regimes
Helmut Breitmeier
Governing Agrobiodiversity
Plant Genetics and Developing Countries
Regine Andersen

The Social Construction of Climate Change
Power, Knowledge, Norms, Discourses
Edited by Mary E. Pettenger

Governing Global Health
Challenge, Response, Innovation
Edited by Andrew Cooper, John Kirton
and Ted Schrecker

Participation for Sustainability in Trade
Edited by Sophie Thoyer and
Benoît Martimort-Asso

Bilateral Ecopolitics
Continuity and Change in Canadian-
American Environmental Relations
Edited by Philippe Le Prestre
and Peter Stoett

Governing Global Desertification
Linking Environmental Degradation,
Poverty and Participation
Edited by Pierre Marc Johnson,
Karel Mayrand and Marc Paquin

Sustainability, Civil Society and
International Governance
Local, North American and
Global Contributions
Edited by John J. Kirton and Peter I. Hajnal